# Cognitive Ecology of Pollination
*Animal Behavior and Floral Evolution*

Important breakthroughs have recently been made in our understanding of the cognitive and sensory abilities of pollinators: how pollinators perceive, memorize, and react to floral signals and rewards; how they work flowers, move among inflorescences, and transport pollen. These new findings have obvious implications for the evolution of floral display and diversity, but most existing publications are scattered across a wide range of journals in very different research traditions. This book brings together for the first time outstanding scholars from many different fields of pollination biology, integrating the work of neuroethologists and evolutionary ecologists to present a multidisciplinary approach. Aimed at graduates and researchers of behavioral and pollination ecology, plant evolutionary biology, and neuroethology, it will also be a useful source of information for anyone interested in a modern view of cognitive and sensory ecology, pollination, and floral evolution.

LARS CHITTKA is Associate Professor at the University of Würzburg, Germany. His research interests focus on the ecology and evolution of insect sensory and cognitive capacities.

JAMES D. THOMSON is Professor and Chair of the Department of Zoology at the University of Toronto, Canada, where he specializes in linking pollinator foraging strategies with floral evolution. He is currently Editor of the *Quarterly Review of Biology* and President of the American Society of Naturalists.

# Cognitive Ecology of Pollination

*Animal Behavior and Floral Evolution*

*Edited by*

### Lars Chittka
*Universität Würzburg*

*and*

### James D. Thomson
*University of Toronto*

CAMBRIDGE UNIVERSITY PRESS
Cambridge, New York, Melbourne, Madrid, Cape Town, Singapore, São Paulo

Cambridge University Press
The Edinburgh Building, Cambridge CB2 2RU, UK

Published in the United States of America by Cambridge University Press, New York

www.cambridge.org
Information on this title: www.cambridge.org/9780521781954



First published 2001
This digitally printed first paperback version 2005

*A catalogue record for this publication is available from the British Library*

*Library of Congress Cataloguing in Publication data*

Cognitive ecology of pollination : animal behavior and floral evolution / edited by Lars
Chittka and James D. Thomson.
    p. cm.
  Includes bibliographical references (p. ).
  ISBN 0 521 78195 7
    1. Pollinators – Ecophysiology. 2. Pollination. I. Chittka, Lars, 1963– II. Thomson,
James D., 1950–
  QK926.C64 2001
  571.8′642 – dc21    00-065072

ISBN-13 978-0-521-78195-4 hardback
ISBN-10 0-521-78195-7 hardback

ISBN-13 978-0-521-01840-1 paperback
ISBN-10 0-521-01840-4 paperback

# Contents


*List of contributors  vii*
*Preface  x*

1  The effect of variation among floral traits on the flower constancy
of pollinators  1
ROBERT J. GEGEAR AND TERENCE M. LAVERTY

2  Behavioral and neural mechanisms of learning and memory as
determinants of flower constancy  21
RANDOLF MENZEL

3  Subjective evaluation and choice behavior by nectar- and pollen-
collecting bees  41
KEITH D. WADDINGTON

4  Honeybee vision and floral displays: from detection to close-up
recognition  61
MARTIN GIURFA AND MIRIAM LEHRER

5  Floral scent, olfaction, and scent-driven foraging behavior  83
ROBERT A. RAGUSO

6  Adaptation, constraint, and chance in the evolution of flower color
and pollinator color vision  106
LARS CHITTKA, JOHANNES SPAETHE, ANNETTE SCHMIDT,
ANJA HICKELSBERGER

7  Foraging and spatial learning in hummingbirds  127
SUSAN D. HEALY AND T. ANDREW HURLY

8  Bats as pollinators: foraging energetics and floral adaptations  148
YORK WINTER AND OTTO VON HELVERSEN

9  Vision and learning in some neglected pollinators: beetles, flies,
moths, and butterflies  171
MARTHA R. WEISS

10  Pollinator individuality: when does it matter?  191
JAMES D. THOMSON AND LARS CHITTKA

11  Effects of predation risk on pollinators and plants  214
REUVEN DUKAS

12  Pollinator preference, frequency dependence, and floral
evolution  237
ANN SMITHSON

13  Pollinator-mediated assortative mating: causes and
consequences  259
KRISTINA NIOVI JONES

14  Behavioural responses of pollinators to variation in floral display
size and their influences on the evolution of floral traits  274
KAZUHARU OHASHI AND TETSUKAZU YAHARA

15  The effects of floral design and display on pollinator economics and
pollen dispersal  297
LAWRENCE D. HARDER, NEAL M. WILLIAMS, CRISPIN Y.
JORDAN AND WILLIAM A. NELSON

16  Pollinator behavior and plant speciation: looking beyond the
"ethological isolation" paradigm  318
NICKOLAS M. WASER

Index  337

# Contributors

LARS CHITTKA
Zoologie II, Biozentrum, Universität Würzburg, Am Hubland, 97074
Würzburg, Germany

REUVEN DUKAS
Department of Biological Sciences, Simon Fraser University, Burnaby,
B.C. V5A 1S6, Canada

ROBERT J. GEGEAR
Department of Zoology, University of Western Ontario, London, Ontario
N6A 5B7, Canada

MARTIN GIURFA
Laboratoire de Cognition Animale, Université Paul Sabatier – Toulouse
III, 11B Route de Narbonne, 31062 Toulouse cedex, France

LAWRENCE D. HARDER
Department of Biological Sciences, University of Calgary, Calgary, Alberta
T2N 1N4, Canada

SUSAN D. HEALY
ICAPB, King's Buildings, Ashworth Laboratory, University of Edinburgh,
West Mains Road, Edinburgh EH9 3JT, Scotland

OTTO VON HELVERSEN
Institut für Zoologie II, Universität Erlangen, Staudtstrasse 5, 91058
Erlangen, Germany

ANJA HICKELSBERGER
Zoologie II, Biozentrum, Universität Würzburg, Am Hubland, 97074
Würzburg, Germany

T. ANDREW HURLY
Department of Biological Sciences, University of Lethbridge, 4401
University Drive, Lethbridge, Alberta T1K 3M4, Canada

KRISTINA NIOVI JONES
Department of Biological Sciences, Wellesley College, Wellesley, MA
02481, USA

CRISPIN Y. JORDAN
Department of Biological Sciences, University of Calgary, Calgary, Alberta
T2N 1N4, Canada

TERENCE M. LAVERTY
Department of Zoology, University of Western Ontario, London, Ontario
N6A 5B7, Canada

MIRIAM LEHRER
Zoological Institute, Department of Neurobiology, Universität Zürich,
Winterthurerstrasse 190, CH 8057 Zürich, Switzerland

RANDOLF MENZEL
Institut für Neurobiologie, Freie Universität Berlin, Königin-Luise-
Strasse 28/30, 14195 Berlin, Germany

WILLIAM A. NELSON
Department of Biological Sciences, University of Calgary, Calgary, Alberta
T2N 1N4, Canada

KAZUHARU OHASHI
Biological Institute, Graduate School of Science, Tohoku University,
Sendai 980-8578, Japan

ROBERT A. RAGUSO
Deptartment of Biological Sciences, University of South Carolina,
Columbia, SC 29208, USA

ANNETTE SCHMIDT
Zoologie II, Biozentrum, Universität Würzburg, Am Hubland, 97074
Würzburg, Germany

ANN SMITHSON
School of Biological Sciences, University of Exeter, Hatherly Laboratories,
Prince of Wales Road, Exeter EX4 4PS, UK

JOHANNES SPAETHE
Zoologie II, Biozentrum, Universität Würzburg, Am Hubland, 97074 Würzburg, Germany

JAMES D. THOMSON
Department of Zoology, University of Toronto, Toronto, Ontario M5S 3G5, Canada

KEITH D. WADDINGTON
Department of Biology, University of Miami, Coral Gables, FL 33124, USA

NICKOLAS M. WASER
Department of Biology, University of California, Riverside, CA 92521, USA

MARTHA R. WEISS
Department of Biology, Georgetown University, Washington DC 20057-1028, USA

NEAL M. WILLIAMS
Department of Biological Sciences, University of Calgary, Calgary, Alberta T2N 1N4, Canada

YORK WINTER
Institut für Zoologie, Universität München LMU, Luisenstrasse 14, 80333 München, Germany

TETSUKAZU YAHARA
Department of Biology, Faculty of Science, Kyushu University, Fukohoka 812-8581, Japan

The idea of making this book arose from a symposium at the XVI International Botanical Congress in St. Louis, USA in August 1999, which brought together some of the contributors of this book. The idea, then, was to inform botanists of important recent developments in pollinator behavior, cognition, and sensory biology. These new findings and perspectives have numerous implications for the evolution of plants and the shaping of plant community structure. Our rationale for such a symposium was that we thought that many botanists are hard-pressed to keep up with the literature concerning pollinator neuroethology and behavioral ecology. Therefore, the field of plant–pollinator interactions is somewhat hobbled by stereotyped, anachronistic, scale-limited, or just simplistic views of how animals really interact with flowering plants.

Our discussions during the symposium (and with other contributors outside the symposium), however, revealed much more profound gaps than just the one between botanists and zoologists. Pollination biology is poised at the boundary between two different traditions, those of proximate and ultimate reasoning in biology. On the one hand, evolutionary ecologists tend to seek *adaptive* explanations for biological characters – how do the observed traits benefit the animal or plant? Physiologists and neuroethologists, on the other hand, prefer to consider the *mechanisms* by which environmental stimuli provoke or modify behavior. Unfortunately, these two groups of scientists have little commerce; they publish in different journals, attend different conferences, and tend to disparage each other's views. This was how the biological world was divided until a few years ago.

In recent times, however, many workers have realized that we cannot fully understand the operation of an animal's senses, learning, and cognition without knowing under what conditions these functions evolved; we cannot understand how flowers (and other biological signals) evolve without understanding mechanisms of information-processing on the receiving side. Attempts to bridge the gap between the two traditions have become fashionable in the past few years. It is in these attempts that the real shortfalls are revealed. Many physiologists will "round up" a paper's discussion by adding a paragraph on the adaptive significance of the traits they describe, where the existence of the trait is usually regarded as sufficient evidence for its adaptive nature. Standard evolutionary tools, such as phylogenetic analyses or measurements of fitness, are generally deemed unnecessary. Evolutionary biologists, on the other hand, do nowadays often discuss behavioral mechanisms, but their understanding of such mechanisms is sometimes rudimentary. The result is that mechanisms, in such papers, seem often tailored to fit *any* kind of observed behavior, rather than being based on what we actually know about them from physiological work. Finally, behavioral ecologists are sometimes guilty of both sins: pan-adaptionism without rigorous tests of evolutionary hypotheses *and* naïveté about neural mechanisms underlying behavior. Because papers in each of these traditions are largely refereed by other workers within the same field, the review process seldom forces authors to adjust their views. This, in turn, antagonizes scientists in the other fields. Each of us knows that our own field of research is a difficult one that demands knowledge of a specialized education and training in certain rigorous ways of thought. When an "outsider" offers a contribution to our field, we are quick to note the imperfections and inadequacies that are virtually inevitable. Perniciously, we may further decide that the inadequacies of outsiders must extend to their performance in their own fields, while they – of course – reach the same conclusion about us. We fear, therefore, that well-intentioned attempts to link ultimate and proximate perspectives, when done clumsily, will further the antagonism between these traditions, rather than smooth it out.

Pollination biology is a field that might serve as a link to tie these fields together, because it involves workers from all of the traditions above, all working on the same or similar experimental subjects. But readers should be warned from the start that our book does not represent the successful reconciliation and fusion of these viewpoints. Rather, we selected authors from different traditions whose work seemed to us most stimulating and innovative in *initiating* the process. Several of our chapters present controversial views that highlight the discrepancy between traditions, but it is our sincere hope that by encountering these views side by side, readers will realize the necessity of *careful* links between proximate and ultimate reasoning.

Our emphasis in selecting authors was to find work that stimulated us intellectually, rather than attempting a complete survey of the field of the sensory, behavioral, and cognitive processes involved in flower visitation, and its implications for floral evolution. The chapters are organized so that we move from the more zoological work to the more botanical (but naturally, a clear distinction is not possible). We start with several chapters on the implications of cognition, memory, and perception for pollinator foraging behavior, then move through several sensory modalities involved in flower detection and recognition (color vision, pattern vision, olfaction, echolocation), at the same time discussing several important classes of pollinators (bees, bats, birds, butterflies, etc.). This is followed by one general chapter on the importance of recognizing pollinators as individuals and another on the influences of predators on pollination systems. The last third of the book has a stronger emphasis on the consequences of pollinator behavior and cognition for the evolution of floral traits, covering frequency dependent selection, assortative mating, speciation, and the influences of floral traits on patterns of pollen movement.

In selecting the contributors for this volume, it was of particular importance for us not to be biased towards age and eminence. We thought that fresh ideas are likely to come from young scientists, and so many of our authors are still in the early stages of their careers. As we received the chapter manuscripts one by one, we became more and more enthusiastic, because we felt that this approach has worked out very well. If this book turns out to be a success, this is due in no small part to the energy and creativity of its contributors, their support in reviewing chapters by other authors, and their patience in dealing with endless suggestions for revisions by two rather censorious editors. We are also extremely grateful to the following external referees: John Allen, Elizabeth A. Bernays, Thomas S. Collett, Catherine L. Craig, Heidi Dobson, Robert Dudley, Ted H. Fleming, Lee Gass, Wayne Getz, Andreas Gumbert, Carlos Greco, Carlos M. Herrera, Katherine Hinman, Almut Kelber, Peter Kevan, Susan J. Mazer, Randall J. Mitchell, Douglass H. Morse, Daniel R. Papaj, Gene Robinson, Flavio Roces, Sharoni Shafir, Sara Shettleworth, and Paul Wilson. Tracey Sanderson and Sarah Jeffery, who handled our manuscript at Cambridge University Press, have been extremely helpful in guiding us through assembling this book. Its completion would have been impossible without Barbara Thomson's meticulous editorial help.

We sincerely hope that the efforts of all the individuals contributing to this work will foster new and innovative work on the interaction of animals and plants, and provide fruitful stimulation for all the biological traditions involved.

Lars Chittka, Würzburg
James D. Thomson, Toronto
August 2000

# 1

# The effect of variation among floral traits on the flower constancy of pollinators

The interaction between floral traits and pollinator behavior has been an important force in the coevolution of plants and their animal pollinators. An element of conflict underlies this interaction because the ideal behavior of the pollinator from the plant's point of view may often diverge from that dictated by the pollinator's own self-interest. Because of their immobility, outcrossed plants require a reliable courier that has a high probability of placing their pollen where it has a chance of fertilizing a conspecific ovule. Pollen finding an inappropriate stigma is effectively wasted, and deposition of heterospecific pollen may block receptive sites on the stigma and reduce seed set (e.g., Waser 1978, 1983; Thomson *et al.* 1981; Campbell & Motten 1985). Thus, plants should benefit if pollinators tend to move sequentially among flowers of the same species, a pattern that an optimally foraging pollinator should rarely adopt unless energetic returns from one plant species regularly exceed those from a mixed diet of some or all of the flower species available. More often, pollinators distribute themselves in an ideal free pattern across resources (Dreisig 1995), thereby minimizing differences in rewards among many different plant species, a pattern that should make generalist foraging the best option.

Yet pollinators often sequentially visit the flowers of one species even though they are bypassing flowers of other available, rewarding plant species (e.g., Grant 1950; Manning 1957; Free 1970; Waser 1983, 1986; Lewis 1989; Goulson & Cory 1993; Laverty 1994*b*). This "flower constant" foraging behavior has been described in many taxa, primarily honeybees (e.g., Wells & Wells 1986; Hill *et al.* 1997), bumble bees (e.g., Free 1970; Heinrich *et al.* 1977), and butterflies (Lewis 1986), but also more recently in solitary bees (Gross 1992), beetles (De Los Mozas Pascual & Domingo 1991), and dipterans (Goulson & Wright 1998).

Chittka *et al.* (1999) recently reviewed the many explanations that have been proposed to account for pollinator flower constancy and suggested that constancy probably has multiple causes. The most popular explanations for floral constancy invoke some sort of limitation on the cognitive abilities of pollinators to process, store, or recall information about multiple flower types at the same time (e.g., Waser 1983, 1986; Lewis 1993; Dukas 1998; Goulson 2000; but see Menzel, this volume). Two main hypotheses concerning the relation between flower constancy and the cognitive abilities of pollinators have been tested experimentally; these are considered below.

### Darwin's hypothesis

The first hypothesis arises from Darwin's (1876) widely quoted statement:

> That insects should visit the flowers of the same species for as long as they can is of great significance to the plant, as it favours cross-fertilization of distinct individuals of the same species; but no one will suppose that insects act in this manner for the good of the plant. The cause probably lies in the insects being thus enabled to work quicker; they have just learnt how to stand in the best position, and how far and in what direction to insert their proboscides.

Darwin's explanation implies that pollinators learn and remember the motor pattern or handling skill associated with flowers of a particular species. Bumble bees (Heinrich 1976, 1979; Laverty 1980, 1994*a*; Laverty & Plowright 1988) and butterflies (Lewis & Lipani 1990) are capable of learning and remembering a variety of different flower-handling skills. Motor patterns needed for working simple flowers with exposed nectar are learned quickly, while more complex flowers with concealed nectar are more difficult to learn. Waser (1983, 1986) interpreted Darwin's statement as meaning that pollinators are constant because they are limited in the number of handling skills that they can remember simultaneously. Lewis (1986) added the idea that learning additional flower-handling skills may interfere with a pollinator's ability to recall a previously learned handling skill. This combined hypothesis, which has been referred to as "Darwin's interference hypothesis" (Woodward & Laverty 1992), suggests that pollinators are constant to the flowers of one or a few plant species to minimize the costs of relearning flower-handling skills after every switch.

Attempts to test Darwin's hypothesis have looked for evidence of increased flower-handling times immediately following a switch between flower types requiring different handling skills in butterflies (Lewis 1986; Goulson *et al.* 1997) and bumble bees (Woodward & Laverty 1992; Laverty 1994*b*; Gegear & Laverty 1995, 1998). The most common design has been to train a test group of individual pollinators to work flowers of species A, then to switch them to learn a second flower type (B), before finally retesting them again on flower type A. If handling times during the retesting period are significantly greater than those recorded during initial training on flower type A, then the increase is attributed to some sort of negative effect from learning a new type, or simply forgetting the motor patterns associated with flower type A with the passage of time. To separate these two possible sources of reduced performance, a control group is run to assess the contribution of forgetting with the passage of time. The difference in retest handling times between the test and control groups is attributed to the effect of switching between flower types requiring different handling skills.

Does switching flowers increase handling times? The answer depends on several factors, including the difficulty and number of new flower-handling skills learned. Data from many studies that measured the increase in flower-handling time attributed to switching (shown as a percentage above the flower handling time for an experienced pollinator) are summarized in Fig. 1.1. Switching between two different handling skills generally involves an increase in the handling time for experienced foragers of only 0–2 s – or a 0%–100% increase in handling time – although one study of butterflies (Lewis 1986) found a 300% increase. In general, these elevated times are still 10–50 times lower than the handling time of naïve individuals learning the skill for the first time. Switching among three or more different flower-handling skills results in much longer handling times, approaching those for naïve individuals, especially if the additional flower types are difficult. These results suggest that switching may weaken or in some cases even erase from memory the motor patterns that bees have learned for handling flowers (see Chittka *et al.* 1999 for discussion).

One complication in interpreting many of these studies is that the animals tested made the switch among different flower handling methods only once. Dukas (1995) found that bumble bees switching among different foraging tasks for the first time showed a 22% reduction in choosing the correct flower color, but this reduction disappeared with practice. Moon (1999) recently demonstrated that bumble bees could

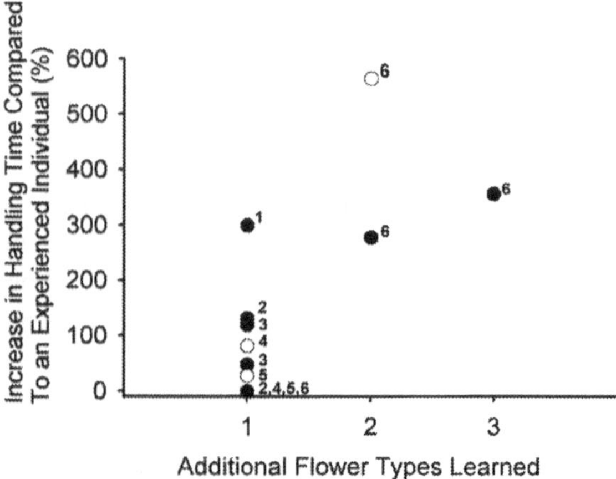

Fig. 1.1 Increase in handling times for flower type A following switches to additional flower types. Increase expressed relative to the mean handling time for experienced foragers. Circles denote the floral complexity of the additional flower type(s) learned (closed = simple; open = complex). Data taken from: (1) Lewis (1986); (2) Goulson *et al.* (1997); (3) Woodward & Laverty (1992); (4) Laverty (1994*b*); (5) Gegear & Laverty (1995); (6) Gegear & Laverty (1998).

learn to switch between two different flower-handling skills without penalty. For bees working three flower types, however, a residual cost of 1.1 s (a 52% increase in handling time) remained, despite repeated practice. These costs may seem small, but could add up over many foraging episodes in the life of an individual.

Are switching costs related to constancy? The key point here is to find a relationship between the magnitude of switching costs and the tendency to move sequentially among like flower types. The larger the penalty for switching, the more the forager would benefit from being constant, all else being equal. Although Lewis (1986) found strong constancy to a single plant species and long relearning times when butterflies switched species, studies with bumble bees have found no consistent link between the magnitude of switching costs and flower constancy (Woodward & Laverty 1992; Laverty 1994*b*; Gegear & Laverty 1998; Moon 1999).

### Search image hypothesis

The second hypothesis is based on the "search image" concept outlined by Tinbergen (1960). Tinbergen argued that in order to increase the efficiency of detecting one prey type, a predator performs "a highly selective sieving operation on the visual stimuli reaching the retina," forming a "searching image" for that prey type. It appears that the perceptual mechanism behind search image effects is related to the "run effect": animals tend to improve their performances by selecting "runs" of one prey type, even though other types are available (Bond & Riley 1991). For example, Pietrewicz & Kamil (1979) found that blue jays (*Cyanocitta cristata*) could find one species of cryptic moth better after practice runs of that species alone, but did not improve if runs involved alternation of two different moth species.

Although many authors have suggested that search images can account for flower constancy (e.g., Levin 1978; Waser 1986; Dukas & Real 1993; Wilson & Stine 1996; Goulson 2000), there is little empirical evidence for the formation of search images in pollinators. In field experiments with flower species that differed in both handling skill and color, Wilson & Stine (1996) found that individual bumble bees visited flowers of similar color but different handling skills (e.g., pink/purple-flowered red clover and self-heal), but not flowers differing in color but with the same handling method (e.g., purple-flowered vetch and white clover). They argued that bees were constant because they formed a search image based on flower color. Thus, as bees sequentially visited the flowers of one species, they became conditioned to the color of that flower, and subsequently tended to visit similarly-colored flowers, even of another plant species.

Most discussions of search image assume that the predator is searching for cryptic prey types (e.g., Dawkins 1971*a*, *b*; Bond & Riley 1991; Reid & Shettleworth 1992); however, Tinbergen did not explicitly state this condition as a component of the hypothesis. Can flowers be cryptic? Goulson (2000) recently proposed that flowers are effectively cryptic when viewed against a background of plant species with similarly colored flowers. In a field experiment, Goulson found that bumble bees took twice as long to find flowers of one yellow-flowered species in an area containing several other yellow-flowered species, compared with their rates in an area where the background flowers were not yellow. However, in this study only flight times to the next nearest flower were assessed, and no data were presented on the relationship between constancy and the floral background mix encountered by pollinators.

### Flower-handling skills or search image for floral features?

Many authors have suggested that hypotheses for flower constancy based on flower handling and those based on search image are not mutually exclusive (Waser 1986; Dukas & Real 1993; Wilson & Stine 1996; Dukas 1998; Goulson 2000). In fact, under natural foraging conditions it is difficult, if not impossible, to separate the two. Every flower has a particular motor pattern associated with a set of sensory cues, such as color, odor, size, and shape. Thus, it is probable that studies using real flowers to test Darwin's hypothesis (supposedly manipulating only flower-handling skill) or the search image hypothesis (supposedly assessing a single floral signal such as color) have been confounded by between-species differences in a variety of other floral traits besides the one of interest. For example, the increased flower-handling times following switches between two flower species observed by Laverty (1994*b*) and Lewis (1986) may have been affected by other uncontrolled floral differences besides handling method such as color and scent.

Similarly, in field experiments testing the search image hypothesis (such as Wilson & Stine 1996), it was concluded that visitors became conditioned to legume flowers of the same color (e.g., vetch and red clover) and ignored differently-colored flowers (white clover) that apparently had the same flower-handling method. However, in this example, corolla tube length may also have been a confounding factor (Laverty 1994*b*): white clover has a short tube relative to the other two legumes, and bees may display some degree of constancy to tube length. These points underscore the advantages of testing hypotheses about effects of floral traits on pollinator behavior under carefully controlled conditions where traits can reasonably be manipulated one at a time.

### Trait variability hypothesis

Given that both motor pattern and sensory stimuli are closely linked together (e.g. Chittka & Thomson 1997), perhaps it is time to take a more comprehensive approach to investigating the effect of floral traits on pollinator behavior. Instead of investigating pollinator choice patterns as a separate response to either handling skill or single traits such as flower color, selective foraging patterns such as flower constancy and learned preferences (an overall bias in choice towards some of available flower types) may be responses to variability over a wide range of floral traits. A key idea in this "trait variability hypothesis", as we have called it, is that pollinators are faced with two fundamentally different types of variation among flower types. First, there can be variation within a single trait (herein referred to as states of a trait). For example, three flower color morphs (blue, purple, or pink) would represent three states (blue, purple, and pink) of the trait (color). Second, flower types may also be defined by variation among several floral traits. For example, flowers from two plant species might show variation in both color (yellow and blue) and size (large and small).

Although the effect of variation within and among floral traits on pollinator behavior has not been previously compared in manipulative experiments, several studies have suggested that when floral characteristics such as color, shape, odor, and handling technique are more distinct, pollinators are more constant (e.g., Bateman 1951; Ostler & Harper 1978; Pleasants 1980; Waser 1986; Dukas & Ellner 1993). In addition, there are many reports that pollinators tend to be inconstant when flowers vary in only one trait, but become more selective if flowers differ in two or more traits (e.g., Waser 1983, 1986; Gross 1992; Gegear & Laverty 1998; Goulson & Wright 1998).

Increases in these two types of floral-trait variation may affect foraging behavior in very different ways if among-trait variability is more difficult than within-trait variability for pollinators to process, remember, or recall. Psychological experiments on humans and pigeons have found differences in responses to within- and among-trait variability, and these have been explained by a concept known as serial and parallel processing (Nakayama & Silverman 1986; Shettleworth 1998; see also Chittka *et al.* 1999). If a target type is presented simultaneously with other non-target types differing in just one feature (e.g., color), then it is processed in parallel with no reduction in the time taken to pick out this target type regardless of the number of other non-target types present. However, if the target is characterized by two or more variable features (e.g., color and shape), then information on each of the two features is processed serially, which takes longer and is less efficient. An analogous mechanism may account for floral constancy in pollinators. The trait variability hypothesis predicts that pollinators should exhibit floral constancy and preference when the number of variable traits increases, because information becomes more difficult to process. However, the same amount of variation in states within a single trait should be much easier for the pollinator to process and so would be expected to produce less selectivity.

We tested these predictions with naïve bumble bees (*Bombus impatiens*) in a series of 10 separate laboratory experiments in which we systematically manipulated the floral variability within and between floral traits in arrays of 80 artificial flowers. Following the methods used by Gegear & Laverty (1998), we first ran discrimination tests on pairwise combinations of the flower types used in each experiment to ensure that the bees recognized the different types. Flowers were constructed from colored Eppendorf centrifuge tubes (1.5 ml), so that floral states and traits could be easily manipulated in a standardized manner. In addition to color, we varied flower size (3 or 6 cm diameter collar around the entrance to the tube), and flower complexity by making it easy (cap on centrifuge tube removed) or difficult (cap blocking most of tube entrance) for bees to crawl into the tube. Table 1.1 summarizes the combinations of within- and among-trait variability, as well as the flower types used in each experiment.

Test bees were pretrained to complete foraging trips on several different pure arrays of single flower types, presented in random order, to ensure that they had fully learned to work each flower type. On the day of testing, each test bee completed one foraging trip on each flower type to be tested. Each bee was then tested on a mixed array of 80 flowers with equal numbers of each flower type arranged equidistantly and containing equal rewards (2 µl of 30% sucrose solution, refilled after each visit). The first 100 flowers visited by each bee in the mixed array were videotaped for later analysis. Separate groups of 10 bees were run in each experiment; 100 bees were tested in total. We measured these response variables: preference of each bee for a particular flower type on the mixed array (tested by comparing the total number of visits to each flower type using a $\chi^2$ test); and, flower constancy of each bee (comparing number of moves to flowers of the same type, "like–like" moves, to the number of moves between unlike flower types in the visit sequence on the mixed array using a $\chi^2$ test). In addition, we quantified flower constancy of all 10 bees in each experiment using Bateman's Index (Bateman 1951). This measure, ranging from $-1$ (inconstancy) to $+1$ (complete constancy), summarizes the tendency of foragers to move sequentially among the same flower types. We used regression analysis to test if the above three variables increased as the number of states within a single trait was increased from 2 to 4 (Experiments 1–7) and also as the number of variable traits was increased from 1 to 3 (Experiments 1–10).

Within experiments, there were no consistent biases in the flower types visited by bees but the percentage of bees showing selective behavior (preference and constancy) varied considerably among different

Table 1.1. *Floral traits, state of each trait and number of flower types used in Experiments 1–10.*

| Experiment | Traits varied | States varied | $N^a$ | Flower types in each experiment[b] |
|---|---|---|---|---|
| 1 | Complexity | Easy (open entry) Difficult (closed entry) | 2 |  |
| 2 | Color | Yellow Blue | 2 | |
| 3 | Size | Small (3 cm diameter) Large (6 cm diameter) | 2 |  |
| 4 | Color | Blue Orange Purple | 3 |  |
| 5 | Color | Yellow Blue Purple | 3 |  |
| 6 | Color | Yellow Blue Orange Purple | 4 |  |
| 7 | Complexity | Easy Difficult | 4 |  |
| 8 | Color | Blue Yellow | 4 |  |
|  | Complexity | Easy Difficult | | |
| 9 | Color | Blue Yellow | 4 | 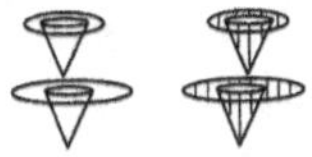 |
|  | Size | Large Small | | |
| 10 | Color | Blue Yellow | 8 | 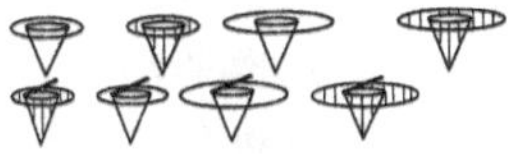 |
|  | Size | Large Small | | |
|  | Complexity | Easy Difficult | | |

*Notes:*
[a] Number of flower types in mixed array.
[b] Flower symbols illustrate different flower types. (Color: clear = blue; striped = yellow; hatched = orange; checkered = purple; Complexity: open = easy; closed = difficult; Size: small = 3 cm; large = 6 cm).

experiments. Bees increased their selectivity as the number of variable traits increased from 1 to 3 in different experiments (Fig. 1.2). All three response variables showed the same trends. As among-trait variability increased, more bees showed preferences for some of the flower types available (Fig. 1.2a, $F = 8.14$, df $= 9$, $p = 0.021$, $r^2 = 0.50$); also, our measure of constancy increased (Fig. 1.2b, $F = 15.88$, df $= 9$, $p = 0.004$, $r^2 = 0.66$) and Bateman's Index approached $+1$, indicating that moves among flowers of the same type were increasingly common (Fig. 1.2c, $F = 31.51$, df $= 9$, $p = 0.0005$, $r^2 = 0.80$).

In contrast, increased variation within states of a single trait (Fig. 1.2, 1–4 states of one trait) did not increase preference (Fig. 1.2a, $F = 0.0006$, df $= 6$, $p = 0.98$, $r^2 = 0.0001$), constancy (Fig. 1.2b, $F = 0.01$, df $= 6$, $p = 0.92$, $r^2 = 0.002$), or Bateman's Index ($F = 0.04$, df $= 6$, $p = 0.57$, $r^2 = 0.07$).

The observed increase in selective foraging behavior is not explained simply by increases in the number of different flower types learned – this is clear from the behavior of bees in experiments testing the same number of flower types. In arrays with 4 flower types, variation within a single trait (4 colors, Experiment 6; floral complexity, Experiment 7) provoked less constancy than the same amount of variation among traits (color and complexity, Experiment 8; color and size, Experiment 9; Fisher's exact test, $p < 0.05$).

Are these patterns of selective foraging behavior consistent with Darwin's hypothesis or the search image hypothesis? Darwin's idea predicts that bees should be constant only when there are costs associated with switching among flower types differing in handling methods. Because bumble bees experience negligible costs associated with switching between two different handling methods (Laverty 1994b; Gegear & Laverty 1998), bees were not expected to display constancy in any experiment. Although bees did forage randomly when flower types differed only in complexity, bees were constant when flowers varied in more than one trait. In addition, bees were constant in several experiments that did not involve variation in complexity. When bees are presented with multiple floral signals, the search image idea predicts that individuals should focus on one floral feature (e.g., yellow color) and restrict their visits to flowers with similar features. Some bees tended to show constancy to color in some experiments, but most did not do so in experiments in which color alone was variable. The observed patterns are not accounted for by either hypothesis.

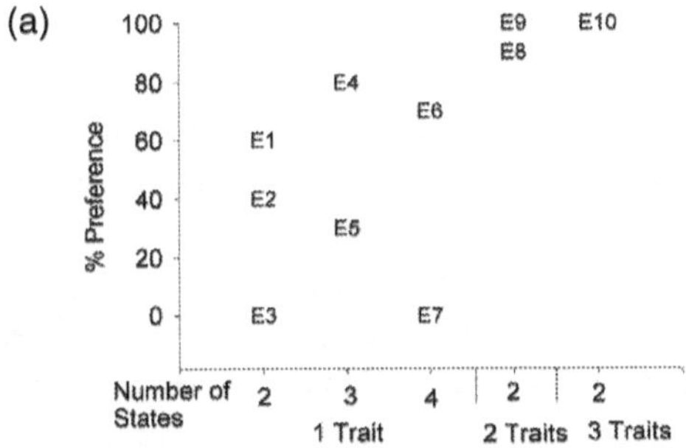


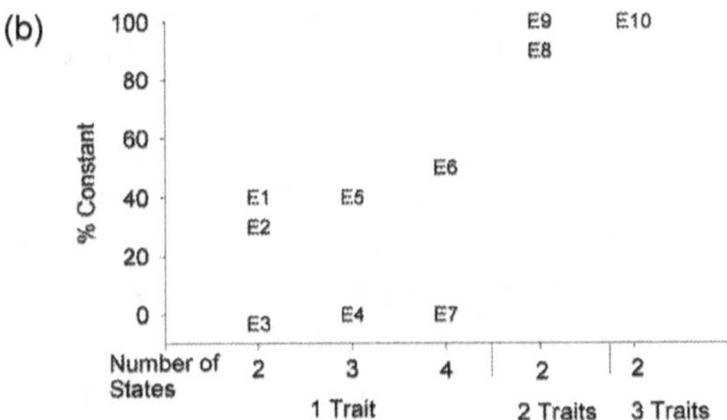


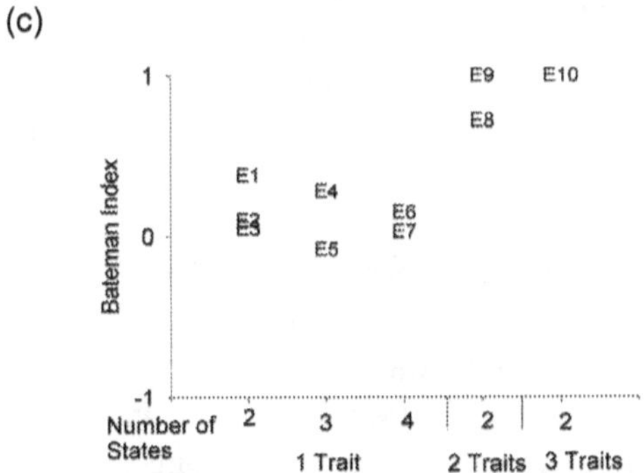


Fig. 1.2. Degree of pollinator (a) preference, (b) constancy (both expressed as a percentage of bees showing the behavior in each experiment), and (c) Bateman's Index values (based on pooled data for the 10 bees in each experiment) for bees foraging on mixed arrays varying within and between different floral traits. Refer to Table 1.1 for the flower state and traits that were manipulated in each experiment.

### Why are pollinators constant?

The tendency of pollinators to move sequentially among the flowers of the same species has attracted attention since Aristotle's observations of honeybee foraging behavior (Grant 1950). Why insects demonstrate this behavior still remains a mystery, and there are probably multiple factors that promote flower constancy (Chittka *et al.* 1999). In the simple experiments reported here, we were able to control many of the factors thought to be important in causing selective foraging behavior (e.g., flower spacing, abundance, tube depth, floral complexity, reward quality and quantity, and incomplete information about those rewards on the part of the pollinator). Under these standardized conditions, we were able to induce both constancy and preference in foragers by increasing the variation among floral traits. What general mechanisms could account for these observations?

All of the hypotheses considered here assume that pollinators are limited in their ability to process, store, or recall information about flowers (for reviews of bee learning see Menzel *et al.* 1974, 1993; Dukas & Real 1993; Gould 1993; Menzel, this volume). As flowers become more and more dissimilar in display, pollinators should become more efficient if they selectively attend to one or a few key traits on which to base decisions, while ignoring variation in other traits (e.g., Manning 1957; Dukas & Ellner 1993; Dukas & Waser 1994; Wilson & Stine 1996). Although this mechanism is plausible, it does not account for the observation that among-trait variation was more effective in promoting constancy and preference than within-trait variation. Many experiments have suggested that when bees displayed constancy and preference, the selectivity was almost invariably based on flower color. But does this mean that they ignored variation in the other traits besides color? In our experiments, bees sometimes formed secondary preferences based on flower size or complexity or both of these traits together; therefore, these bees did process information on traits besides color.

Once information on floral signals associated with rewarding flowers is stored in short-term memory, it may enter long-term memory through consolidation (Menzel *et al.* 1974; see Menzel, this volume). Honeybees can learn to associate several colors with reward (Menzel 1969), but if a flower type has not been visited for a critical amount of time, the probability that the next visit will be to the same flower type diminishes (Greggers & Menzel 1993). Studies of honeybee choices on artificial flowers (Marden & Waddington 1981; Greggers & Menzel 1993) and bumble bee foragers in the field (Chittka *et al.* 1997) suggest that whether a bee is constant or switches to a different flower type may depend partly on the time (distance) after departing a particular type until the forager encounters another target flower of the same type. If a bee encounters a match with its current flower target within a short time window (3 s of flight), it has a high probability of visiting the same flower type and constancy will be observed; if no matching target is encountered within 4–5 s, a switch of flower types becomes much more likely. These findings suggest that bees are searching for a remembered image as they leave a flower, but that this image decays within a few seconds if the forager does not detect a similar target. This may explain why honeybees are notoriously constant, and why variation within a single trait like color often seems to be sufficient to induce constancy to one color on closely spaced arrays of artificial flowers (e.g., Wells *et al.* 1983, Wells & Wells 1985), even if alternate colors had a greater reward (Hill *et al.* 1997). This sort of mechanism fits less well with observations of bumble bees foraging on mixed arrays. They have been reported to switch among flower types differing in single traits such as color, size, shape, and scents, whereas honeybees on the same arrays were constant to one flower type (e.g., Manning 1957). In experiments discussed here, most bumble bees foraged randomly on arrays with variation within single traits. Because flowers were close together in arrays (within 10 cm), foragers would typically have encountered a matching target well within the 3 s window, so it is difficult to reconcile the observed behavior with the above target-matching rule. Comparative studies of both honeybees and bumble bees foraging on the same experimental arrays confirm that there are consistent differences in their learning abilities and foraging patterns (R. J. Gegear & T. M. Laverty, unpublished data).

## Selective foraging behavior in pollinators: implications for floral diversity

An obvious characteristic of many natural plant assemblages is the astounding variation of flower types differing in color, shape, scent, size, complexity, etc. Traditionally, this diversity has been seen as driven by the advantages of floral specialization associated with distinct pollinator groups, a process giving rise to "syndromes of pollination" as species diverge through adaptation to the sensory and morphological features of their most effective pollinators (Stebbins 1970; Faegri & van der Pijl 1971; Proctor & Yeo 1973). Paradoxically, however, floral diversity remains high even when most plants in a community are pollinated by the same pollinator group (e.g. Heinrich 1975). Furthermore, field observations usually indicate that most plant species receive visits from a variety of pollinator species and *vice versa* (Heinrich 1975, 1976; Waser 1983, 1998; Ollerton 1996; Chittka *et al.*, this volume).

An additional explanation is that floral diversity is primarily a means of promoting selectivity by individual pollinators in their choices of flowers, thereby increasing the efficiency of pollination (Heinrich 1975). In this view, floral diversity may not necessarily represent adaptation to specific pollinator species, but rather could be largely an incidental outcome of the typical behavioral response of individual pollinators to variation in floral traits. How do pollinators respond to floral variation? Our results suggest that, other things being equal, floral variation within a single trait should be less effective at promoting selective foraging behavior in individual pollinators than variability in two or more floral traits. It is tempting to consider these trends with reference to divergence among co-flowering plants in the same area. One of the key components in sympatric divergence is the effectiveness of natural floral variation and pollinator selectivity as a potential isolating mechanism (Chittka *et al.* 1999). These topics are reviewed below.

### Intraspecific variation

Intraspecific floral variation in single traits, particularly in flower color, has been well documented (Kay 1978). In some cases, pollinators seem indifferent to the variation (e.g., Darwin 1876; Manning 1957; Waser 1983; Goulson & Wright 1998). However, other studies report preferential foraging towards different color morphs (e.g., Levin 1972; Kay 1976; Heinrich *et al.* 1977; Waser & Price 1981; Stanton 1987; see Smithson, this volume) or scent morphs (Galen & Kevan 1980; Galen 1985) of the same species. Preferences for particular morphs may vary from site to site, among different pollinator groups, and at different times of the year. Overall, there seems to be no consistent pattern in these studies, and it is unlikely that such variation could lead to reproductive isolation of different morphs (Waser, this volume). Possibly, examples that have documented selective foraging behavior on different morphs of a single floral trait have overlooked less obvious variation in other traits besides the one of interest, but this needs to be examined in future studies.

### Interspecific variation

Much evidence suggests that pollinators do become more selective as flowers become more dissimilar in their traits (e.g., Grant 1950, 1994; Pleasants 1980; Waser 1986; Chittka *et al.*, this volume; Jones, this volume). Examples of cases where flowers differ in two or more floral traits often appear to involve separate races or species. When hybridization is rare despite interfertility of floral forms, workers have often identified at least two traits that supposedly account for the behavioral selectivity shown by the pollinators (e.g., Mather 1947; Grant 1950; McNaughton & Harper 1960; Levin 1972; Jones 1978). Bradshaw *et al.* (1995) reported that reproductive isolation in two interfertile species of monkeyflowers (the *Mimulus lewisii–cardinalis* complex) was likely based on differences in quantitative trait loci for eight floral traits, including color, nectar reward, and flower shape. Hybrids between these two species are never found in the wild, perhaps because the two combinations of traits are each preferred by bumble bees or hummingbirds (Schemske & Bradshaw 1999). However, since both types of pollinators have been recorded on each species (Sutherland & Vickery 1993; Waser 1998), the isolation may also be explained by strong constancy shown by individuals visiting flowers differing in multiple traits. This explanation is consistent with observations of bumble bees visiting mixed arrays of both species: individual bees were constant to flowers of one species or the other at a time (R. J. Gegear & T. M. Laverty, unpublished data). More recently, Stout *et al.* (1998) tested bees on arrays of pairwise combinations of flowers differing in their floral complexity. Bees tended to be more constant to the flowers in the array if the flowers also differed in other traits (such as shape and size) besides the traits (handling method) that were manipulated.

Collectively, these studies support the idea that pollinators become more selective when flower types differ in multiple traits, and that assortative movements of individual pollinators could potentially provide effective isolation of cross-fertile forms (see Jones, this volume). Future studies should examine the importance of floral-trait covariation on the selective behavior of pollinators, and also the genetic mechanisms governing the expression of floral traits.

## Floral diversity in communities

An attractive proposition is that pollinator behavior, through the benefits of constancy, has selected for divergence of floral traits among co-occurring outcrossed plants. Plant species that competed with each other because they shared pollinators that were inconstant could be "moved" by natural selection to a more isolated location (phenotype) in the space defined by floral traits and sensorimotor learning capabilities of pollinators. This floral-trait niche could represent many dimensions, as long as they interacted to influence sensorimotor learning. Gumbert *et al.* (1999) recently asked whether flower colors (as defined according to properties of bee color vision) of co-flowering species showed evidence of divergent structure, compared to a random model. At two of five sites, rare plants were more distinct than expected by chance, but common plants at all sites had flower colors that were not distinguishable from chance. Though many reasons could account for not finding strong evidence of divergence in flower colors (see review by Chittka *et al.* 1999), it is also possible that divergence would not be evident within a single floral trait. Rather, co-flowering species may be distinct when viewed over several floral traits (e.g., color, scent, complexity) because variation in several traits seems most effective in promoting a strong constancy response in pollinators.

Studies looking for structure in floral signals have often focused on single traits such as color, but there is a least one data set that examined community-level patterns of several floral traits among outcrossed species. Ostler & Harper (1978) analyzed floral features of co-occurring (not necessarily co-flowering) plant species in 25 plant communities. Floral-trait diversity was strongly correlated with the number of co-occurring outcrossed species. Flower-color diversity (as assessed by human eyes) in 14 open communities was positively associated with the number of co-occurring species. More important, other floral traits associated with flower-handling methods also showed the same trends. The frequency of restrictive corolla tubes and flowers with bilateral symmetry (which require more elaborate flower-handling techniques) both increased with the diversity of animal-pollinated flowers. That both these traits show the same trend suggests that co-occurring plants are isolated in sensorimotor space. Not only do they vary in several traits, they vary in combinations of traits that are well suited to induce flower selectivity in individual pollinators. Multivariate analyses of floral-trait diversity in plant communities may detect non-randomness that would not necessarily be evident from analyses of a single floral trait such as color.

## Acknowledgements

We thank Lars Chittka, Reuven Dukas, James Thomson, and an anonymous reviewer for their helpful comments on the manuscript. This work was supported by an NSERC operating grant to T.M. Laverty.

## References

Bateman AJ (1951) The taxonomic discrimination of bees. *Heredity* 5:271–278

Bond AB & Riley DA (1991) Searching image in the pigeon: a test of three hypothetical mechanisms. *Ethology* 87:203–224

Bradshaw HDJ, Wilbert SM, Otto KG & Schemske DW (1995) Genetic mapping of floral traits associated with reproductive isolation in monkeyflowers (*Mimulus*). *Nature* 376:762–765

Campbell DR & Motten AF (1985) The mechanism of competition for pollination between two forest herbs. *Ecology* 66:554–563

Chittka L & Thomson JD (1997) Sensory-motor learning and its relevance for task specialization in bumble bees. *Behav Ecol Sociobiol* 41:385–398

Chittka L, Gumbert A & Kunze J (1997) Foraging dynamics of bumblebees: correlates of movements between and within plant species. *Behav Ecol* 8:239–249

Chittka L, Thomson JD & Waser NM (1999) Flower constancy, insect psychology, and plant evolution. *Naturwiss* 86:361–377

Darwin C (1876) *On the effects of cross and self-fertilization in the vegetable kingdom*. John Murray, London

Dawkins M (1971a) Perceptual changes in chicks: another look at the 'search image' concept. *Anim Behav* 19:566–574

Dawkins M (1971b) Shifts of "attention" in chicks during feeding. *Anim Behav* 19:575–582

De Los Mozos Pascual M & Domingo LM (1991) Flower constancy in *Heliotaurus ruficollis* (Fabricius 1781), Coleoptera, Alleculidae. *Elytron (Barc)* 5:9–12

Dreisig H (1995) Ideal free distribution of foraging bumblebees. *Oikos* 72:161–172

Dukas R (1995) Transfer and interference learning in bumble bees. *Anim Behav* 49:1481–1490

Dukas R (1998) Constraints on information processing and their effects on behavior. In: Dukas R (ed) *Cognitive ecology*, pp 89–128. University of Chicago Press, Chicago

Dukas R & Ellner S (1993) Information processing and prey detection. *Ecology* 74:1337–1346

Dukas R & Real LA (1993) Cognition in bees: from stimulus reception to behavioral change. In: Papaj DR & Lewis AC (eds) *Insect learning: ecological and evolutionary perspectives*, pp 343–374. Chapman & Hall, New York

Dukas R & Waser NM (1994) Categorization of food types enhances foraging performance of bumblebees. *Anim Behav* 48:637–644

Faegri K & van der Pijl L (1971) *Principles of pollination ecology*, 2nd edn. Pergamon Press, Oxford

Free JB (1970) The flower constancy of bumble bees. *J Anim Ecol* 39:395–402

Galen C (1985) Regulation of seed set in *Polemonium viscosum*: floral scents, pollination and resources. *Ecology* 66:792–797

Galen C & Kevan PG (1980) Scent and color, floral polymorphisms and pollination biology in *Polemonium viscosum* Nutt. *Am Midl Natur* 104:281–289

Gegear RJ & Laverty TM (1995) Effect of flower complexity on relearning flower handling skills in bumble bees. *Can J Zool* 73:2052–2058

Gegear RJ & Laverty TM (1998) How many flower types can bumble bees forage on at the same time? *Can J Zool* 76:1358–1365

Gould JL (1993) Ethological and comparative perspectives on honey bee learning. In: Papaj DR & Lewis AC (eds) *Insect learning: ecological and evolutionary perspectives*, pp 18–50. Chapman & Hall, New York

Goulson D (2000) Are insects flower constant because they use search images to find flowers? *Oikos* 88:547–552

Goulson D & Cory JS (1993) Flower constancy and learning in the foraging behavior of the green-veined white butterfly, *Pieris napi*. *Ecol Entomol* 18:315–320

Goulson D & Wright NP (1998) Flower constancy in the hoverflies *Episyrphus balteatus* (Degeer) and *Syrphus ribesii* (L.) (Syrphidae). *Behav Ecol* 9:213–219

Goulson D, Stout JC & Hawson SA (1997) Can flower constancy in nectaring butterflies be explained by Darwin's interference hypothesis? *Oecologia* 112:225–231

Grant V (1950) The flower constancy of bees. *Bot Rev* 16:379–398

Grant V (1994) Modes and origins of mechanical and ethological isolation in angiosperms. *Proc Natl Acad Sci USA* 91:3–10

Greggers U & Menzel R (1993) Memory dynamics and foraging strategies of honey bees. *Behav Ecol Sociobiol* 32:17–29

Gross CL (1992) Floral traits and pollinator constancy: foraging by native bees among three sympatric legumes. *Austral J Ecol* 17:67–74

Gumbert A, Kunze J & Chittka L (1999) Floral colour diversity in plant communities, bee colour space and a null model. *Proc R Soc Lond B* 272:1711–1716

Heinrich B (1975) Bee flowers: a hypothesis on flower variety and blooming time. *Evolution* 29:325–334

Heinrich B (1976) The foraging specializations of individual bumblebees. *Ecol Monogr* 46:105–128

Heinrich B (1979) "Majoring" and "minoring" by foraging bumblebees, *Bombus vagans*: an experimental analysis. *Ecology* 60:245–255

Heinrich B, Mudge PR & Deringis PG (1977) Laboratory analysis of flower constancy in foraging bumblebees: *Bombus ternarius* and *B. terricola*. *Behav Ecol Sociobiol* 2:247–265

Hill PSM, Wells PH & Wells H (1997) Spontaneous flower constancy and learning in honey bees as a function of color. *Anim Behav* 54:615–627

Jones CE (1978) Pollinator constancy as a pre-pollination isolating mechanism between sympatric species of *Cercidium*. *Evolution* 32:189–198

Kay QQN (1976) Preferential pollination of yellow-flowered morphs of *Raphanus raphanistrum* by *Pieris* and *Eristalis* spp. *Nature* 261:230–232

Kay QQN (1978) The role of preferential and assortative pollination in the maintenance of flower color polymorphisms. In: Richards AJ (ed) *The pollination of flowers by insects*, pp 175–190. Academic Press, London

Laverty TM (1980) The flower-visiting behavior of bumble bees: floral complexity and learning. *Can J Zool* 58:1324–1335

Laverty TM (1994a) Bumble bee learning and flower morphology. *Anim Behav* 47:531–545

Laverty TM (1994b) Costs to foraging bumble bees to switching plant species. *Can J Zool* 72:43–47

Laverty TM & Plowright RC (1988) Flower handling by bumble bees: a comparison of specialists and generalists. *Anim Behav* 36:733–740

Levin DA (1972) The adaptiveness of corolla-color variants in experimental and natural populations of *Phlox drummondii. Am Nat* 106:57–70

Levin DA (1978) Pollinator behavior and the breeding structure of plant populations. In: Richards AJ (ed) *The pollination of flowers by insects*, pp 133–150. Academic Press, London

Lewis AC (1986) Memory constraints and flower choice in *Pieris rapae. Science* 232:863–865

Lewis AC (1989) Flower visit constancy in *Pieris rapae*, the cabbage butterfly. *J Anim Ecol* 58:1–13

Lewis AC (1993) Learning and the evolution of resources: pollinators and flower morphology. In: Papaj DR & Lewis AC (eds) *Insect learning: ecological and evolutionary perspectives*, pp 219–242. Chapman & Hall, New York

Lewis AC & Lipani GA (1990) Learning and flower use in butterflies: hypotheses from honey bees. In: Bernays EA (ed) *Insect–plant interactions*, pp 95–110. CRC Press, Boca Raton, FL

Manning A (1957) Some evolutionary aspects of the flower constancy of bees. *Proc R Phys Soc Edinb* 25:67–71

Marden JH & Waddington KD (1981) Floral choices by honeybees in relation to the relative distances to flowers. *Physiol Entomol* 6:431–435

Mather K (1947) Species crosses in *Antirrhinum*. I. Genetic isolation of the species *majus, glutinosum* and *orontium. Heredity* 1:175–186

McNaughton IH & Harper JL (1960) The comparative biology of closely related species living in the same area. I. External breeding barriers between *Papever* species. *New Phytol* 59:15–26

Menzel R (1969) Das Gedaechtnis der Honigbiene für Spektralfarben. II. Umlernen und mehrfachlernen. *Z vergl Physiol* 63:290–309

Menzel R, Erber J & Mashur J (1974) Learning and memory in the honey bee. In: Browne LB (ed) *Experimental analysis of insect behavior*, pp 195–217. Springer-Verlag, New York

Menzel R, Greggers U & Hammer M (1993) Functional organization of appetitive learning and memory in a generalist pollinator, the honey bee. In: Papaj DR & Lewis AC (eds) *Insect learning: ecological and evolutionary perspectives*, pp 79–125. Chapman & Hall, New York

Moon PDK (1999) Effects of practice on the recall of flower handling skills and foraging behavior in bumble bees (*Bombus impatiens*). MSc thesis, University of Western Ontario, London, Canada

Nakayama K & Silverman GH (1986) Serial and parallel processing of visual feature conjunctions. *Nature* 320:264–265

Ollerton J (1996) Reconciling ecological processes with phylogenetic patterns: the apparent paradox of plant–pollinator systems. *J Ecol* 84:550–560

Ostler WK & Harper KT (1978) Floral ecology in relation to plant species diversity in the Wasatch Mountains of Utah and Idaho. *Ecology* 59:848–861

Pietrewicz AT & Kamil AC (1979) Search image formation in the blue jay (*Cyanocitta cristata*). *Science* 204: 1332–1333.

Pleasants JM (1980) Competition for bumblebee pollinators in Rocky Mountain plant communities. *Ecology* 61:1446–1459

Proctor M & Yeo P (1973) *The pollination of flowers*. Collins, London

Reid PJ & Shettleworth SJ (1992) Detection of cryptic prey: search image or search rate? *J Exp Psych: Anim Behav Proc* 18:273–286

Schemske DW & Bradshaw HD (1999) Pollinator preference and the evolution of floral traits in monkeyflowers (*Mimulus*). *Proc Natl Acad Sci USA* 96:11910–11915

Shettleworth SJ (1998) *Cognition, evolution, and behavior*. Oxford University Press, Oxford

Stanton ML (1987) Reproductive biology of petal color variants in wild populations of *Raphanus sativus*. I. Pollinator response to color morphs. *Am J Bot* 74:178–187

Stebbins GL (1970) Adaptive reproduction of reproductive characteristics in angiosperms. I. Pollination mechanisms. *Ann Rev Ecol Syst* 1:307–326

Stout JC, Allen JA & Goulson D (1998) The influence of relative plant density and floral morphological complexity on the behaviour of bumblebees. *Oecologia* 17:543–550

Sutherland SD & Vickery RK Jr (1993) On the relative importance of floral color, shape and nectar rewards in attracting pollinators to *Mimulus*. *Great Basin Nat* 53:107–117

Thomson JD, Andrews BJ & Plowright RC (1981) The effect of a foreign pollen on ovule development in *Diervilla lonicera* (Caprifoliaceae). *New Phytol* 90:777–783

Tinbergen L (1960) The natural control of insects in pine woods. I. Factors influencing the intensity of predation by songbirds. *Arch Neerland Zool* 13:265–343

Waser NM (1978) Interspecific pollen transfer and competition between co-occurring plant species. *Oecologia* 36:223–236

Waser NM (1983) The adaptive nature of flower traits: ideas and evidence. In: Real LA (ed) *Pollination biology*, pp 241–285. Academic Press, Orlando, FL

Waser NM (1986) Flower constancy: definition, cause and measurement. *Am Nat* 127:593–603

Waser NM (1998) Pollination, angiosperm speciation, and the nature of species boundaries. *Oikos* 81:198–201

Waser NM & Price MV (1981) Pollinator choice and stabilizing selection for flower color in *Delphinium nelsonii*. *Evolution* 35:376–390

Wells PH & Wells H (1985) Ethological isolation of plants. 2. Odour selection by honeybees. *J Apic Res* 24:86–92

Wells PH & Wells H (1986) Optimal diet, minimal uncertainty and individual constancy in the foraging of honey bees, *Apis mellifera*. *J Anim Ecol* 55:375–384

Wells H, Wells PH & Smith DM (1983) Ethological isolation of plants. 1. Colour selection by honeybees. *J Apic Res* 22:33–44

Wilson P & Stine M (1996) Floral constancy in bumble bees: handling efficiency or perceptual conditioning? *Oecologia* 106:493–499

Woodward GL & Laverty TM (1992) Recall of flower handling skills by bumble bees: a test of Darwin's interference hypothesis. *Anim Behav* 44:1045–1051

2

# Behavioral and neural mechanisms of learning and memory as determinants of flower constancy

Flowers are unreliable, widely distributed food sources, normally offering minute rewards. Flowers of the same kind tend to bloom in close proximity, because plants of the same species growing in patches often bloom simultaneously, or a single plant has many blossoms. Thus, a patch of flowers of the same kind has a location in space and exists for some time, perhaps longer than the lifespan of an insect pollinator. A typical habitat consists of several to many patches of flowers, some of the same species, some of others; pollinators must choose between them.

Hymenopteran pollinators visit flowers to provide food for themselves and their brood. They frequently travel long distances between the nest site and the flower patches, carrying pollen and nectar. Since they must visit many flowers per foraging bout, they need to decide between different flowers in quick succession. Both innate preferences and experience guide the decision-making process (Menzel 1985). Since most of the approach flights are either return visits to a plant or first visits to nearby ones, pollinators are guided mainly by their memories of the location of productive flowers and the particular features of the flowers (signals, manipulatory properties, reward conditions) that the insects learned during previous visits.

Insects' learning capacity and richness of memories are usually underestimated, but studies of learning and memory in honeybees (under both natural and laboratory conditions) demonstrate that learning is fast, and comprises various levels of cognitive processing, such as generalization, categorization, concept formation, configuration, and context-dependency (Menzel & Giurfa 2001). Memory is rich, highly dynamic, and long-lasting (Menzel 1999).

Here I take up the case for the decision-making process being guided by navigational memories and memories of signals from the flowers themselves. Specifically, I shall argue that the components of the pollinators' navigational memory are intimately connected with memories of the flower signals, leading to a unique neural representation of localized and qualified objects (nest site, feeding places with particular properties, landmarks passed, etc.). Patches of flowers are localized in space, and bees navigate between loci in a goal-directed fashion. They establish locus-specific memories, and thus their navigational capacities are a major component in returning to a flower, identifying it as a productive one, and handling it efficiently.

Learning all these features of a flower – location, signals, construction – establishes composite memories, whose impact on choice behavior is continuously updated, both with reference to new experience and to elapsing time. Most importantly, memory is not a unique and stable entity of information storage – not in bees nor in any other animal (Milner *et al.* 1998) – but rather a dynamic process establishes different and sequential forms of memory phases, which are then transformed by consolidation processes. The central argument put forward here is that the contents and dynamics of the memory phases are the major factors controlling choice behavior, and thus flower constancy.

Our case study is the honeybee. There is no reason to believe that the honeybee is in any way special in its cognitive capacities, because the main requirement, namely, goal-directed navigation between nest site and feeding places, must be met by any hymenopteran pollinator species. In this sense, honeybees can be studied as a representative species of hymenopteran pollinators, including both social and solitary bees.

### Localization on a rough scale: the structure of navigational memories

Foraging bees embark on feeding flights and return to the hive using sun compass information (von Frisch 1965; Wehner & Menzel 1990), visual distance estimation (Esch & Burns 1995; Srinivasan *et al.* 1996), path integration (Wehner 1992), and visual landmarks (von Frisch 1965; Menzel *et al.* 1996; Collett & Zeil 1998). These sources of information are tightly interconnected: compass directions are derived from both extended landmarks (von Frisch 1965, Dyer & Gould 1981) and from home vectors associated with local landmarks (Menzel *et al.* 1998), establishing a memory for the flight route between the hive and a frequently visited feeding site.

Besides this specialized route memory (SRM), bees leaving the hive for the first time do not fly straight to a feeding site, but, rather, perform elaborate orientation and learning flights (von Frisch 1965) of increasing duration and distance (Becker 1958; Vollbehr 1975; Capaldi & Dyer 1999; Capaldi *et al.* 2000). It was recently shown (Menzel *et al.* 2000) that the spatial memory established during these learning flights leads to a general landscape memory (GLM), coding and storing the layout of landmarks in a geometric sense with the hive at the center.

To uncover the structure of GLM, it was necessary to perform experiments in which bees were not trained along a route, because route-trained animals apply the vector memory for the route when released at an unexpected place, thereby giving the impression that they are unable to localize the release site relative to the intended goal (Wehner & Menzel 1990). We therefore tested bees that had not established a SRM. We chose a moving feeder (called the "variable feeder") which circled around the hive at close range (5–10 m) several times a day. Bees trained under these conditions are called V-bees. These V-bees were compared to bees trained in the usual way to a constant feeder at a greater distance to the hive (C-bees). Both groups of bees were released at five sites 350 m away from the hive. For C-bees one of these sites was their familiar training site. To measure navigational performance, both vanishing bearings at the release site and flight time were recorded.

It was found that C-bees follow their compass memory at the release site as expected. They take a long time to return to the hive, particularly when the initial flight route carries them further away from the hive, but eventually all of the bees arrived at the hive, indicating that they refer to some other form of spatial memory when their active memory about the flight vector has vanished. V-bees, on the other hand, showed a weak tendency to fly into the 180° sector toward the hive from any of the five release sites and, most importantly, arrived at the hive after a brief flight time, a flight time that was not significantly longer than the flight time of C-bees along their trained route (Menzel *et al.* 2000).

These results indicate that bees do indeed possess a form of geometric representation of the landmark layout when they refer to GLM, but not when they refer to SRM. Since no natural feeding spots were available during the test period, bees must have established GLM during their orientation flights. GLM is suppressed by SRM, as indicated by the fact that the C-bees first follow their active flight memory; however, SRM does not erase GLM, since, if SRM did not lead back to the hive, C-bees were able to activate GLM from a remote store and use it for navigating. Otherwise we would not have observed C-bees returning to the hive.

The map-like organization of GLM proves a hitherto unexpected dimension of navigational capacity in a pollinating insect. Using the harmonic radar tracking technique (Riley *et al.* 1996), we have recently shown that bees referring to GLM do not only return to the hive on direct flights over distances of several hundreds of meters, but may also choose to fly to a feeding site first. This indicates that the structure of GLM is not confined to spatial relationships between the central spot (hive) and landmarks, but, rather, any location within GLM can be chosen as a goal from any other location. The neural structure of GLM might be that of a map-like representation of the landscape and thus indicative for a "cognitive or mental map". Such a claim has been made by Gould (1986) on the basis of vanishing bearings of C-bees taken at the release site. Gould's observations could not be verified in any of our studies or those of other researchers (Wehner & Menzel 1990): route-trained animals always applied their SRM and flew in the wrong direction. The vanishing bearings at the release sites were the only data available to Gould, and it is still unclear how he arrived at the conclusion that bees refer to geometric structure of spatial memory.

Bees forage in a known landscape whose geometric structure is stored in their spatial memory. The locations of rewarding sites are characterized by their particular features and are memorized accordingly. Bees learn the local features (signals, localization relative to landmarks, reward conditions) of two to four feeding sites, and behave accordingly: they choose the correct color at the correct time and place (Menzel *et al.* 1999; Lehrer 1999) or the correct color pattern at the correct step in a sequence (Collett 1992); they choose the correct odor at a particular time (Koltermann 1971); they indicate the correct direction and distance to one of two feeding sites according to time of the day (von Frisch 1965, table 37); and, they match the frequency of their visits to the reward quantities of at least four feeding sites (Greggers & Menzel 1993).

Furthermore, bees have the capacity to switch their motivation according to recent experience and activate remote memory according to the motivational change. Take the following experiment as an example of the flexible use of location-related information. Bees were trained to two sites, one in the morning and one in the afternoon. When captured in the morning at the hive heading out to the feeding site and released at the afternoon site, or captured in the afternoon heading out to the afternoon feeding site and released at the morning site, they flew back directly to the hive from either site, indicating that the landmarks characterizing each site are able to retrieve a remote memory (here, the homeward flight vector; Menzel *et al.* 1996). Furthermore, when bees were released halfway between the morning and afternoon sites, at a site that resembled landmark constellations characteristic of both the morning and the afternoon sites, 50% of them flew directly toward the hive, a flight direction that they did not show at any other site and that must have resulted from the retrieval of both site-specific memories.

When these data were published, the complexity of the integration process was enigmatic, and we argued that it might be explained as an automatic process of path integration on a large scale, or as a sensorimotor routine of fast sequential reference to the landmark constellations, or as an integration process at the level of two separate memories. On the basis of the results reported above on the use of SRM and GLM we can now interpret these results more specifically. Since the bees were tested at a moment when GLM should still have been depressed by the dominant SRMs established at the two feeding sites, the results also indicate a flexible use of SRMs, and an integration of such memories if more than one is activated. Under such conditions it may also be possible that rivaling SRMs decrease their control over flight behavior, so that GLM is no longer depressed by the dominant SRM. In such a case, the novel flight direction of the bees may indicate a reference to GLM, and in that case they would have localized their release site and steered toward the intended goal (the hive) along the shorter route.

The role of a flower's location for finding and choosing it again in a foraging trip could be a function of the structure of the landscape, the kinds of landmarks, and the vegetation density. Plants flowering in a landscape with dense subtropical and temperate vegetation tend to appear in closer patches, and these patches are thus less well-characterized by their different location in the general landscape memory (GLM). Such flowers may need to provide signals that allow spotting a patch over greater distances. The achromatic signal produced by the green receptors of the compound eye allows detection over further distances than that produced by the color vision system (Giurfa & Lehrer this volume). We thus asked how achromatic and color signals of flowers are correlated with habitat structure. In a comparison of visual signals provided by Mediterranean and desert flowers in Israel, Menzel *et al.* (1997) found that the achromatic signal is more pronounced in the densely grouped Mediterranean flowers than in the sparsely distributed desert flowers, whereas the color signal does not differ between the species in these two habitats. It is possible that bees mainly use their spatial memory to spot sparsely growing desert plants. Desert plants may thus rely less on their own green-contrast signals for the intermediate range of detection than densely blooming plants in the Mediterranean habitat do. The color signals of both kinds of plants should depend less on habitat features, because this signal may be needed for the flying insect's proper posture when approaching the flower for fast and effective handling, irrespective of how the plant was spotted. And indeed the color signals are not different in these two habitats. The color signal, together with the shape and pattern, may also more reliably indicate the nutritional status of the flower, a feature that should also be independent of the habitat.

This interpretation is based on two arguments. (1) Plants are the evolutionarily adaptive units, whereas the habitat's features are the constraints. Flowers are selected to be repeatedly located, identified, and recognized within the conditions provided by the habitat. If the habitat allows easy localization (e.g., in the desert because of the low growth density), the pollinator's navigational system may need less support from further-ranging flower signals (e.g., the green signal). (2) Spatial memory is intimately connected with associative learning processes at the feeding site. It is, therefore, likely that the different sets of external stimuli to which these navigational tasks refer are elements of a rich spatial memory with "qualified" and localized components. The "qualification" relates to the localization in the GLM and the goal's specific features (e.g., visual and olfactory stimuli, flower mechanics, reward properties). The concept of a rich and unique navigational memory composed of interrelated memory items underlying the task of navigation between nest site and feeding sites supports the view that flower signal evolution should depend on all the components guiding pollinator navigation. This is a testable hypothesis for further ecophysiological studies.

### Localization on a small scale: choice sequences and memory dynamics

A foraging bout is structured in time (Fig. 2.1). Because flowers mostly occur in patches, intrapatch choices follow each other quickly and are more likely to hit on the same kind of flower. Interpatch choices are more spread out in time, and are likely to expose bees to flowers of other

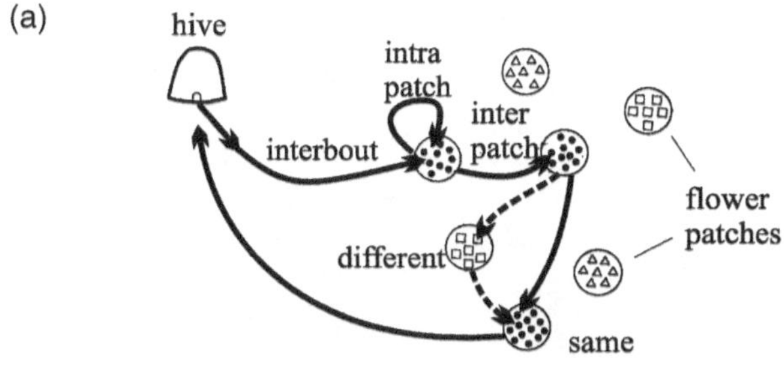


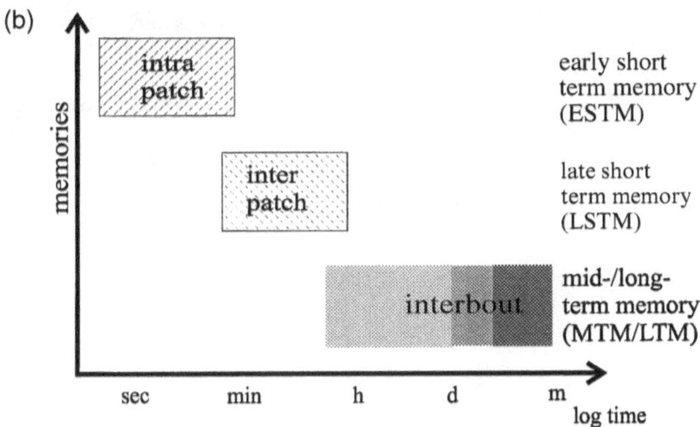


Fig. 2.1. Foraging cycle in honeybees. (a) Flowers usually appear in patches of similar flowers. A bee departing from the hive will arrive at such a patch and make intrapatch choices at short intervals. Interpatch choices follow each other at longer intervals and require a decision between similar and different flowers. Intervals between bouts range among many minutes, hours, and days (Menzel 1985). (b) Working hypothesis of different memories as defined by the sequences of events during a natural foraging cycle. For more detailed explanation see text.

species, forcing them to make decisions between "same" and "different" flowers. Intervals of visits to individual flowers were counted for four different plant species, and it was found that the most frequent intervals are in the range of a few seconds (Menzel 1987: Fig. 1b). There will certainly be a great scattering of such intervals, depending on plant species, habitat, and many other ecological conditions, but as long as we do not have more complete data, I consider that intrapatch choice intervals are typically a few seconds. Choice intervals between different patches have not been

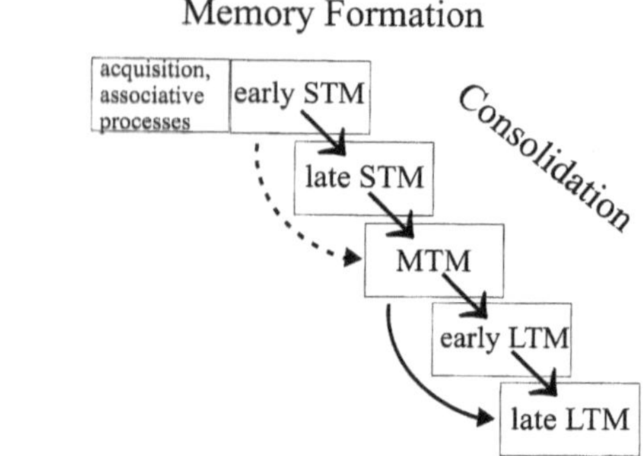



Fig. 2.2. Models of memory storage and memory retrieval as applied to foraging behavior in honeybees. (a) Memory storage is initiated by the associative acquisition process, which leads to an early form of STM lasting in the range of 1 min. The memory stages that follow (a late form of STM, MTM, and early and late forms of LTM) are predominantly arranged in sequence, but parallel processing has been shown for the transition from MTM to both forms of LTM (dotted and solid arrows). The memory stages are characterized by their time courses, their differential control of behavior, and their physiological substrates as described in the text.

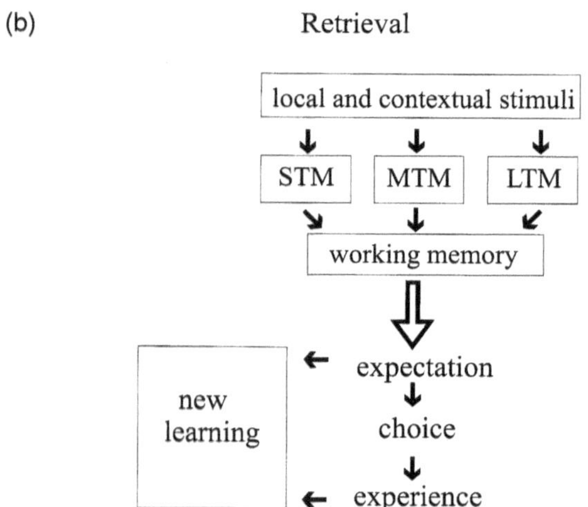



Fig. 2.2 (*cont.*). (b) Memory retrieval is initiated by local and contextual stimuli, which are believed to feed into each memory store directly for the activation of stimulus-specific memory traces. The contents of these traces are integrated into working memory, a memory phase controlling behavior via a neural stage addressed as "expectation". This stage is a necessary assumption from experiments showing that learning follows the difference rule as formulated by Rescorla & Wagner (1972) (see Greggers & Menzel 1993). As a consequence, retrieval is intimately combined with new learning, which leads to an updating of the memory contents according to the mismatch between the expected and the experienced outcome of behavior.

estimated and may vary to a large degree, but they are likely to last longer on average than within-patch choices. The distribution of interbout intervals from training experiments with bees on artificial feeders had a median around 4–5 min (Menzel 1987, Fig. 1a). Much longer intervals are expected during bad weather conditions, e.g., Lindauer (1963) reported that overwintering bees visited the feeding place from the previous autumn at their first flight in the spring.

Different memories may determine choice behavior at these different intervals. Evidence for different memory phases in the honeybee comes from behavioral, neurophysiological, and biochemical studies (Menzel & Müller 1996; Menzel 1999). The concept emerging from these results assumes five sequential stages during the process of memory formation (Fig. 2.2a). Consolidation from early to late memory stages is time- and event-dependent, meaning that both elapsing time and new experience during the process of consolidation define the speed of transfer between the memory phases. Most importantly, the consolidation process changes the content of memory, a general property of memory processing in animals (including invertebrates and humans; Müller & Pilzecker 1900; Milner *et al.* 1998).

I shall first give a short characterization of the mechanistic basis of the five memory stages and relate them to behavioral measures of retention as revealed by simple forms of associative learning. Then I shall try to incorporate this information into a model of sequential decision-making as it relates to choice behavior of a foraging bee. At this point it will become important to consider two aspects of memory: memory formation and memory retrieval (Fig. 2.2a, b). During foraging, both processes are intimately connected, and it is extremely difficult to assign any particular character of the choice behavior to one process or the other. The central theme here will be the concept of working memory, a form of memory that controls ongoing behavior and retrieves its information from all memory stores.

## Memory phases

### Associative induction and early short-term memory

An associative learning trial involves the pairing of the stimuli to be learned (conditioned stimuli, CS) with the rewarding stimulus (e.g., sucrose solution; unconditioned stimulus, US). Olfactory conditioning, for example, leads to associative induction and an early form of short-

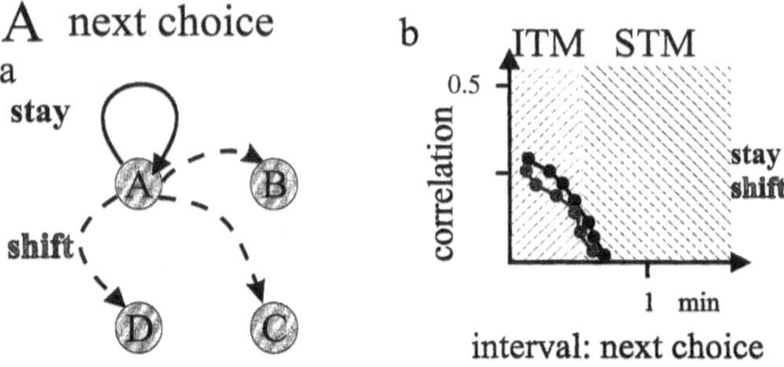



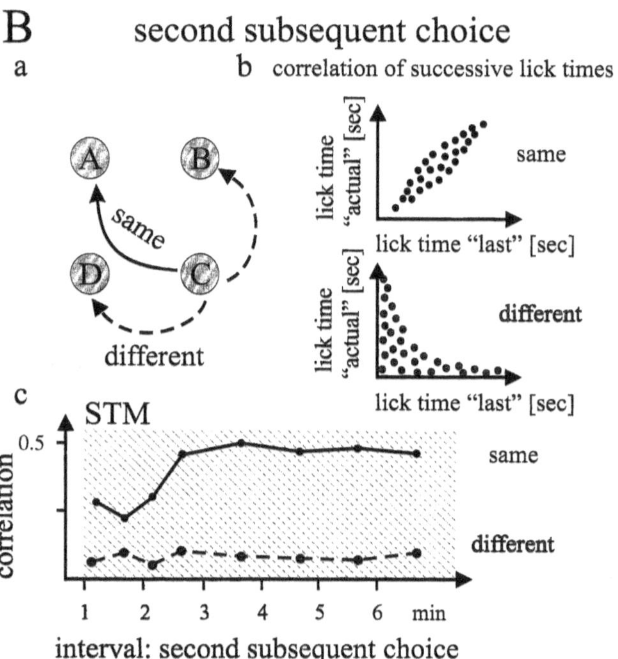



Fig. 2.3. Temporal dynamics of choice behavior in single honeybees foraging in a patch of four artificial feeders (A, B, C, D). Each feeder provided the same flow rate of sucrose solution. The flow rate was so low that the bee visited each of the feeders. Since the flow rate was similar in all feeders, the frequency of visits was the same for each feeder. Two kinds of subsequent choices were performed: **A**: next choices called "stay" and "shift" flights, and B: second subsequent choices – here they are called "same" and "different" choices, depending on whether the bee flies back to the same feeder after visiting another feeder or flies to a different feeder. One of the four possible cases is shown in a and b. Reward memory for the last (**A**) or second-to-last (**B**) feeder visited was evaluated by correlating lick time at the last or second-to-last

Fig. 2.3 (*cont.*)
feeder visited and the feeder for which reward memory is being evaluated
(actual feeder). There is a high correlation when the same feeder is visited, and
no correlation when a different feeder is visited (see Greggers & Mauelshagen
1997). Correlation (as a measure of memory, and thus expectancy during the
test visit) depends on the interval between visits. **A**: Next choices follow each
other quickly (<1 min). "Stay" and "shift" flights show the same rapidly
decaying time dependence of correlation of the respective lick times. This
indicates that at short intervals during eSTM, reward memory is unspecific for
the feeder visited. **B**: Second subsequent choices follow each other at intervals
of 1–7 min (thus during the lSTM time window). The time course of choices
differs for "same" and "different" choices. At different feeders, bees do not
expect to get the same reward as during the second-to-last visit at any of the
three other feeders; at the same feeder, however, they develop an increasing
(consolidating) memory for reward at that feeder. (Data for this figure come
from unpublished work by U. Greggers.)

term memory (eSTM) over the time course of a few seconds (Bitterman *et
al.* 1983; Menzel 1990). This memory is strongly dominated by appetitive
arousal induced by the US. Arousal sensitizes animals to a broad range of
stimuli generally related to the arousal-inducing events, e.g., the unexpected experience of food. eSTM is rather unspecific and imprecise. For
example, a bee conditioned to an odor by one trial will respond to other
stimuli (a mechanical stimulus to the antenna or to other odor stimuli)
much more strongly during the first 30 s than later (Menzel 1990; Smith
1991).

Furthermore, in an experiment with free-flying bees choosing among
four feeders of continuous but low sucrose solution flow, Greggers &
Mauelshagen (1995) found that the lick time during each visit is a measure
of what the animals recall about the reward at a particular feeder. Thus,
the correlation between actual and last lick time can be taken as a measure
of reward memory (Fig. 2.3). If sequential visits over short intervals
(<1 min) are considered, memory is unspecific for the four feeders, and
decays quickly. The early form of STM is, therefore, characterized by
general arousal, which decays within one minute. eSTM covers the time
window during which bees can expect to be exposed to the same stimuli.
No specific choices need to be performed at this time, and general arousal
(depending on the strength of the US) will suffice to control whether the
animal stays in the patch or postpones choices for a later time. Affirmative
information in this time window will inform the animal whether specific
memory storage is worthwhile, and this will lead to quicker formation of
longer-lasting forms of memory (see below).

At the level of cellular and neural processes, stimulus association is reflected in the convergence of excitation of the CS and US pathways. The neuroanatomical convergence sites of these pathways are known for olfactory learning in bees (antennal lobe, lip region of mushroom body calyces, lateral protocerebrum; Hammer 1997), and the putative primary transmitters and second messenger pathways have been identified (Menzel & Müller 1996).

### Late short-term memory (lSTM)

The transition to the selective associative memory trace (consolidation) during lSTM is a rather slow process after a single learning trial, lasting up to several minutes. It is quicker after multiple learning trials (Menzel 1968; Erber 1975*a*, *b*; review by Menzel 1999). Thus, consolidation of the stimulus-specific associative memory trace is both time- and event-dependent. Odor-conditioned animals show increasing retention for intervals >3 min, and during this consolidation period retention becomes more specific for the learned odor. Free-flying bees foraging in the patch of four feeders mentioned above (Greggers & Mauelshagen 1995; see Fig. 2.3) behave identically: retention for the reward quality of each feeder rises after a minimum around 2 min and increases over the next minutes, becoming more and more specific for each feeder. Therefore, consolidation during lSTM is a process that establishes a more precise memory. This memory is more resistant to new information (Menzel 1979, 1990), and is no longer susceptible to amnestic treatment (Menzel 1968; Erber 1975*a*, *b*). Transition from eSTM to lSTM is accelerated by multiple learning trials and learning trials in quick succession.

The behavioral relevance of these findings for foraging behavior under natural conditions may be related to the temporal separation between intra- and interpatch visits, and the different memories established for later use. First, memory needs to be highly specific after leaving a patch, because distinctions need to be made between similar and different flowers. Second, such a specific memory trace should also be established after a single learning trial, because in some rare cases a single flower may offer a very high amount of reward. Third, discovering a rewarding flower in a different patch means that the local cues just learned are now presented in a different context (localization within GLM; see above). Consolidation is the process that allocates different memories to different stores, enabling the bee to store many different memories according to the contextual cues related to the separate experiences. How many memories may reside in working memory will be discussed below.

### Mid-term memory (MTM)

At the beginning of MTM, behavior is controlled by consolidated, highly specific memory. At this stage, memory is more resistant to extinction, conflicting information, and elapsing time, and some context-dependent information may have already been stored. Under natural conditions, bees have usually returned to the hive and departed on a new foraging bout usually within the time window of MTM. Upon arrival at the feeding area, memory for flower cues does not reside in STM any longer, but needs to be retrieved from a more permanent store and put into working memory. Therefore, MTM is a memory stage clearly disconnected from a continuous stream of STMs into working memories, which regulate foraging behavior during the quick successions of intra- and interpatch choices (see below).

MTM is physiologically characterized by the continuing activation of a particular second messenger pathway, the protein kinase C pathway (Grünbaum & Müller 1998). It was concluded that both the sensory neuropil (the antennal lobe, in the case of olfactory learning) and the mushroom bodies are involved in the memory trace. The mushroom body provides the information that relates the memory traces in the primary sensory neuropils to contextual stimuli across modalities. The memories necessary to guide the bee back to the feeding place after returning to the hive involve many different behavioral faculties (e.g., compass orientation to celestial and landmark cues, information about the time of day, sequential landmark appearance, social encounters, and information about the colony's needs). These behavioral faculties can be integrated only across sensory modalities, and are most likely related to mushroom-body function (Menzel *et al.* 1994; Rybak & Menzel 1998).

### Long-term memory (LTM)

LTM requires multiple learning trials, indicating that specific information, which can be extracted only from multiple experiences (reliability of signals, context dependence), characterizes its contents. Two forms of LTM are distinguished based on physiological characteristics – early LTM (eLTM, between several hours and 2 days) and late LTM (lLTM, $\geq 3$ days). Only lLTM depends on protein synthesis, and thus structural changes in the wiring of neurons appear to store memory lasting longer than 3 days

(Grünbaum & Müller 1998). The physiological basis of eLTM is not yet well understood. Since protein-synthesis blockade does not interfere with eLTM formation, and the signaling pathway of protein kinase C is constitutively activated, it is concluded that covalent transformations within the existing neural pathways store eLTM.

The biological circumstances of two forms of LTM may be related to the distinction between those forms of learning that usually lead to life-long memories (e.g., visual and olfactory cues characterizing the home colony) and those that are stable but need updating on a regular basis (e.g., visual and olfactory cues of feeding places). However, lifelong memories can also be formed for floral cues (e.g., color information; see Menzel 1968, Lindauer 1963), and standard lifelong memories (e.g., localization of the colony) can also be changed by new experience (e.g., swarming).

### Sequential choice behavior during foraging: memory dynamics at work

Memories guide choice behavior. Anyone who has trained bees to a feeding station knows that bees choose the signals associated with reward even after long intervals: days, weeks, or – as Lindauer (1963) reported – several months. Therefore, bees activate remote memories stored in long-term form when motivational and contextual conditions are favorable. Similarly, memories kept in MTM and STM will contribute. The active form of all these types of memory is usually called working memory. Working memory is understood to provide the animal with specific expectations about the signals, reward conditions, and manipulatory requirements of a food source for which a bee strives. The choice resulting from an expectation will lead to new experience and thus to new learning, which – in the manner discussed above – will add new information to memory (Fig. 2.2b).

Working memory will be continuously updated by STM during the quick succession of flower visits during a foraging trip, and, therefore, the dynamics of STM will be most important for the actual status of working memory. The content of working memory is certainly limited, as it is in all animals (Baddeley 1986). It is, therefore, likely that the intervals between successive choices and the sequence of experiences will control the expectation, or non-surprisingness as proposed by Wagner (1978), at any moment of the choice process. Indeed, expectation about reward properties changes over time (Fig. 2.3). As already pointed out, very short intervals (<1 min) lead to high but rather unspecific expectations, intermediate intervals (1–2 min) to low and unspecific expectations, and long intervals to specific expectations. These observations apply to expectations both for reward properties and learned signals, because the same time dependencies were found in olfactory conditioning experiments and in tests with free-flying bees trained to color signals (Menzel 1999). Some indications that favor this interpretation come from observations by Chittka *et al.* (1997), who recorded the frequency of intervals between stay and shift flights of bumble bees foraging on more than two plant species. Stay flights appear at shorter intervals than shift flights, indicating that immediate choices are dominated by the most recent and the most effective STM, but reference to more remote memories needs more time, or, to interpret it from another perspective, longer intervals release working memory from the dominant memory of the last visit and allow for contributions from earlier, consolidated memories. The time-scale in the observations of Chittka *et al.* (range of 1–20 s) differs from that in the experiments with bees foraging in a patch of four artificial feeders mentioned above (Fig. 2.3) (Greggers & Menzel 1993; Mauelshagen & Greggers 1993; Greggers & Mauelshagen 1995), in which the average retrieval between successive visits is in the 1-min range. However, since handling time ranges from 20–40 s in the Greggers *et al.* experiments, the actual intervals in flight time are rather similar. I conclude, therefore, that the temporal dynamic of STM contributes substantially to working memory, and that initial and later phases (eSTM and lSTM) are an important determinant of working memory.

Capacity of working memory can also be estimated from the experiments with artificial feeders. As Fig. 2.3B indicates, the expectancy during revisits to the same feeder after second subsequent choices is significantly different from expectancies expressed in visits to different feeders. This finding applies to any of the four feeders. Therefore, bees store the reward properties of at least four different feeders in working memory. The same result was found for eight feeders, indicating that the reward properties of eight feeders can also be stored in eight feeder-specific memories. A larger number of feeders has not yet been tested. The capacity of working memory is, therefore, at least eight items, and the time-range for all these specific working memories lies above 6 min (Fig. 3B).

## The framework of a mechanistic model of flower choice

It is well documented that memory in both animals and humans is processed in temporal stages. The cellular substrates appear surprisingly similar both among different species (*Aplysia, Drosophila*, mouse, chick, man) and among different forms of memory (memories consciously addressable and those under automatic control, appetitive and aversive memories, and emotional and non-emotional ones; Milner *et al.* 1998; Rosenzweig 1998). This has led to the assumption that the memory formation process is determined by its underlying cellular machinery, and that similar time courses for the respective stages are indicative of general mechanisms rather than species-specific and task-specific adaptations. However, studies on memory stages have focused primarily on their neural and cellular substrates, and have not yet asked the question of how these stages are adapted to the needs of an animal behaving in natural surroundings. In particular, very little attention has been paid to the dynamics of natural sequences of behavioral events that simultaneously create new memory and need to be controlled by memory. The notion presented here is that the cellular machinery may not be the defining factor, but, rather, the systems' requirements for the installation, sequence, and character of memory stages. In particular, I favor the view that the similarities in memory stages discovered so far reflect basic and general requirements of the continuous process of concurrently learning, retrieving, storing, and applying information (Menzel 1999).

Species- and task-specific adaptations are to be expected, and these may be the deciding parameters for the dynamics and significance of memory stages. As pointed out above, we need to separate memory states during memory formation and during memory retrieval (Fig. 2.2). During choice performance, memory retrieval guides the next choice, but it is important to keep in mind that any experience will always induce a learning process, which in turn leads to memory formation and alteration of the content of all memory stages. In fast behavioral sequences, such as during foraging within and between flower patches, STM of the last encounter will first feed strongly into working memory, but with time elapsing, the memory from former experiences will gain by a consolidation process. Specifying properties of a food source, such as local signals and contextual cues, will become increasingly important during consolidation. These highly specific memories are stored via multiple experiences in MTM and LTM, and their contribution to working memory will make expectations rather specific.

A particular aspect of memory specificity is the combination of different memories established independently in LTM. One of these memories, general landscape memory (GLM), has been discussed here. The hypothesis is put forward that food sources are represented along with their properties (signals, rewards, mechanics) in GLM, and these can be chosen by directed flights between them, even over long distances.

The capacity of such a compound LTM is unknown, but it may be safe to assume that multiple locations can be stored that represent different loci in GLM. The spatial resolution of such loci and their maximum number will have to be addressed in future experiments. Future experiments must also evaluate the upper limits of working memory with respect to both the capacity of stored items and the timespan. The ranges found so far for bees (at least eight items, over at least 6 min for a honeybee foraging in a patch of artificial feeders) are already quite impressive when compared with other animals (Baddeley 1986).

Another issue refers to the question of which memory stages (with their accompanying dynamics) are evolutionarily adapted to the task to be solved, namely efficient foraging in an unreliable and scattered food market. Although a mechanistic model need not refer to ultimate causation, it is tempting to speculate in what sense the structure and dynamics of memories might be shaped by evolution. I noted in the introductory section of this chapter that the patchiness of food sources poses different demands on the tasks to be solved in sequence. The decision to stay in the patch is mainly a motivational one controlled by the amount (or quality) of food gained as compared to the amount (or quality) expected. Thus, rather unspecific behavioral control as induced by food arousal is a dominant characteristic of choices occurring in quick succession. I would expect that the dynamic of decay in arousal might reflect the spacing (as measured in flight time) between chosen patches.

The concept of consolidation of associative events includes the notion that different memory items are consolidated separately (Müller & Pilzecker 1900), a notion that has been substantiated for the honeybee. Different memory items are characterized and later retrieved specifically by their contextual cues, which should be mainly those defining different locations in GLM. Thus, lSTM should be adapted to the spacing between patches. It might be an interesting question whether species adapted to different intrapatch distances or with very different flight speeds developed different dynamics of lSTM.

Flower constancy of hymenopteran pollinators results from choice behavior, which is at any moment guided by the memory of former experience. The richness, duration, complexity, and dynamics of memory have been underestimated, and have only recently become clear. Although we still have to learn a large amount about the structures, mechanisms, and contents of the various forms of memory, we certainly can no longer assum that major components of the choice processes are dictated by the limited capacity or duration of memory. Rather, it is the dynamics of the memory stages and their transitions that allow for highly flexible choice behavior and thus for flower constancy.

## References


Baddeley AD (1986) *Working memory*. Oxford University Press, Oxford

Becker L (1958) Untersuchungen über das Heimfindevermögen der Bienen. *Z vergl Physiol* 41:1–25

Bitterman ME, Menzel R, Fietz A & Schäfer S (1983) Classical conditioning of proboscis extension in honeybees (*Apis mellifera*). *J Comp Psychol* 97:107–119

Capaldi EA & Dyer FC (1999) The role of orientation flights on homing performance in honeybees. *J Exp Biol* 202:1655–1666

Capaldi EA, Smith AD, Osborne JL, Fahrbach SE, Farris SM, Reynolds DR, Edwards AS, Martin A, Robinson GE, Poppy GM & Riley JR (2000) Ontogeny of orientation flight in the honeybee revealed by harmonic radar. *Nature* 403:537–540

Chittka L, Gumbert A & Kunze J (1997) Foraging dynamics of bumble bees: correlates of movements within and between plant species. *Behav Ecol* 8:239–249

Collett TS (1992) Landmark learning and guidance in insects. *Phil Trans R Soc Lond B* 337:295–303

Collett TS & Zeil J (1998) Places and landmarks: an arthropod perspective. In: Healy S (ed) *Spatial representation in animals*, pp 18–53. Oxford University Press, Oxford

Dyer FC & Gould JL (1981) Honey bee orientation: a backup system for cloudy days. *Science* 214:1041–1042

Erber J (1975a) The dynamics of learning in the honeybee (*Apis mellifera carnica*). I. The time dependence of the choice reaction. *J Comp Physiol* 99:231–242

Erber J (1975b) The dynamics of learning in the honeybee (*Apis mellifera carnica*). II. Principles of information processing. *J Comp Physiol* 99:243–255

Esch HE & Burns JE (1995) Honeybees use optic flow to measure the distance of a food source. *Naturwiss* 82:38–40

Frisch K von (1965) *Tanzsprache und Orientierung der Bienen*. Springer-Verlag, Heidelberg

Gould JL (1986) The locale map of honey bees: do insects have cognitive maps? *Science* 232:861–863

Greggers U & Mauelshagen J (1995) The temporal composition of task-specific and reward related memory components in free-flying honeybees. In: Elsner N &

Menzel R (eds) *Learning and memory, Proceedings of the 23rd Göttingen Neurobiology Conference*, p 13. Thieme Verlag, Stuttgart

Greggers U & Mauelshagen J (1997) Matching behavior of honeybees in a multiple-choice situation: the differential effect of environmental stimuli on the choice process. *Anim Learn Behav* 25:458–472

Greggers U & Menzel R (1993) Memory dynamics and foraging strategies of honeybees. *Behav Ecol Sociobiol* 32:17–29

Grünbaum L & Müller U (1998) Induction of a specific olfactory memory leads to a long-lasting activation of protein kinase C in the antennal lobe of the honeybee. *J Neurosci* 18:4384–4392

Hammer M (1997) The neural basis of associative reward learning in honeybees. *Trends Neurosci* 20:245–252

Koltermann R (1971) 24-Std-Periodik in der Langzeiterinnerung an Duft- und Farbsignale bei der Honigbiene. *Z vergl Physiol* 75:49–68

Lehrer M (1999) Dorsoventral asymmetry of colour discrimination in bees. *J Comp Physiol A* 184:195–206

Lindauer M (1963) Allgemeine Sinnesphysiologie. Orientierung im Raum. *Fortschr Zool* 16:58–140

Mauelshagen J & Greggers U (1993) Experimental access to associative learning in honeybees. *Apidol* 24:249–266

Menzel R (1968) Das Gedächtnis der Honigbiene für Spektralfarben. I. Kurzzeitiges und langzeitiges Behalten. *Z vergl Physiol* 60:82–102

Menzel R (1979) Behavioral access to short-term memory in bees. *Nature* 281:368–369

Menzel R (1985) Learning in honey bees in an ecological and behavioral context. In: Hölldobler B & Lindauer M (eds) *Experimental behavioral ecology*, pp 55–74. Gustav Fischer Verlag, Stuttgart

Menzel R (1987) Memory traces in honeybees. In: Menzel R & Mercer AR (eds) *Neurobiology and behavior of honeybees*, pp 310–325. Springer-Verlag, Berlin

Menzel R (1990) Learning, memory, and "cognition" in honey bees. In: Kesner RP & Olton DS (eds) *Neurobiology of comparative cognition*, pp 237–292. Erlbaum, Hillsdale, NJ

Menzel R (1999) Memory dynamics in the honeybee. *J Comp Physiol A* 185:323–340

Menzel R & Giurfa M (2001) Cognitive architecture of a minibrain: the honeybee. *Trends Cognite Sci*, in press.

Menzel R & Müller U (1996) Learning and memory in honeybees: from behavior to neural substrates. *Ann Rev Neurosci* 19:379–404

Menzel R, Durst C, Erber J, Eichmüller S, Hammer M, Hildebrandt H, Mauelshagen J, Müller U, Rosenboom H, Rybak J, Schäfer S & Scheidler A (1994) The mushroom bodies in the honeybee: from molecules to behavior. In: Schildberger K & Elsner N (eds) *Neural basis of behavioral adaptations, Fortschritte der Zoologie* vol. 39, pp 81–102. Gustav Fischer Verlag, Stuttgart

Menzel R, Geiger K, Chittka L, Joerges J, Kunze J & Müller U (1996) The knowledge base of bee navigation. *J Exp Biol* 199:141–146

Menzel R, Gumbert A, Kunze J, Shmida A & Vorobyev M (1997) Pollinators' strategies in finding flowers. *Israel J Plant Sci* 45:141–156

Menzel R, Geiger K, Müller U, Jörges J & Chittka L (1998) Bees travel novel homeward routes by integrating separately acquired vector memories. *Anim Behav* 55:139–152

Menzel R, Giurfa M, Gerber B & Hellstern F (1999) Elementary and configural forms of memory in an insect: the honeybee. In: Friederici AD & Menzel R (eds) *Learning: rule extraction and representation*, pp 259–282. Walter de Gruyter, Berlin

Menzel R, Brandt R, Gumbert A, Komischke B & Kunze J (2000) Two spatial memories for honeybee navigation. *Proc R Soc Lond B* 267:961–968

Milner B, Squire LR & Kandel ER (1998) Cognitive neuroscience and the study of memory. *Neuron* 20:445–468

Müller GE & Pilzecker A (1900) Experimentelle Beiträge zur Lehre vom Gedächtnis. *Z Psychol* 1:1–288

Rescorla RA & Wagner AR (1972) A theory of classical conditioning: variations in the effectiveness of reinforcement and non-reinforcement. In: Black AH & Prokasy WF (eds) *Classical conditioning II: Current research and theory*, pp 64–99. Appleton–Century–Crofts, New York

Riley JR, Smith AD, Reynolds DR, Edwards AS, Osborne JL, Williams IH, Carreck NL, & Poppy GM (1996) Tracking bees with harmonic radar. *Nature* 379:29–30

Rosenzweig MR (1998) Some historical background of topics in this conference. *Neurobiol Learn & Mem* 70:3–13

Rybak J & Menzel R (1998) Integrative properties of the Pe1-neuron, a unique mushroom body output neuron. *Learn & Mem* 5:133–145

Smith BH (1991) The olfactory memory of the honeybee *Apis mellifera*. I. Odorant modulation of short- and intermediate-term memory after single-trial conditioning. *J Exp Biol* 161:367–382

Srinivasan MV, Zhang SW, Lehrer M & Collett TS (1996) Honeybee navigation *en route* to the goal: visual flight control and odometry. *J Exp Biol* 199:237–244

Vollbehr J (1975) Zur Orientierung junger Honigbienen bei ihrem ersten Orientierungsflug. *Zool Jb Physiol* 79:33–69

Wagner AR (1978) Expectancies and the priming of STM. In: Hulse SH, Fowler H, & Honig WK (eds) *Cognitive processes in animal behavior*, pp 224–251. Erlbaum, Hillsdale, NJ

Wehner R (1992) Arthropods. In: Papi F (ed) *Animal homing*, pp 45–144. Chapman & Hall, London

Wehner R & Menzel R (1990) Do insects have cognitive maps? *Ann Rev Neurosci* 13:403–414

3

# Subjective evaluation and choice behavior by nectar- and pollen-collecting bees

During the last 30 years, animal behaviorists have become serious players in the quest to understand the interaction between plants and their flower-visiting, foraging pollinators (Waddington 1983, 1997; Barth 1985). Flower-visiting bees, flies, butterflies, and beetles are the sole agents for reproduction in many species of plants. Through the larder of pollen and nectar they provide, plants also affect the foraging success and reproductive output of these insects. The pollinator and the plant, each of separate evolutionary lineages, are in a long-term game where each is dependent on the other and each affects the evolution of the other (Selten & Shmida 1991).

On a local scale, in a field of flowers, a forager such as a nectar-collecting bee makes thousands of sequential decisions during a foraging trip. These decisions are reflected in the choice of flowers visited. These decisions determine: which flowers receive visits and which do not; who mates with whom; the distance between mating plants; the transfer of intra- or interspecific pollen; and the amount of self-pollination and out-crossing. The decisions also affect the bee's success on its foraging foray. Through experience, the bee makes associations between different kinds of flowers (e.g., species) and the rewards they provide, and it seeks out the flowers with the greatest net rewards. Animal behaviorists have played an important role in learning how pollinators make these choices among flowers.

Although general patterns of pollinator foraging behavior have been found, variation among individual foragers has not been well studied. Individuals observed in foraging experiments often differ in their behavior even when given the same problem (e.g., choosing among flowers in the same patch; Waddington & Holden 1979). Inter-individual variation might be due to differences in experience, such that each forager has a slightly different experience in the same patch and makes decisions based on different information (see Thomson & Chittka, this volume). Sampling error could produce important differences in experience among individuals, especially if the sample of previous flower visits used to make future decisions is small. However, genetic differences among individuals may also contribute to behavioral variation. We need studies designed specifically to examine the underlying causes of variation among individuals in their assessment of information and choice behavior.

Most published work on the genetics of foraging behavior describes honeybees. Genotypic variability has been shown for many important traits (from Page *et al.* 1995): (1) the decision to forage for pollen or nectar (Calderone & Page 1992; Guzman-Novoa & Gary 1993); (2) behavioral plasticity associated with switching foraging resources (Fewell & Page 1993); (3) nectar and pollen load sizes (Milne *et al.* 1986; Deng 1996); (4) round-trip time and foraging activity (Guzman-Novoa & Gary 1993); (5) rate of foraging trip initiations (Deng 1996); (6) the age at which individuals initiate foraging (Calderone & Page 1988; Deng 1996), and the duration of the foraging career (Deng 1996). These constitute most components of individual foraging behavior, and some could affect the effectiveness, quantity, and quality of pollination.

Equipped with an understanding of how bees choose among flowers and an understanding of causes of individual variability in choice, we can better evaluate how plants may evolve to "manipulate" the bee's behavior so as to enhance their reproductive success. On this larger scale, both in time and space, the evolution of floral morphology and color, flowering phenology, and patterns of nectar and pollen presentation are likely influenced in no small way by the choice behavior of their floral visitors (Grant 1949; Macior 1970; Faegri & van der Pijl 1979). The salience of flowers in the pollinator's olfactory and visual fields, the spatial patterning of flowers, and the quality and quantity of the food are in part the result of selection for the reliable services of pollinators and the manipulation of their behavior.

In this chapter, I focus on the decision-making process of foraging bees with the goal of better understanding their choice behavior. This decision-making process includes the evaluation of relevant information used by the bees to make decisions which result in choice. Particularly relevant to their success on each foraging trip – and ultimately their fitness –

is information on the quality and quantity of their food. I will especially consider bees' evaluations of nectar concentration and pollen quality in making choices. How are nectar concentration and pollen quality evaluated, and what choices are made based on these evaluations? I will also explore some studies of genetic variation for these evaluations and the choice of food. Because most work has been conducted on bumble bees and honeybees, genera of the same family Apidae, this chapter will reflect that bias.

## The decision-making process

Individuals just beginning a foraging career must learn which flowers are likely to provide the most pollen or nectar, and which flowers are not profitable to visit. During this early stage, colors, shapes, and odors of flowers are associated with profits; then this information is used to make decisions in the future. Foragers must decide where and when to look for food, and which food to search out, pollen or nectar. Perhaps before each takeoff from a flower, but certainly before each landing, a bee decides which flower to visit. Should it visit another flower on the same plant, or a more distant flower of the same color and odor (same species), or a flower of a different species? The outcome of these innumerable decisions, the amount of food collected, likely affects the bee's reproductive fitness. Recent studies have aimed at better understanding this decision-making process in order to understand the parameters that affect bees' choices. Both proximate (e.g., von Frisch 1967; Waddington & Holden 1979; Waddington 1983; Waddington & Gottlieb 1990; Greggers & Menzel 1993; Raveret-Richter & Waddington 1993; Shafir 1994) and ultimate (e.g., Possingham *et al.* 1990) causation of choice have been investigated.

Most studies of pollinators' flower choices address the relationship between gains and costs associated with foraging, and the pattern of choice behavior (Pyke 1984). The aim of these studies is to understand the direct relationship between objective information (the actual volume of nectar, the actual quality of pollen, the actual time to access the nectar inside the flower, etc.) and choice behavior. The implicit assumption made in these studies is that objective information is the direct guide to choice behavior. That is, choice behavior is guided directly by the actual (absolute) concentration or volume of nectar, the actual time to fly between flowers, the actual time to handle flowers, and so on. In fact, to do otherwise is assumed by most behavioral ecologists to be irrational behavior (Shafir 1994).

However, the cognitive psychologist will note that the direct guide to behavior is the subjective evaluation of the objective information (Rachlin 1989). Ideally, one would like to study subjective evaluations of objective information and their relationship to flower choice behavior. Studies done in my laboratory examine this relationship in the honeybee.

### Subjective evaluations by honeybees, and genotypic variation – the dance

It is possible to study independently various stages of human beings' decision-making behavior. Choice behavior in the marketplace can be directly observed. The kind of investment instruments traded under various conditions can be determined. Furthermore, a person's subjective evaluation of the objective alternatives and other available information can be accessed independently. Questionnaires are used in the laboratory to determine these relationships. Thus, the relationships between the subjective evaluations of the options and the person's choice behavior can be understood. We do not have a well-developed protocol for tapping into non-human animals' evaluation of objective information. Recently, we developed a protocol for examining the relationships between objective costs of foraging, objective rates of energy intake, and intakes, and the subjective evaluation of these variables by honeybees.

Nectar and costs

Aspects of honeybees' dance are well known to be correlated with the distance and direction of food from the hive (von Frisch 1967). The bees use their perception of distance and direction to perform the dance in a certain way. I conducted experiments designed to find whether the honeybees' in-hive dance could be used to measure their evaluation of recently experienced foraging costs and gains (Fig. 3.1).

Von Frisch (1967) found that the "vigor" of the dance changed with the concentration of sugar solution imbibed. When bees had collected highly concentrated solution from a feeder, their dance was more excited than after collecting a weak solution. I expanded his studies to quantify "vigor" (Waddington 1982). Bees foraged back and forth between two artificial flowers that contained sugar solution (nectar). I manipulated gains by varying concentration and costs by varying the distance between the two flowers. The rate of the "round dance" increased as concentration

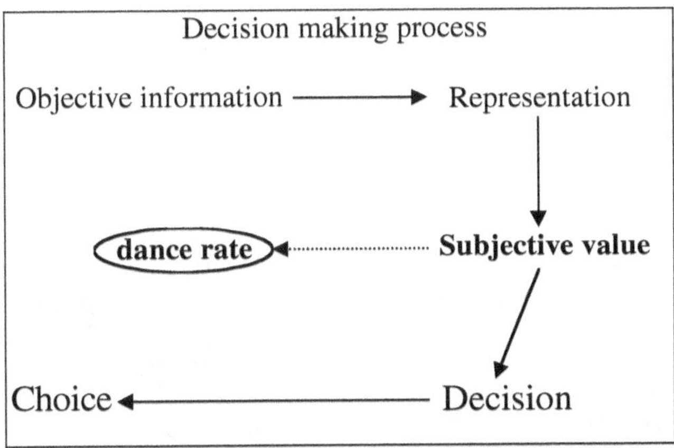


Fig. 3.1. Diagram of the honeybee's decision-making process. The rate of
180° turns of the honeybee's round dance is used to quantify the subjective
evaluation of objective information such as nectar concentration. The
subjective values are used by the bees as the direct guide to choice behavior.

increased (this reflects von Frisch's "vigor"), and the rate of the dance
decreased as flight costs increased. Rate of the dance was quantified as the
number of 180° changes in direction per minute. Thus, bees assessed and
integrated both foraging costs and gains, then revealed their evaluation
in the dance.

We since have used rate of the dance to quantify the functional relationship between costs and sucrose concentration and the bees' assessments, or subjective evaluations, of them. We discovered the following
relationships. First, the relationship between sugar concentration and
subjective concentration is non-linear (Waddington & Kirchner 1992); the
dance rate increases with increasing concentration, but at a decreasing
rate (Fig. 3.2). This is qualitatively consistent with the Weber–Fechner
law of perception that generally describes the non-linear relationships
between perception and the magnitude of stimuli (Carterette & Friedman
1974; see discussion in Perez & Waddington 1996).

Second, costs are subjectively weighted in relation to gains
(Waddington 1985). That is, some incremental change in gains (e.g., joules
obtained per flower visit) results in a lesser change in the dance rate than
the same incremental change in cost (joules expended per flower visit for
flight and handling).

More recently, we investigated whether bees make absolute or relative
assessments of objective information (Raveret-Richter & Waddington

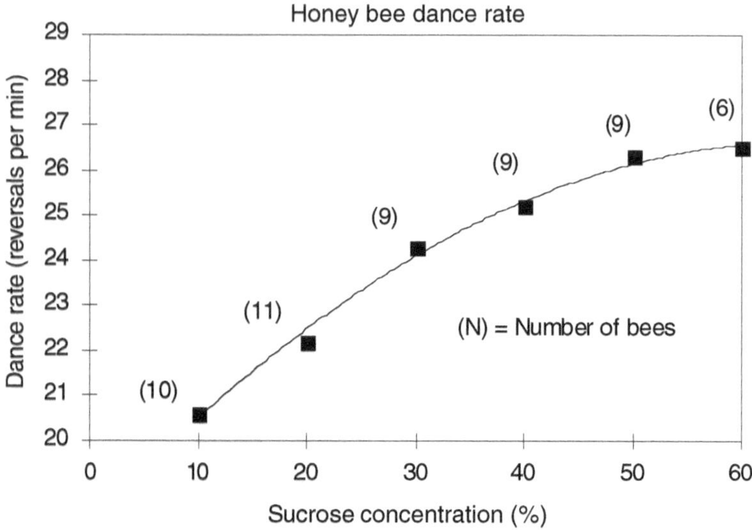



Fig. 3.2. Relationship between dance rate and sucrose concentration. (After Waddington & Kirchner 1992, Fig. 1A. Reprinted by permission of Blackwell Wissenschafts-Verlag, Berlin.)

1993). Most foraging models assume the former (Shafir 1994). We allowed bees to forage at artificial flowers for solutions of various sugar concentrations. The concentrations were offered to the same bees sequentially in increasing order of concentration (i.e., 10%, 20%, ..., 60%; weight of solute per weight of solution) on the first day of foraging, and in decreasing order, starting at 60%, on the second day. We then analyzed the dance after each foraging trip. The question was whether the bees' assessment, as indicated by dance rate, was always the same whether visiting a particular concentration of sugar solution in increasing or decreasing order (indicating absolute assessment) or different (indicating a relative evaluation). The bees' assessments of a concentration clearly varied with context (Fig. 3.3). For example, the dance rate after feeding on the 40% solution placed in an increasing sequence was higher (15 reversals/min) than when placed in a decreasing sequence of concentrations (7 reversals/min). The results suggest that losses (decreasing sequence) are subjectively weighted in relation to increases because of the different magnitudes of change in dance rate with increases and decreases in concentration. Analogous phenomena are well known in human beings (Tversky & Kahneman 1981). In order to understand the meaning of these relationships to the attractiveness of flowers to pollinators in the field, it will be

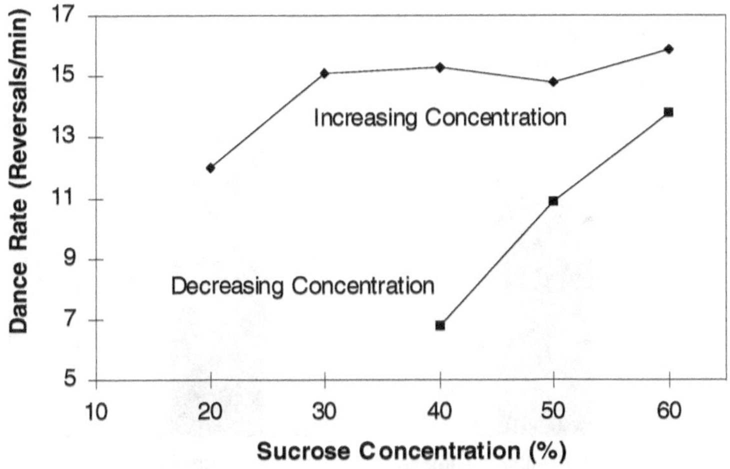


Fig. 3.3. Relationship between dance rate and sucrose concentration under two conditions: when each concentration except 10% was experienced after a 10% lower concentration (increasing concentration) and when each concentration except 60% was experienced after a 10% higher concentration (decreasing concentration). The results show that concentrations are assessed differently under different conditions. (After Raveret-Richter & Waddington 1993, Fig. 1. Reprinted by permission of Academic Press.)

necessary to study over what scale the phenomenon holds. For example, departure decisions from inflorescences may be affected. Vertical inflorescences sometimes have more nectar in the bottom male flowers than in the upper female flowers (Waddington 1981). Bees tend to arrive at the inflorescence on a lower flower and then move upward (Pyke 1979). Perhaps a bee's weighted assessment of decreasing nectar volume as it moves upward on single inflorescences would hasten its departure for another inflorescence.

### Pollen

Little is known about bees' assessments of pollen quality and quantity and choice (Rasheed & Harder 1997*a*, *b*). The amount of pollen available in individual flowers can be detected by bumble bees (Buchmann & Cane 1989; Harder 1990). Waddington *et al.* (1998) checked to see if the dance rate varied with pollen quality as it does with nectar quality. Pollen loads were collected from honeybees, dried, and ground to a powder. The pollen was given to bees in a petri dish in two ways: pure pollen or pollen mixed with alpha-cellulose powder (1:1 by volume). Alpha-cellulose powder does not have nutritional value to bees; thus, the mixture is lower in quality than pure pollen. The dance rate was higher after bees

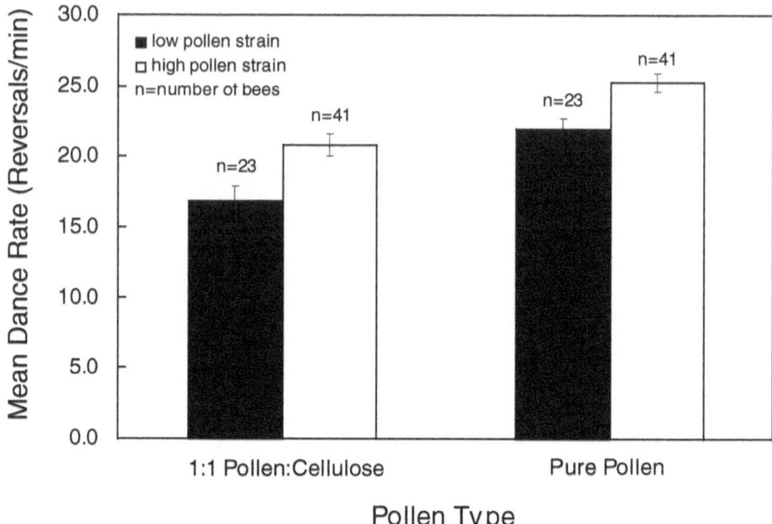



Fig. 3.4. Relationship between pollen quality and dance rate of honeybees from two genotypic strains. (After Waddington *et al.* 1998, Fig. 1A. Reprinted by permission of Academic Press.)

foraged for pure pollen than after they foraged for the lower-quality pollen–cellulose mixture (Fig. 3.4). These results suggest that the dance can indeed be used in the future to make more detailed studies of honeybees' evaluation of pollen quality and perhaps quantity.

Genotypic variation has been found in the dance rate, and therefore in subjective evaluation, in honeybees. We studied dance behavior of bees that had been subjected to colony-level selection for variation in collecting and storing pollen. Colonies of the high-pollen strain filled the comb with pollen, while bees from the low-pollen strain stored little pollen. High-strain bees preferentially foraged for pollen and bees from the low-pollen strain preferentially foraged for nectar (Page *et al.* 1995). Bees from the two genotypic strains were given dishes of pure pollen and a pollen and alpha-cellulose mixture (as described above). Pollen foragers from the high-pollen strain had higher dance rates (Waddington 1982) for both types of pollen than did bees from the low-pollen strain; this suggests a genotypic basis for the subjective evaluation of pollen quality (Fig. 3.4) (Waddington *et al.* 1998).

In summary, the honeybee's dance may give insight into the perceptual processes of foraging honeybees and, by extension, such processes in other taxa of pollen and nectar foragers.

### Subjective evaluations by honeybees and genotypic variation – proboscis extension response

The proboscis extension response (PER) is routinely used to study classical conditioning in honeybees (Bitterman *et al.* 1983). Page *et al.* (1998) suggested that the PER test could also be used as a window into honeybees' perceptions of sugar concentration. They collected bees and harnessed each bee in a metal tube for the PER test. The antennae, which have sugar receptors, were touched with a droplet of sugar solution; a record was made of whether the bee responded by extending its proboscis. Page *et al.* found that the probability of proboscis extension increased with increasing sugar concentration. This is interpreted as follows. There is a threshold sugar concentration that will elicit proboscis extension and imbibition of the solution. Concentrations below that threshold do not elicit a response; those above threshold do elicit a response. A honeybee's threshold can be a measure of its subjective evaluation of the solution.

As judged by the PER test, the perception of sucrose is not constant, but is affected by previous experience. T. Pankiw, KD Waddington & RE Page (unpublished data) demonstrated that the evaluation depends on the concentration of sugar previously imbibed and on the bee's nutritional status. Bees recently fed lower concentrations of sugar or starved were more responsive than bees that had fed on high concentrations or were well fed. The former bees made a higher subjective evaluation of sucrose than those fed a higher concentration. The response also was affected by the amount of fluid, water, or nectar in the crop at the time of testing. Empty bees evaluated sucrose concentration more highly than filled bees.

The studies of the PER by Page *et al.* (1998) and by Pankiw *et al.* (unpublished data) also suggest genotypic variance for evaluation of nectar concentration. Bees from the high-pollen strain were more likely than bees from the low-pollen strain (selection described above) to respond to each concentration touched to the antennae.

### Choice behavior: expected rewards

### Some models and concerns about currency

Models of food choice based on energetics (e.g., Charnov 1976) initially focused on responses made in relation to the long-term expected rate of net energy gain [(gain − cost)/time] of alternative foods. Because of a presumed positive relationship between long-term rate of intake and fitness, animals are expected to prefer alternatives with the highest intake rate. Some experiments suggest that the long-term rate of net energy gain provides a reasonable currency and time-scale for studying choice behavior based on expected payoffs (Stephens & Krebs 1986). In some situations other than individual food choice behavior, the ratio [(gain – cost)/cost] seems to better predict the behavior of honeybees (Schmid-Hempel *et al.* 1985; Seeley 1986; Wolf & Schmid-Hempel 1990).

Other researchers, however, have suggested that bees maximize short-term rather than long-term rates of energy gain. Waddington & Holden (1979) developed an optimality model that assumed bees choose the best flower (highest gain/time) among all nearby flowers on each move. The model successfully predicted honeybees' choice behavior under the limited set of conditions tested. Bumble bees also appear to maximize short-term rate of net energy gain (Real *et al.* 1982; Harder & Real 1987). Real *et al.* (1990) point out that bumble bees may be forced to base foraging decisions on a few recent visits, because they are constrained by the amount of information they can remember from previous visits. Although some evidence suggests that bees maximize in the short-term, this important problem deserves further investigation.

A problem with this general modeling approach used to predict choice behavior is that the model can fail at different places in the modeling process. If predicted and observed choice behavior differ, it may be because an incorrect maximized currency is assumed, or it may be that one of the operational assumptions is incorrect. These possibilities are very difficult to tease apart. How, for example, does one, outside of the choice model itself, independently test the validity of the assumed currency?

### Choice – subjective evaluation of objective information

If the goal is simply to predict bees' choice behavior between two or more different kinds of flowers with different expected rewards (e.g., nectar concentration), it should be possible to make the predictions based on the functional relationships between subjective evaluations and objective information (Fig. 3.1). Since animals are presumed to make decisions based on their evaluations, it is these relationships that should be most revealing in predicting choices.

#### Nectar concentration

The strength of preference for the more rewarding of two types of flowers that differ in associated expected concentration of nectar is expected to vary depending on the magnitude of the concentration difference. The more similar the concentrations, the more likely it is that the difference

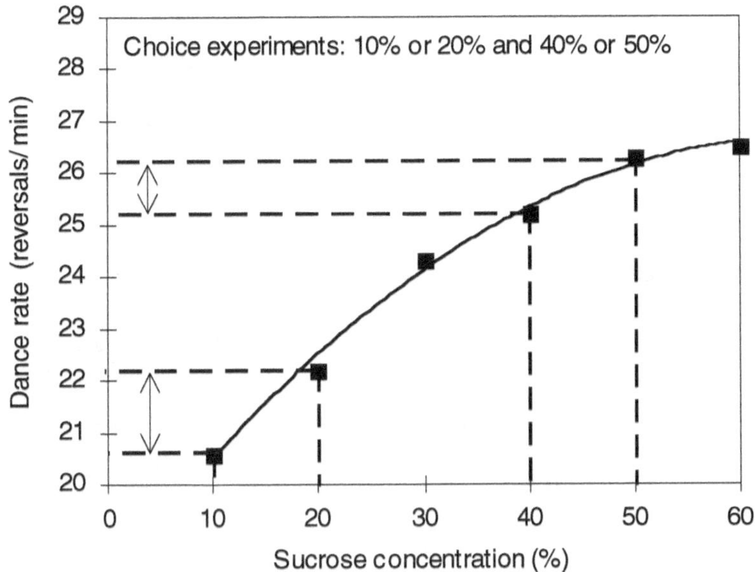


Fig. 3.5. Honeybee's dance rate in relation to concentration, taken from Fig. 3.2. The perceived values of concentration are shown by projecting concentration onto the perceptual scale (*y* axis). On the perceptual scale the difference between 10% and 20% is greater than the difference between 40% and 50%; thus, the preference for 20% over 10% is predicted to be stronger than the preference for 50% over 40%.

will not be perceived, or that the perceived difference will not elicit a preference. Thus, a bee might prefer a 40% solution to 30%, but might be indifferent to, or prefer less strongly, the 5% difference of 40% over 35%. Now, because the relationship is non-linear between dance rate (subjective evaluation) and sucrose concentration, the strength of preference for any given difference in concentration should vary along the concentration scale. For example, it is expected that the preference for 50% over 40% will be less than the preference for 20% over 10%, because the perceived difference between 20% and 10% is greater than the perceived difference between 50% and 40% (Fig. 3.5).

We examined this prediction (Fig. 3.6). Bumble bees (*Bombus impatiens*) were given a choice between two types (colors) of artificial flowers that differed in sugar concentration (K. D. Waddington, S. Lamenta & M. Jordan, unpublished data). Two sets of bees were tested. One set always obtained 1 μl of 10% sucrose from one type of flower (e.g., blue) and 20% from the other type (e.g., yellow). The other set of bees obtained 40% from one type

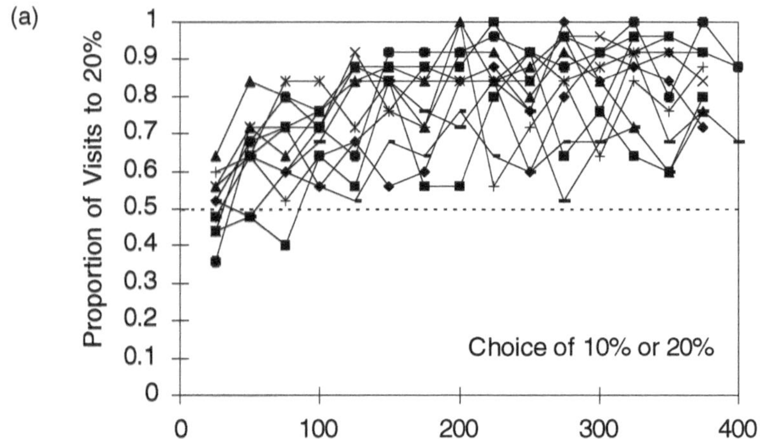


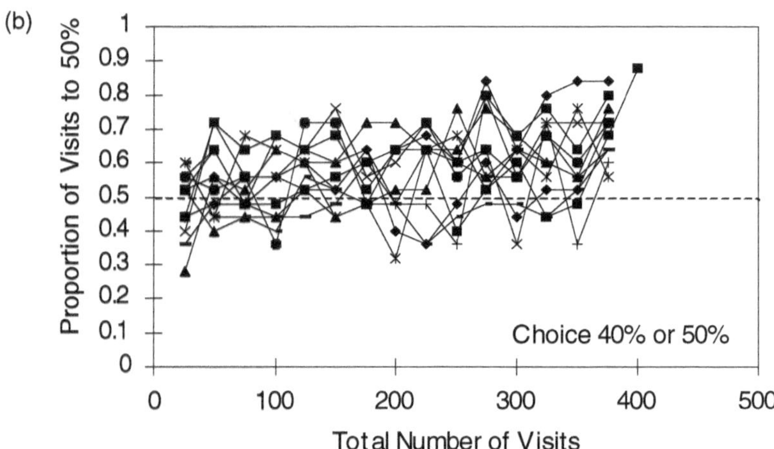


Fig. 3.6. Proportion of visits to the flower type, of two types, containing the higher concentration of 1 μl of sucrose solution. The foraging history of each *Bombus impatiens* worker is connected by a line. (a) One type provided 10%, the other 20%; (b) one type provided 40%, the other 50%. The dotted line in each graph shows the behavior of indifferent bees.

and 50% from the other. Experienced bumble bees (those with 250 or more visits) tended to prefer the 20% over the 10% ($\bar{x} = 0.82$, SD = 0.090, $n = 12$ bees) more strongly than they preferred the 50% over the 40% ($\bar{x} = 0.61$, S.D. = 0.071, n = 12 bees), the results qualitatively predicted by the perceptual scale (Fig. 3.6).

Waddington & Gottlieb (1990) investigated the choice behavior of honeybees by giving them a binary choice between tubular blue and yellow artificial flowers. The flowers contained different volumes of sucrose solution. The degree of preference for the flower type having more solution depended on the relative difference in volume, rather than absolute difference between the flower types. Such behavior is consistent with decisions based on a non-linear relationship between subjective evaluation and objective nectar volume.

### Pollen

Foragers make decisions and choices based on the quality and quantity of pollen. Harder's (1990) work indicated that bumble bees make more flower visits per inflorescence visit when the flowers contain more pollen; they also revisit pollen-rich flowers more frequently. In some way, the bees gather information on the volume of pollen they have harvested, then base decisions on it. Buchmann & Cane (1989) also found clear empirical evidence that bumble bees and bees of the genus *Ptiloglossa* assessed the amount of pollen collected from individual *Solanum* flowers. Schmidt (1982, 1984) demonstrated that honeybees prefer pollen from certain plant species over others; these choices may be based on odor or taste of the pollen. Schmidt & Johnson's (1984) work suggests choices may be based on the pH and the percentage of crude protein in the pollen.

K.D. Waddington, C.M. Nelson & R.E. Page (unpublished data) quantified honeybees' choices between pollens of different quality to see if the choices were consistent with a non-linear subjective evaluation of pollen quality. We manipulated pollen quality by diluting pure pollen with various proportions of alpha-cellulose (vol./vol.) (see above). Bees were presented with choices between two dishes of pollen: pure pollen and a 1:1 dilution, pure pollen and a 3:1 (pollen:cellulose) dilution, or a 1:1 dilution and a 3:1 dilution. The difference in quality between the 1:1 and 3:1 mixtures is the same as that between the 3:1 mixture and pure pollen. The number of foragers counted at each dish of pollen of a pair was used to gauge indifference (equal number of foragers) or a preference for one dish of the pair. The bees were indifferent between pure pollen and the 3:1 mixture, but they preferred the 3:1 mixture (and pure pollen) over the 1:1 mixture. The results further confirm that bees base their choices on pollen quality. The choices observed suggest that the honeybees evaluated pure pollen and the 3:1 mixture equally but the 3:1 mixture was evaluated as greater than the 1:1 mixture; these choices are consistent with the hypothesis that evaluation of pollen quality is a non-linear function of pollen quality.

All these results are consistent with the hypothesis that bees base choices on subjective evaluations that are non-linearly related to objective nectar volume or concentration and pollen quality.

### Choice behavior: variation in nectar concentration

Within and among species of plants, flowers vary in the concentration and volume of their nectar. The question is: do bees make choices based on these sources of variation? For example, if two types of flowers provide the same expected nectar concentration, but the predictability of the concentration is high in one and low in the other, does it matter? Behavioral ecologists have recognized the effects of variability on choice, and incorporated reward variation into optimality models of foraging behavior. Response to variation, called "risk sensitivity," has become central to studies on choice behavior (Caraco 1980; Stephens & Charnov 1982). Risk sensitivity has been examined experimentally using many kinds of animals, including several species of nectarivorous invertebrates and vertebrates (see Kacelnik & Bateson 1996 for references).

Elsewhere, summaries can be found of risk-sensitive foraging models (Stephens & Paton 1986). However, Harder & Real (1987) began a discussion of the mechanisms underlying risk-sensitive foraging behavior which has been reviewed elsewhere (Perez & Waddington 1996; Waddington 1997). Here I will restrict the discussion by relating some recent data on choice behavior in relation to variation in nectar concentration to the ideas derived from measurement of subjective evaluation of concentration in honeybees.

We have studied response to variation in sugar concentration in three genera of bees: *Apis mellifera* (honeybees; Banschbach & Waddington 1994), *Xylocopa micans* (carpenter bees; Perez & Waddington 1996), *Bombus impatiens* and *Bombus fervidus* (bumble bees; Waddington 1995, K.D. Waddington, S. Lamenta & M. Jordan, unpublished data). Bees were given a choice between two types (colors) of artificial flowers which were equivalent in expected concentrations of sugar solution, but the types differed in the distributions of concentrations provided. The experimental paradigm follows the early study of response to variation in volume (Waddington *et al.* 1981). The bee always found the same concentration in the low-variance type and found a random sequence of two concentrations in the high-variance type. We obtained two results. Honeybees, *Bombus fervidus*, and carpenter bees were indifferent to (showed no preference between) the two types when the low-variance type provided 20%

sucrose and the high-variance provided 10% or 30% equally frequently in random sequence. However, when *Bombus impatiens* experienced a mean concentration of 30% and the high variance flower provided 10% or 50%, the bees preferred the low-variance flower (30%, $p = 1.0$), i.e., they became risk-averse (Fig. 3.7a).

Both indifference and the preference for the low-variance flower are consistent with the rule: visit the flower type with the higher perceived expected concentration. In Fig. 3.7b, concentrations are projected onto the perception axis using the relationship between concentration and honeybee dance rates (Fig. 3.2). A constant 30% has a higher perceived expected value than the lottery of 10% and 50% ($p = 0.5$); bees appear to be maximizing perceived expected concentration.

In the studies of three other species of bees, a constant 20% was not preferred to the variable 10% and 30% presented equally frequently. A return to Fig. 3.7b, provides an explanation for the indifference. If these concentrations (10%, 20%, and 30%) were each projected onto the perceptual axis, the perceived expected concentrations would be nearly equivalent (dance rate about 22.5) for the high- and low-variance flowers, because the evaluation is linearly related to concentrations over the range of 10%–30% (Perez & Waddington 1996).

These are remarkable matches between the relationship describing perception of concentration and response to variance in concentration, especially considering that I used the honeybee's scale to predict choice in three other species of bees. Of course, the actual relationship describing perception over the whole dynamic range of concentration in bumble bees and carpenter bees may differ from that of honeybees.

### Summary

I suggest that we can link a bee's choice behavior to that bee's subjective evaluations of the quantities and qualities of pollen and nectar. These evaluations can be quantified – independently of choice behavior – using the honeybee's round dance. This technique is analogous to asking human beings to place the magnitude of a stimulus on a scale (Carterette & Friedman 1974). Thus far, we have found that the value perceived by bees is a non-linear function of nectar concentration. The non-linear relationship is generally described by the Weber–Fechner law of perception, which is a very general property of animals; the weighting of losses

(a)

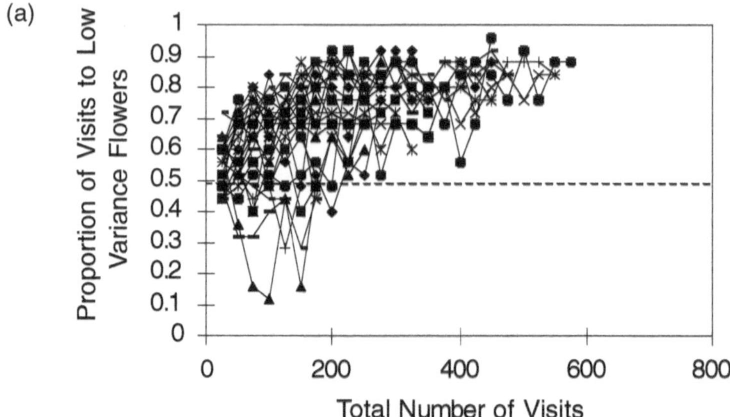


(b)

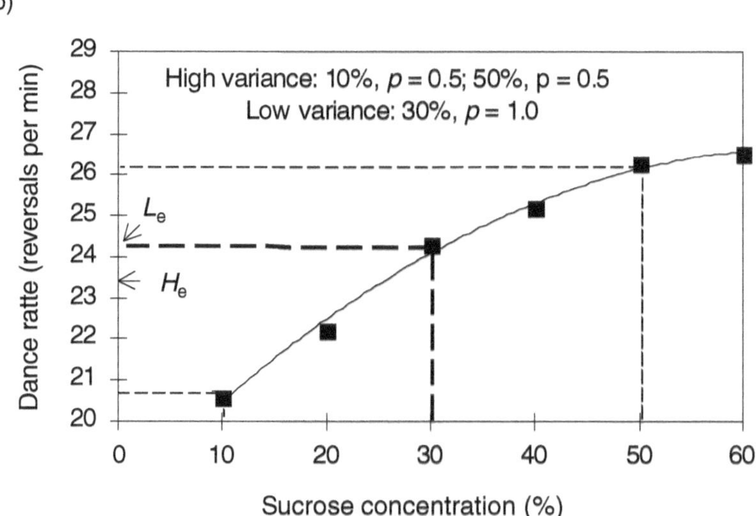


Fig. 3.7. (a) Proportion of visits by 24 *Bombus impatiens* workers to the low-variance flowers that provided 1 μl of 30% sucrose with $p = 1.0$. The high-variance flowers provided 10% or 50% in random sequence with $p = 0.5$. The dotted line shows the behavior of risk-indifferent bees. (b) As an estimate of the bumble bees' subjective evaluation of nectar concentration, I have reproduced the graph from Fig. 3.2 on the relationship between dance rate and concentration for the honeybee. The perceived concentrations are shown for the high- and low-variance flowers by projecting concentration onto the perceptual scale ($y$ axis). The arrows indicate the perceived expected concentration for the high ($H_e$) and low ($L_e$) variance flowers. The subjective evaluations of concentrations may differ in detail for bumble bees and honeybees.

to gains is known for human beings (Tversky & Kahneman 1981). This suggests that in a competitive market for pollinator service, there is a decelerating payoff to plants (in terms of attracting pollinators) for increasing the quality of their nectar. The same likely applies to volume. We also found that losses (declining concentration) and energetic costs are perceptually weighted relative to gains. Concerning the latter, a change in floral morphology that increases or decreases handling cost only slightly can have important effects on the attractiveness of the flower to pollinators. I think that we can expect these patterns of behavior to apply generally to other pollen- and nectar-foraging taxa.

The perceptual scale based on the honeybee dance successfully predicted, in laboratory studies using artificial flowers, choice behavior in relation to expected sucrose concentration and variance in sucrose concentration. Preference for concentration depends on the magnitude of difference between the concentrations and magnitudes of the concentrations. Response to variance in our experiments depended on the magnitude of the expected concentrations and of the magnitude in variance of the high-variance flower. Choices among flowers differing in expected nectar volume and variance in volume, and pollen quality, all are explained by non-linear perceptual relationships; however, there is variation among bees in their perception. Our studies show that genotypic variation is one source of variation in perception. Genotypic variation in honeybees' choice behavior is consistent with their differing perceptual scales indicated by the dance. Finally, the proboscis extension response; (PER) is just beginning to be used to study perception of sucrose concentration. Thus far, PER results suggest genotypic variation in response and experiential factors which modulate the response; PER also has promise for studying risk-sensitive foraging behavior (Shafir *et al.* 1999). This technique will be especially useful for exploring perceptual differences among individual bees.

### Acknowledgments

I thank Dr Peter Vanderborght of the International Society for Horticultural Science for giving me permission to reprint parts of Waddington (1997) here. I also thank Academic Press and Blackwell Wissenschafts-Verlag for permitting me to reproduce figures. Dr C. Mindy Nelson made helpful comments on the manuscript.

## References

Banschbach VS & Waddington KD (1994) Risk-sensitive foraging in honeybees: no consensus among individuals and no effect of colony honey stores. *Anim Behav* 7:933–941

Barth FG (1985) *Insects and flowers: the biology of a partnership.* Princeton University Press, Princeton, NJ

Bitterman ME, Menzel R, Fietz A & Schafer S (1983) Classical conditioning of proboscis extension in honeybees (*Apis mellifera*). *J Comp Psychol* 97:107–119

Buchmann SL & Cane JH (1989) Bees assess pollen returns while sonicating *Solanum* flowers. *Oecologia* 81:289–294

Calderone N & Page RE (1988) Genotypic variability in age polyethism and task specialization in the honeybee, *Apis mellifera* (Hymenoptera; Apidae). *Behav Ecol Sociobiol* 22:17–25

Calderone N & Page RE (1992) Effects of interactions among genetically diverse nestmates on task specialization by foraging honeybees (*Apis mellifera*). *Behav Ecol Sociobiol* 30:219–226

Caraco T (1980) On foraging time allocation in a stochastic environment. *Ecology* 61:119–128

Carterette EC & Friedman MP (1974) *Handbook of perception*, vol. 2. Academic Press, New York

Charnov E (1976) Optimal foraging: attack strategy of a mantid. *Am Nat* 110:141–151

Deng G (1996) Foraging performance of honeybees (*Apis mellifera*). PhD thesis, University of Miami, FL

Faegri K & van der Pijl, L (1979) *The principles of pollination ecology*, 3rd edn. Pergamon Press, New York

Fewell JH & Page RE (1993) Genotypic variation in foraging responses to environmental stimuli by honeybees, *Apis mellifera*. *Experientia* 49:1106–1112

Frisch K von (1967) *Dance language and orientation of bees*. Belknap Press, Cambridge, MA

Grant V (1949) Pollination systems as isolating mechanisms in flowering plants. *Evolution* 3:82–97

Greggers U & Menzel R (1993) Memory dynamics and foraging strategies of honeybees. *Behav Ecol Sociobiol* 32:17–29

Guzman-Novoa E & Gary NE (1993) Genotypic variability of components of foraging behaviour in honeybees (Hymenoptera; Apidae) *J Econ Entomol* 86:715–721

Harder LD (1990) Behavioral responses by bumble bees to variation in pollen availability. *Oecologia* 85:41–47

Harder L & Real LA (1987) Why are bumble bees risk averse? *Ecology* 68:1104–1108

Kacelnik A & Bateson M (1996) Risky theories – the effects of variance on foraging decisions. *Am Zool* 36:402–434

Macior LW (1970) The pollination ecology of *Pedicularis* in Colorado. *Am J Bot* 57:716–728

Milne CP, Hellmich RL & Pries KJ (1986) Corbicular size in workers from honeybee lines selected for high or low pollen hoarding. *J Apic Res* 25:50–52

Page RE, Waddington KD, Hunt GJ and Fondrk MK (1995) Genetic determinants of honeybee foraging behaviour. *Anim Behav* 50:1617–1625

Page RE, Erber J & Fondrk MK (1998) The effect of genotype on response thresholds to sucrose and foraging behavior of honeybees (*Apis mellifera* L.). *J Comp Physiol A* 182:489–500

Perez SP & Waddington KD (1996) Carpenter bee (*Xylocopa micans*) risk-indifference and a review of nectarivore risk-sensitivity studies. *Am Zool* 36:435–446

Possingham HP, Houston AI & McNamara JM (1990) Risk-averse foraging in bees: a comment on the model of Harder and Real. *Ecology* 71:1622–1624

Pyke GH (1979) Optimal foraging in bumblebees: rule of movement between flowers within inflorescences. *Anim Behav* 27:1167–1181

Pyke GH (1984) Optimal foraging theory: a critical review. *Ann Rev Ecol Syst* 15:523–575

Rachlin H (1989) *Judgement, decision, and choice.* WH Freeman, New York

Rasheed SA & Harder LD (1997*a*) Foraging currencies for non-energetic resources: pollen collection by bumblebees. *Anim Behav* 54:911–926

Rasheed SA & Harder LD (1997*b*) Economic motivation for plant species preferences of pollen-collecting bumble bees. *Ecol Entomol* 22: 209–219

Raveret-Richter M & Waddington KD (1993) Past foraging experience influences honeybee dance behaviour. *Anim Behav* 46:123–128

Real L, Ott J & Silverfine E (1982) On the trade-off between the mean and the variance in foraging: effects of spatial distribution and color preference. *Ecology* 63:1617–1623

Real LA, Ellner S & Harder LD (1990) Short-term energy maximization and risk-aversion in bumblebees: comments on Possingham *et al. Ecology* 71:1625–1628

Schmid-Hempel P, Kacelnik A & Houston AI (1985) Honey bees maximize efficiency by not filling their crop. *Behav Ecol Sociobiol* 17:61–66

Schmidt JO (1982) Pollen foraging preferences of honeybees. *Southwest Nat* 7:255–259

Schmidt JO (1984) Feeding preferences of *Apis mellifera* L. (Hymenoptera: Apidae): individual versus mixed pollen species. *J Kans Entmol Soc* 57:323–327

Schmidt JO & Johnson BE (1984) Pollen feeding preference of *Apis mellifera*, a polylectic bee. *Southwest* Nat 9:41–47

Seeley T (1986) Social foraging by honeybees: how colonies allocate foragers among patches of flowers. *Behav Ecol Sociobiol* 19:343–354

Selten R & Shmida A (1991) Pollinator foraging and flower competition in a game equilibrium model. In: Selten R (ed) *Game theory in behavioural sciences*, pp 195–256. Springer-Verlag, Berlin

Shafir S (1994) Intransitivity of preferences in honeybees: support for "comparative" evaluation of foraging options. *Anim Behav* 48:55–67

Shafir S, Wiegmann DD, Smith BH & Real L (1999) Risk-sensitive foraging: choice behaviour of honeybees in response to variability in volume of reward. *Anim Behav* 57:1055–1061

Stephens DW & Charnov E (1982) Optimal foraging: some simple stochastic models. *Behav Ecol Sociobiol* 10:251–263

Stephens DW & Krebs JR (1986) *Foraging theory.* Princeton University Press, Princeton, NJ

Stephens DW & Paton SR (1986) How constant is the constant of risk-aversion? *Anim Behav* 34:1659–1667

Tversky A & Kahneman D (1981) The framing of decisions and psychology of choice. *Science* 211:453–458

Waddington KD (1981) Factors influencing pollen flow in bumblebee-pollinated *Delphinium virescens. Oikos* 37:153–159

Waddington KD (1982) Honeybee foraging profitability and round dance correlates. *J Comp Physiol* 148:279–301

Waddington KD (1983) Foraging behavior of pollinators. In: Real L (ed) *Pollination biology*, pp 213–239. Academic Press, NY

Waddington KD (1985) Cost–intake information used in foraging. *J Insect Physiol* 31:891–897

Waddington KD (1995) Bumblebees do not respond to variance in nectar concentration. *Ethology* 101:33–38

Waddington KD (1997) Foraging behavior of nectarivores and pollen collectors. *Acta Hort* 437:175–191

Waddington KD & Gottlieb N (1990) Actual vs. perceived profitability: a study of floral-choice of honeybees. *J Insect Behav* 3:429–441

Waddington KD & Holden LR (1979) Optimal foraging: on flower selection by bees. *Am Nat* 114:179–196

Waddington KD & Kirchner W (1992) Acoustical and behavioral correlates of profitability of food sources in honeybee round dances. *Ethology* 92:1–6

Waddington KD, Allen T & Heinrich B (1981) Floral preferences of bumblebees (*Bombus edwardsii*) in relation to intermittent versus continuous rewards. *Anim Behav* 29:779–784

Waddington, KD, Nelson CM & Page RE (1998) Effects of pollen quality and genotype on the dance of foraging honeybees. *Anim Behav* 56:35–39

Wolf T & Schmid-Hempel P (1990) On the integration of individual foraging strategies with colony ergonomics in social insects: nectar collection in honeybees. *Behav Ecol Sociobiol* 27:103–111

4

___

# Honeybee vision and floral displays: from detection to close-up recognition

In a social insect such as the honeybee, the survival of the colony depends on the success of its foragers. The bee optimizes its foraging success by returning to flowers of the species at which it has previously found food. This so-called flower constancy (see Chittka *et al.* 1999 for references) is based on the bee's capacity to learn and memorize specific flower signals (Menzel *et al.* 1993; Menzel & Müller 1996; Menzel 1999 and this volume) and to discriminate among different species by their different signals.

A bee returning to the feeding site in search of a flower, be it natural or artificial, must first detect the target from a distance. Once the flower has been detected, the bee will approach it up to a distance at which it is able to recognize whether or not the flower is similar to that stored in memory. Among the different sensory cues used, visual cues are of fundamental importance. In the rich market of coexisting and competing flower species, flower colors, shapes, and patterns are the visual cues that allow bees to recognize and discriminate profitable species.

Here we review studies concerned with the bee's use of visual signals for detecting and recognizing food sources. In the first part of the chapter, we examine the role of the bee's color vision in these tasks. In the second part, we look at the role of several spatial parameters contained in achromatic (black-and-white) stimuli. In both cases, we ask whether or not the visual cues that serve the bee for detecting a target at some distance are identical with those that serve for recognizing it at a closer range. We further ask whether the bee's use of particular visual cues in particular situations can be modified through experience. Flexibility of behavior based on learning may indicate some cognitive capacities on the part of the animal.

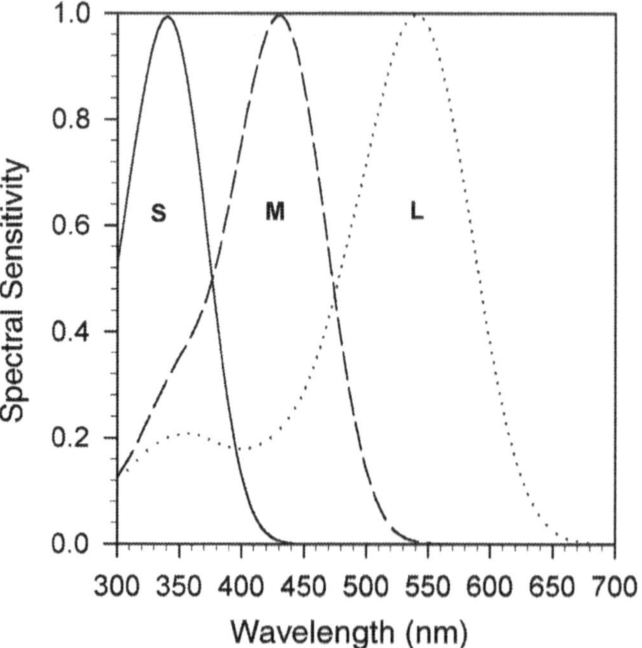

Fig. 4.1. The spectral sensitivity curves of the three photoreceptor types of the honeybee, *Apis mellifera*: S (UV receptor, solid line), M (blue receptor, dashed line), and L (green receptor, dotted line).

## Detection and recognition of colored targets

### The bee's color vision system

Behavioral (von Frisch 1914; Daumer 1956; Menzel 1967; Helversen 1972) and electrophysiological studies (Autrum & Zwehl 1964; Menzel & Blakers 1976; Peitsch *et al.* 1992) have shown that the honeybee possesses trichromatic color vision, with three spectral types of photoreceptors. Their sensitivity peaks are at 344 nm in the short-wave (ultra violet) region of the spectrum (S receptor), 436 nm in the middle-wave (blue) region (M receptor), and 544 nm in the long-wave (green) region of the spectrum (L receptor), respectively (Fig. 4.1).

An individual receptor can determine the amount of light it absorbs, but not the spectral composition of that light. Because the eye has three spectral types of photoreceptors, however, two juxtaposed color areas (a target and its background) will produce against each other three types of contrast: (1) Color contrast. Color is encoded in the comparison among the inputs arriving from the three spectral types of receptor, i.e., the three receptor types interact at a higher neural level. Several models of these interactions have been proposed, but are not discussed in detail here (for reviews, see Menzel & Backhaus 1991; Vorobyev & Brandt 1997). The output of the interactions is the perceptual color distance (i.e., the animal-subjective color difference) between stimuli. In the studies reviewed here, the calculations of color contrast are based on the neural interactions postulated in the color-opponent model proposed by Backhaus (1991) (see Giurfa *et al.* 1997 for formulas). (2) Receptor-specific contrasts, defined as the difference in excitations produced by two stimuli in each of the three receptor types, S (UV), M (blue), and L (green). We thus differentiate among "UV contrast," "blue contrast," and "green contrast". Because only one receptor type is involved in each case, receptor-specific contrasts are achromatic, i.e., they do not give rise to the perception of a color. (3) Intensity contrast, defined as the difference between the two stimuli in the sum of excitations that they evoke in all three receptor types. Summing the excitations prevents the comparison among the individual inputs. Therefore, intensity contrast is achromatic. In color discrimination tasks, bees make no use of intensity differences (Daumer 1956; Helversen 1972; Backhaus 1991; Menzel & Backhaus 1991).

### The detection phase: the role of chromatic and achromatic cues and the minimum detectable angle

Traditionally, flower colors have been considered to constitute long-distance signals that enable detection of the flower from afar (von Frisch 1967; Kevan & Baker 1983; Chittka & Menzel 1992). However, the distance at which a flower is first detected cannot measure the detectability of its *color*, because that distance depends on the size of the flower. Therefore, the relevant measure of detectability of a distant target is the visual angle that it subtends on the insect's eye. This angle depends on both distance and size.

To find the minimum visual angle for detection of a colored target, bees were trained to collect a food reward at a colored disc presented against an achromatic (gray) background (Giurfa *et al.* 1996*b*). The experimental set-up consisted of a Y-shaped, dual-arm apparatus (Fig. 4.2). In one of the arms, termed positive, a colored disc was presented against the gray background. A bee entering that arm received a reward of sucrose solution when it arrived at the colored disc. The alternative arm, termed negative, displayed the gray background only, with no reward. Thus, bees

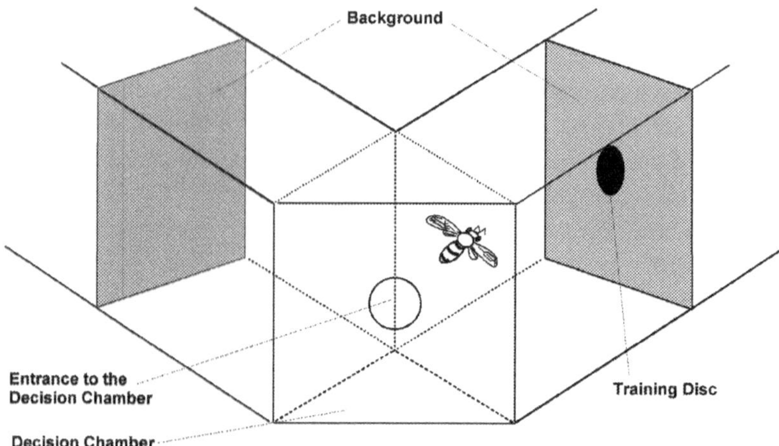



Fig. 4.2. View of the Y-maze apparatus. The decision between the arm with the training disc and that with the background alone could be made by the bee only after it has entered the decision chamber, from which the back walls of both arms could be viewed simultaneously. The decision point was defined as the center of the decision chamber. A choice of either arm was scored when the bee crossed the imaginary line between the decision chamber and one of the arms. The visual angle subtended by the rewarding disc at the bee's eye was calculated from the distance of the disc from the decision point, and the diameter of the disc. For more details, see Giurfa *et al.* (1996*b*).

were trained to distinguish between the presence and the absence of a colored spot.

Six different colored discs were used in six different experiments. The visual angle subtended by the disc was varied by changing the disc diameter or the distance between the back wall and the decision point. Because changing the disc's distance had the same effect as changing its diameter (see also Lehrer & Bischof 1995), the results were pooled. The minimum visual angle at which a given stimulus is detectable was defined as the angle at which the bees chose the positive arm with a frequency of 60%.

The six stimuli presented comparable amounts of color contrast against the gray background. The results, however, differed among the six experiments (Fig. 4.3). Four of the stimuli rendered a threshold of detectability at a minimum visual angle of 5°. These stimuli produced, in addition to color contrast, (achromatic) green contrast against the background. The other two stimuli were not detected unless they subtended a visual angle of at least 15°. In these stimuli, green contrast was absent.

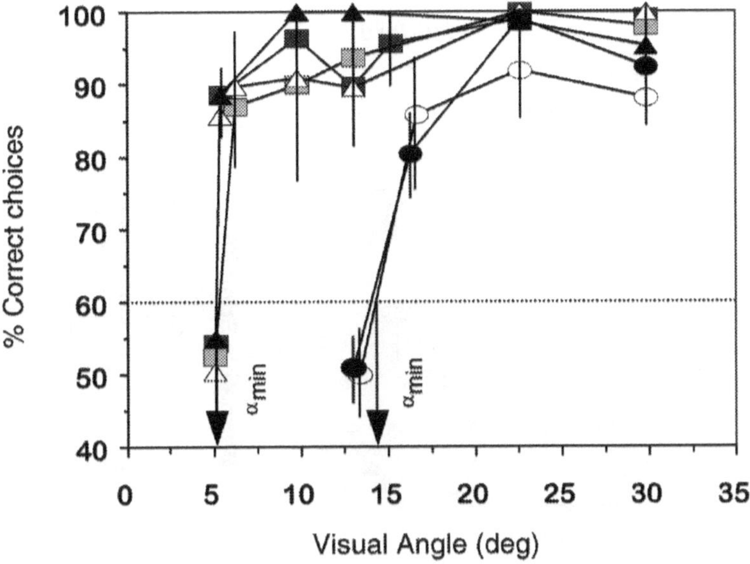

Fig. 4.3. Percentage of correct choices as a function of the visual angle subtended by different color discs presented against an achromatic (gray) background. Each symbol corresponds to a different color stimulus. A 95% confidence interval is depicted for each data point. The dotted horizontal line represents the 60% choice level used to determine the minimum visual angle ($\alpha_{min}$) that the disc must subtend at the eye in order to be detected.

Variation in UV and blue contrast did not affect detection (Giurfa *et al.* 1996b).

These results suggest that achromatic and chromatic cues are used in succession during the approach flight: from afar, detection requires the presence of green contrast (an achromatic cue), whereas, as the bee comes closer, it uses chromatic (color) information.

### Detection versus recognition: alternative use of achromatic and chromatic cues

The results above (Giurfa *et al.* 1996b) suggested the alternative use of achromatic and chromatic cues in the detection and recognition of colored targets. However, the stimuli that were detected at an angle of 5° (Fig. 4.3) contained both chromatic and achromatic (green) contrast, whereas those that were only detected at a visual angle of 15° contained chromatic contrast, but no green contrast. The data do not let us conclude that green contrast cannot be used when the stimulus subtends a large angle at the

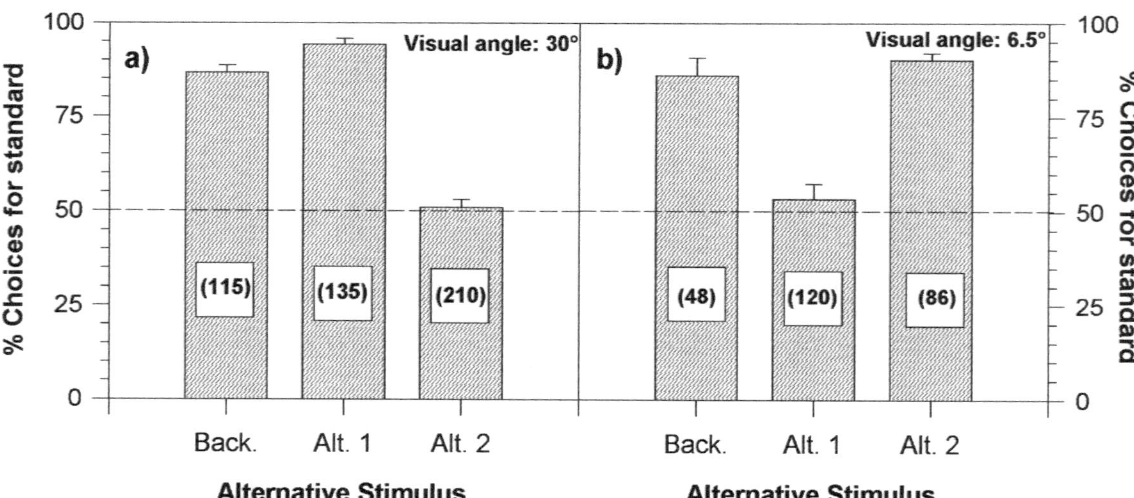


Fig. 4.4. Percentage of choices for the rewarding standard color stimulus in tests against the background alone ("Back.") and against each of two alternative color stimuli ("Alt. 1" and "Alt. 2"). The dashed line at 50% indicates random choice. Values in parentheses depict the total number of choices recorded in each test situation. The visual angle subtended by the stimuli was (a) 30° or (b) 6.5°. With either visual angle, bees discriminated between the arm containing the standard colored disc and the arm containing the background alone. However, discrimination of the standard from each of the two alternatives, 1 and 2, differed significantly between the two experiments.

eye. Nor can we conclude that green contrast is used exclusively at small visual angles. Before such conclusions can be drawn, one must investigate the independent roles of chromatic and achromatic cues at small as well as at large visual angles.

To this end, individual bees were rewarded at a colored disc (henceforth termed "standard") that produced both chromatic and achromatic (green) contrast against the gray background (Giurfa *et al.* 1997). Bees were first trained to discriminate the standard from the background alone, as described above. The standard was then presented alternately against two different stimuli. Alternative 1 differed from the standard in color contrast, but it contained a similar amount of green contrast. Alternative 2 differed in the amount of green contrast, but not in the chromatic contrast that it produced against the background (for details of the colors used, see Giurfa *et al.* 1997). The visual angle subtended by the stimuli was 30° in one experiment and 6.5° in another.

Bees trained and tested with stimuli subtending 30° detected the standard against the background alone, and chose it correctly when it was presented against Alternative 1, i.e., they used chromatic cues (Fig. 4.4a). At the same visual angle, bees chose the standard at random when it was tested against Alternative 2, i.e., the amount of green contrast played no role.

When the visual angle subtended by the stimuli was 6.5°, bees still detected the standard against the background alone (Fig. 4.4b). In the tests against the alternative stimuli, however, their performance was reversed compared to the previous experiment. They now discriminated the standard from Alternative 2, which differed from it in green contrast but not in chromatic contrast, and were incapable of discriminating the standard from Alternative 1, which differed from it in chromatic contrast, but not in green contrast.

Thus, bees use either chromatic or achromatic cues, depending on the visual angle subtended by the stimuli. At small visual angles, i.e., at a distance at which detection first occurs, bees rely on the achromatic signal provided by the green contrast, and not on chromatic information. At large visual angles, they rely on chromatic information. At these large angles, achromatic cues are ignored, even when they are, in principle, available to the visual system. These conclusions are in accordance with the results of earlier studies on color discrimination in bees (for selected references, see e.g., Menzel & Backhaus 1991), all of which were conducted using colored stimuli viewed at a very close range and thus subtending large angles at the bee's eye.

### Bees discriminate among different amounts of green contrast

Most natural flowers produce green contrast against the background, because all natural backgrounds (foliage, soil, rocks) are green or gray, whereas animal-adapted flowers are hardly ever green. If the bee's performance in the detection phase were to depend only on the presence of green contrast, then bees arriving at the feeding site would be expected to steer towards any flower. Only an ability to learn a particular amount of green contrast would ensure that the bee steers towards the particular flower that it has memorized on a previous visit.

The question of whether or not bees possess this ability was investigated by training bees to a standard stimulus that was later tested against three alternative stimuli, all of which differed from it in the amount of color contrast (Giurfa *et al.* 1997). At large visual angles, bees preferred the standard against each of the three alternatives, i.e., their choices were based on color discrimination. At small visual angles, however, the bees discriminated the standard from the two alternatives that presented either a higher or a lower amount of green contrast, but not from the alternative that matched the standard in the amount of green contrast. Therefore, when a rewarding stimulus containing green contrast is viewed at a small visual angle, bees not only detect it, they also recognize the particular amount of green contrast contained in it. Bees can thus discriminate the learned stimulus from other stimuli as soon as they can detect it.

### The use of achromatic spatial cues contained in black-and-white stimuli

Black, white, and gray stimuli evoke equal amounts of excitation in all three spectral types of photoreceptor. The comparison among the three photoreceptor excitations (see section "The bee's color vision system," above) thus renders no differences among them, and therefore no color is perceived. Such achromatic stimuli can be discriminated only on the basis of intensity differences (intensity defined as the sum of the three excitations).

The role of intensity contrast in resolution and discrimination of achromatic stimuli has been demonstrated in many studies (see Wehner 1981 for review). Even in object-detection tasks, intensity contrast is crucial when achromatic stimuli are used (Lehrer & Bischof 1995). Most of the studies on pattern discrimination in the honeybee were conducted using black-and-white stimuli, because these stimuli produce the highest possible amount of intensity contrast. We shall restrict our attention, as we did in our first section, to studies that compare performance at different visual angles. Experimental results allow such a comparison with respect to three spatial parameters: (1) pattern disruption (spatial frequency); (2) contour orientation (spatial alignment); and (3) symmetry.

### Pattern disruption and its role in pattern-discrimination tasks

Natural flowers differ in their degree of outline disruption, i.e., in the amount of edge per unit area (also referred to as "contour density"). This parameter is a good candidate to serve the bee in pattern-discrimination tasks. For example, a flower with six petals will appear more disrupted to the eye of a flying bee than a flower of the same area with only three petals. Indeed, all of the earlier workers on bee spatial vision have agreed that the degree of disruption constitutes the main spatial parameter that bees use for pattern discrimination (Hertz 1929, 1933; Zerrahn 1933; Wolf & Zerrahn-Wolf 1935; Free 1970; Anderson 1977). It was suggested that the bee measures contour density in terms of the average frequency of intensity fluctuations (on- and off-stimulation, also termed "flicker") that the photoreceptors experience: more disrupted patterns produce a higher temporal frequency of intensity fluctuations than do less disrupted patterns.

In most of the early studies, investigators presented the patterns on a horizontal plane, recording the bee's choices at a very close distance from the patterns. When given a choice among novel patterns, bees preferred highly disrupted patterns over less disrupted ones (for references, see von Frisch 1967, Wehner 1981). However, patterns presented on a horizontal plane may change their appearance from one visit to the next, depending on the direction of the bee's approach. In this situation, contour density, a cue that is largely independent of the bee's direction of approach, is likely to be more reliable than any other spatial parameter.

To demonstrate the bee's use of spatial parameters other than disruption it was therefore necessary to present patterns on vertical planes, thus forcing the bees to arrive at the pattern from a constant direction. Even in the vertical plane, though, contour density was found to constitute an effective discrimination cue. Bees previously trained to a sectored disc of a particular spatial frequency discriminated that disc, in subsequent tests, from each of a series of novel discs that differed from it in spatial frequency. The greater the difference in disruption between the trained disc and the novel disc, the better the discrimination (Wehner 1981).

Presenting patterns in a vertical plane allows an experimenter to examine the bee's choice behavior towards disrupted patterns even at some distance from the patterns. To this end, Lehrer *et al.* (1995) used an experimental set-up consisting of 12 identical arms opening into a central arena from which bees had access to any of the arms (Fig. 4.5a). During training, six checkerboard patterns (Fig. 4.5b) were presented, one at a time in a quasi-random sequence, on the back wall of one the arms, the access to the reward box being through a tube penetrating the center of the pattern. The six checkerboards were randomized with respect to contour density, so bees could not memorize any particular spatial frequency. In subsequent tests, each of the 12 arms had a novel pattern placed on its back wall. In each test, a set of four patterns was used, each pattern being repeated in three different arms. The criterion for a choice was a bee entering an arm, i.e., 30 cm from the pattern. Regardless of the type of pattern, the lowest spatial frequency was the most attractive one (Fig. 4.5c, d), which is in contrast to the results of the earlier workers, who found a spontaneous preference for high-frequency patterns (see above).

The results indicate that, when patterns are viewed at a close range, bees prefer high spatial frequency, regardless of whether the stimuli are presented on a horizontal or vertical plane. At greater ranges, however, global cues – those provided by the pattern as a whole – dominate (e.g., Zhang *et al.* 1992; Lehrer *et al.* 1995; Horridge 1997). In this case, low spatial frequency imparts more accurate shape information than does high frequency.

### Pattern recognition by means of a template

Although it makes no difference whether patterns are presented on a horizontal or a vertical plane for discrimination of pattern disruption (see previous section), the mode of presentation is crucial when cues other than disruption are involved. Using patterns presented on a vertical plane, Wehner (1972) trained bees to a half-white and half-black disc, with the edge between the white and the black halves oriented at 45° with respect to the vertical. In subsequent dual-choice tests, bees discriminated

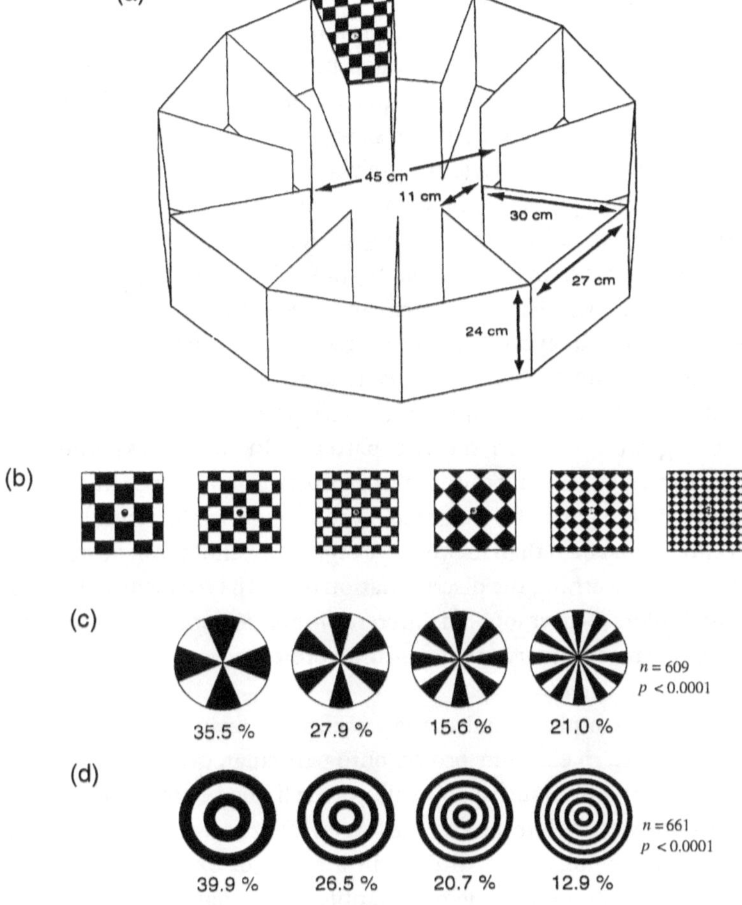



Fig. 4.5. Discrimination among patterns of different spatial frequencies viewed at a distance. (a) The apparatus consisted of 12 arms opening into a central arena through which bees had access to any of the arms. (b) Bees were trained to six different checkerboard patterns presented, one at a time, on the back wall of one of the arms. In the tests, the trained bees had to choose between four novel patterns, each presented on the back wall of three different arms. Choices were recorded at each arm's entrance. (c) and (d) The four patterns offered in each test differ in frequency, but not in type. Percentage of choices is given under each pattern. n depicts the total number of choices. Values of $p$ are results of $\chi^2$ tests comparing the test results with the distribution of choices expected under random-choice conditions. (Modified from Lehrer *et al.* 1995.)

the rewarding disc from each of a series of test discs rotated about their centers at various angles. The larger the angle of rotation, the better the discrimination. However, a comparison between results obtained with clockwise and counterclockwise rotation of the test patterns showed that the cue used in that discrimination task was not the spatial orientation of the edges, but rather the distribution of black and white areas in the visual field. These findings led to the formulation of the so-called "template theory" (see reviews by Srinivasan 1994; Heisenberg 1995), which proposed that bees store an eidetic ("photographic") image of the rewarding pattern in such a way that the pattern can be recognized only when its elements project on the eye in the same retinal positions that the bee viewed them during training. Discriminating a memorized pattern from a novel pattern is then based on the amount of overlapping and non-overlapping contrasting areas between the two patterns. In further experiments, Wehner (1972) showed that the discrepancy between the template and the currently viewed pattern is weighted more strongly in the ventral part of the frontal visual field than in other eye regions. A similar conclusion was later drawn concerning the discrimination of spatial frequencies (Lehrer 1997) and colors (Lehrer 1999) in different frontal regions of the bee's eye: best performance is observed when stimuli project onto the ventral part of the frontal visual field.

The use of a template requires the presence of a fixation point at which the various pattern elements project onto particular, constant regions of the eye. When viewed, such a point allows the flying bee to position itself in a constant way with respect to the target. The occurrence of a fixation reaction was demonstrated directly in a series of experiments using, among other techniques, cinematographic recordings (Wehner & Flatt 1977).

The patterns used by Wehner were rather large, subtending 130° at the fixation point. It is very likely that the bees were able to discriminate between the previously rewarding pattern and the novel one when they were still some distance from the target, i.e., during the approach flight. Therefore, although a bee's choice was recorded as the bee touched the hole in the center of one of the patterns (where it expected the reward), the observed preference for the rewarding disc need not be based on a space-invariant eidetic memory. More direct evidence for the use of a template would be presenting the bees with patterns that cannot be resolved before the bee has arrived at the fixation point.

In a recent study, Giurfa *et al.* (1995*a*) used four-petal model flowers

(7 cm diameter) whose shape could not be resolved before the bee approached them at very close distance (5 cm). Bees were trained to a "flower" of a constant spatial alignment that was then tested against the same flower displaying other orientations. The bees' choices confirmed Wehner's conclusion that the rewarding pattern is stored as a template in which the lower half of the visual field is weighted more than the rest of the visual field (Wehner 1972). Furthermore, even when bees can resolve the patterns at some distance, as in Wehner's experiments, their choice behavior is based mainly on the image viewed during the fixation phase (see also Wehner & Flatt 1977). Some patterns, such as radially symmetrical sectored discs, offer a fixation point even at some distance (Horridge 1999), but, in all cases, the presence of a fixation point is a prerequisite for template learning.

### The role of pattern orientation

To determine whether or not pattern orientation could be learned and used by bees as an independent feature, bees must be prevented from forming a template. Hateren *et al.* (1990) studied the role of pattern orientation using a Y-maze similar to that shown in Fig. 4.2. The patterns were linear gratings in which the distribution of black and white areas was randomized in both the positive and the negative grating throughout the training, keeping the spatial alignment of the patterns constant. This training procedure prevented the bees from memorizing a particular distribution of black and white areas. Each bee's first decision was scored as it entered one of the two arms, i.e., decisions were evaluated at some distance from the patterns. In those and in a series of similar experiments (see review by Srinivasan 1994), bees were shown to learn the orientation of contours and to use this parameter in subsequent discrimination tasks even when they were presented with novel patterns.

In other experiments, again using the Y-maze, Giger & Srinivasan (1995) trained bees to discriminate between two gratings that differed either in the orientation of bars (horizontal versus vertical) or in the distribution of black and white areas. Their results show that, when patterns contain strong directional cues, as linear gratings indeed do, bees ignore the template and use the contour orientation for accomplishing the discrimination. They use eidetic memory only when they are forced to do so, e.g., when the two patterns to be discriminated differ in the distribution of areas, but not in the spatial orientation of the contours (Giger & Srinivasan 1995).

In earlier experiments that recorded the bee's choices very close to the patterns, discrimination of orientation of linear gratings was excellent, even though the distribution of the black and the white areas was kept constant (Wehner 1971; Lehrer *et al.* 1985). In the light of the results obtained by Giger & Srinivasan (1995) described above, the results obtained in those earlier studies need not be based on eidetic memory. There might be no difference between close-range and long-range performance in the context of the discrimination of orientation when patterns contain strong directional cues. When directional information is weak, as was, for example, the case in the half-black, half-white patterns used by Wehner (1972), or in the four-petals patterns used by Giurfa *et al.* (1995*a*), then bees form a template and use the distribution of areas to accomplish pattern discrimination, regardless of whether decisions are made at a close range (Wehner 1972; Giurfa *et al.* 1995*a*) or at a distance (Horridge 1999).

### The role of symmetry

Symmetry is a visual cue available in almost all floral patterns (Neal *et al.* 1998). In recent years, results of several behavioral studies revealed that bees perceive symmetry and use this parameter in pattern-discrimination tasks.

To investigate the bee's long-range appreciation of symmetry, Lehrer *et al.* (1995) used the 12-arm experimental set-up described above (see Fig. 4.5a), training the bees, again, to the randomized checkerboard patterns. In subsequent tests, bees were required to choose between four patterns, each pattern being repeated in three different arms. When the four patterns differed in type, bees expressed a significant preference for radially symmetrical sectored patterns over all other types of pattern (Fig. 4.6a). When the four test patterns contained a constant number of bars that differed in arrangement, bees displayed a preference for radially symmetrical arrangements over less symmetrical or asymmetrical ones (Fig. 4.6b). And when bilaterally symmetrical patterns were presented, bees preferred patterns with a vertical axis of symmetry to patterns with a horizontal axis of symmetry (Fig. 4.6c). These and further results (Lehrer *et al.* 1995) suggest that bees prefer patterns that resemble natural flowers.

In another study, Giurfa *et al.* (1996*a*) examined the bees' capacity to learn bilateral symmetry using very small patterns (diameter 7 cm) and evaluating the bees' decisions at a very close distance (5 cm) from the patterns. During training, three patterns were presented simultaneously on

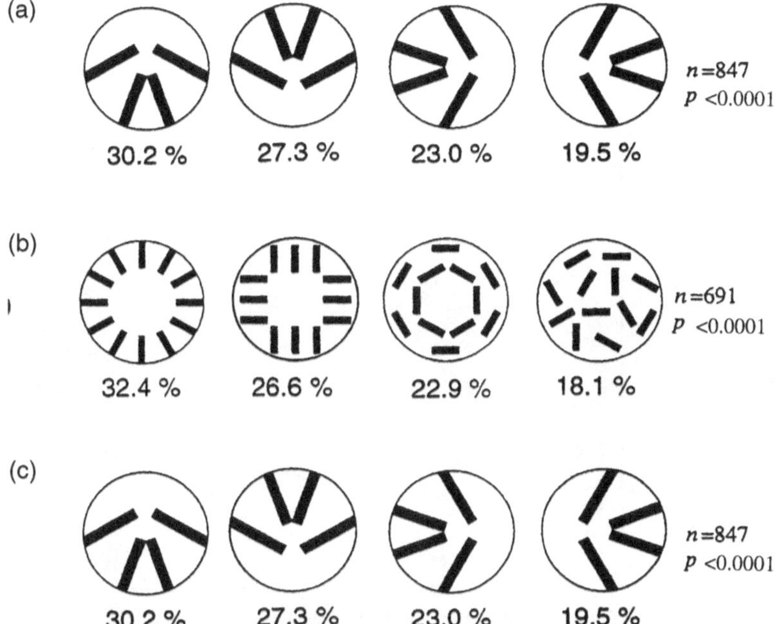



Fig. 4.6. As in Fig. 4.5, but the cues to be discriminated are: (a) the type of pattern; (b) the degree of symmetry; and (c) the orientation of the axis of bilateral symmetry. (Data from Lehrer *et al.* 1995.)

a vertical plane, one positive (rewarding), the other two not. When the positive pattern was symmetrical, the two negative patterns were asymmetrical (Fig. 4.7a), and vice versa. Throughout the training, the patterns were randomized with respect to their shapes (Fig. 4.7a), so that the bees could not form a template of the rewarded pattern and could rely on no cue other than symmetry. In the tests, the trained bees were given a multiple choice among 12 novel patterns, six of which were symmetrical, the other six asymmetrical (Fig. 4.7b).

The results (Fig. 4.7c) show that bees learned to prefer either the symmetrical or the asymmetrical test patterns, depending on whether they had previously been trained to the former or to the latter pattern category. The trained bees generalized the parameters "symmetry" and "asymmetry", respectively, to the novel patterns presented in the tests. However, bees trained to the symmetrical patterns performed significantly better in the discrimination task than did bees trained to the asymmetrical patterns (Giurfa *et al.* 1996*a*). Such a bias may be either innate or acquired through previous experience of the insects with symmetrical flowers in

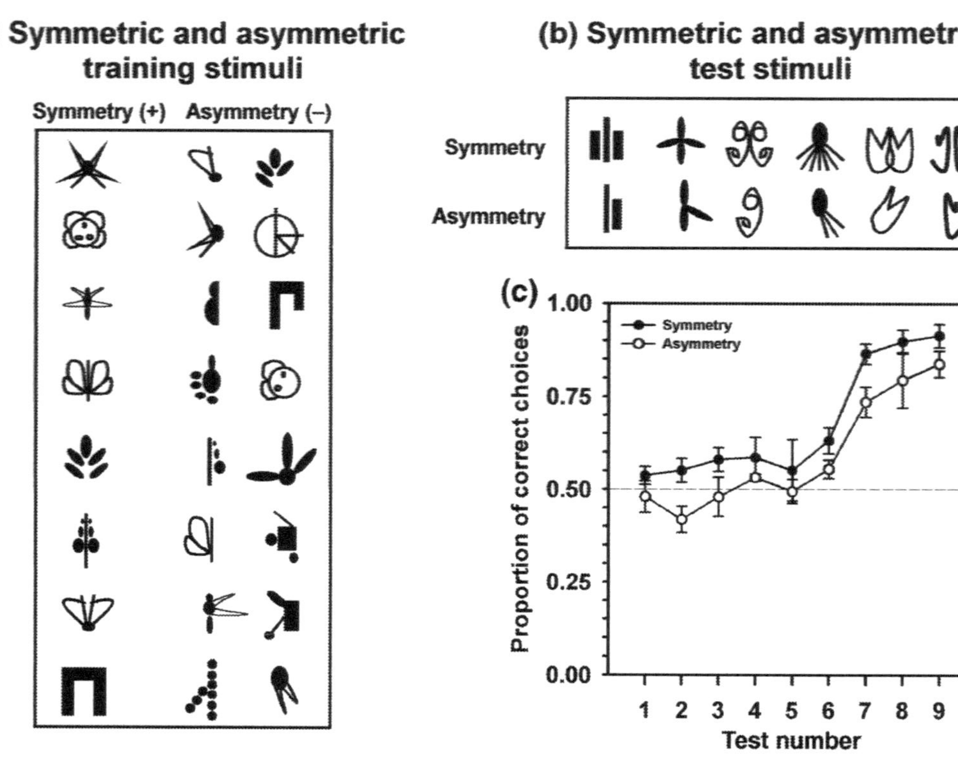


Fig. 4.7. Categorization of bilateral symmetry by honeybees. (a) Example of a succession of triads used to train honeybees to symmetry. The three stimuli differed in shape among triads, but each triad consisted of one symmetrical, rewarding stimulus (+) and two asymmetrical non-rewarding stimuli (−) presented simultaneously. Training with each triad was interspersed with multiple-choice tests. (b) Symmetrical and asymmetrical novel test stimuli were presented; none had rewards. (c) Choice frequency for the trained feature in bees trained for symmetry (black circles) and for asymmetry (empty circles). 0.5 means random choice between the two categories of novel stimuli. From test 7 onwards, bees transferred the information appropriately to the novel stimuli, thus demonstrating a capacity to detect, learn, and abstract symmetry or asymmetry as an independent visual pattern feature (modified from Giurfa *et al.* 1996a).

the field. Since the bees did not show a particular bias towards symmetry or asymmetry at the beginning of the training procedure (see Fig. 4.7c, tests 1–6), it must be concluded that the formation of categories ("symmetry" and "asymmetry") observed from test 7 on was a result of the training procedure used in those experiments.

Horridge (1996), again training bees to bilaterally symmetrical stimuli, used large patterns presented in the Y-maze set-up at a distance of 27 cm from the decision point. Thus, unlike the experiments of Giurfa *et al.* (1996*a*), Horridge's experiments tested symmetry detection at a farther range. In different training experiments, the rewarding patterns had either a vertical or horizontal plane of symmetry. Bees learned the orientation of the symmetry axis in either case. They discriminated the learned orientation from other orientations even when the test patterns were novel to them. Thus, bees generalize the orientation of the axis of symmetry much the same way in which they generalize the orientation of visible contrasting edges (Srinivasan 1994), although the symmetry axis does not constitute a visible edge between two contrasting areas.

Thus, in the case of symmetry perception, there are no fundamental differences between performance at a close range and that at a further range. The only condition that must be met is that the bee sees the whole pattern; the symmetry of large patterns is perceived at a larger distance than is the symmetry of small patterns. Lehrer (1999) proposed that radial symmetry (but not bilateral symmetry) might also be recognized in large patterns viewed at a close distance, on the basis of the presence of a number of neighboring radiating pattern elements.

### Final remarks

### Experience-based flexibility of behavior and cognitive capacities

Pattern recognition in the bee is based on several capacities, the use of which depends on the experimental procedures. Forming a template requires constant spatial relations between eyes and patterns. This condition is met only when a fixation point is available to the bee, and when the distribution of areas contained in the rewarding pattern is kept constant throughout the training. In natural flowers, a fixation point is usually provided by the site of reward in the center of the flower.

Bees can, however, learn to extract a particular cue and use it in the discrimination task, even when they are presented with novel patterns, as seen in several examples above. Using artificial stimuli, there are two ways that bees can be made to generalize a particular cue. One is to train bees using two patterns, one rewarding and one not, that differ from each other only in the one parameter to be learned. In this type of experiment, bees learn to "pay attention" to the parameter in which the two patterns differ. Selective attention is a capacity than can be considered cognitive, at least to some extent (Goldstone 1998). In the present review, some examples of this capacity are the detection of presence or absence of a colored disc against a contrasting background, the discrimination between two colored stimuli that differ in amount of green contrast, and the discrimination between a horizontal and a vertical axis in patterns that do not differ in symmetry.

A second method is to train bees to a set of training patterns, all rewarding, that have one parameter in common but are randomized with respect to all other parameters. This procedure prevents the forming of an association between any of the variable features and the reward, leaving the animal with only one useful cue, namely the one that is preserved in all of the training stimuli. A measure of the animal's capacity to learn that particular cue is the degree to which the animal generalizes it to novel stimuli. This methodology has been traditionally used in testing the cognitive abilities of some vertebrates (Harlow 1949). The first experiments of this type with honeybees were performed by Lehrer *et al.* (1988) to study distance estimation by size-independent cues. Two other examples cited above are the generalization of orientation (Hateren *et al.* 1990) and of symmetry (Giurfa *et al.* 1996*a*). It is particularly this type of performance that reveals a cognitive capacity, because it involves some type of "categorization" (Pearce 1994). Categorization is defined as the capacity to discriminate on the basis of a feature common to all members of a set of stimuli, and to generalize that feature to novel stimuli. These requirements are met in the experiments on range discrimination (Lehrer *et al.* 1988), on the use of pattern orientation (Hateren *et al.* 1990), and on symmetry perception (Giurfa *et al.* 1996*a*) by honeybees.

### Genetically fixed capacities

The flexibility evinced by honeybees in visual tasks reviewed in this chapter is clearly adaptive. In natural conditions, the appearance of a particular flower species that the bee has visited on a previous foraging trip may change slightly from one visit to the next. For instance, the color may differ slightly among individuals of the same species. Pattern parameters and the spatial orientation of flowers may vary, even within the same plant, depending on genetic and environmental factors. Further, not all bilaterally symmetrical flowers are vertically oriented; the axis of symmetry may be subjected to changes in spatial alignment. Behavioral flexibility thus allows the insect to cope with a changing environment in which visual cues display a certain degree of variability.

Still, bees do use particular parameters more efficiently than others, even in the absence of previous experience and even despite experimental manipulations. For example, pattern disruption is more effective than shape, symmetrical patterns are more attractive than asymmetrical patterns, black shapes on a white background are more effective than white shapes on a black background, and color is at any time a more effective signal than any spatial parameter (Menzel 1985; Gould & Gould 1988). Thus, some innate, genetically fixed behavioral programs must be involved in the bee's choice behavior. Such programs help the bees discover natural flowers even on their very first foraging trips.

In tasks involving color and spatial vision, innate preferences may be weakened through training, but no training procedure can cause them to disappear completely. Bees can be trained to asymmetrical patterns, although they are better at symmetrical ones (Giurfa *et al.* 1996*a*); they can be trained to low frequencies at short distances, although they prefer high frequencies (Lehrer 1997); they can learn a horizontal axis of symmetry, although they prefer a vertical one (Horridge 1996); and, they can even be trained to a white shape on a black background, although they prefer black shapes on a white background (Wehner 1972). However, in these cases, learning is slow, and the frequencies of correct choices seldom reach the high value that they do in training experiments that support the bee's innate tendencies. From the ecological point of view, this makes sense: natural flowers display exactly those parameters that bees tend to prefer (Giurfa *et al.* 1995*b*; Lehrer *et al.* 1995; Møller 1995; Møller & Eriksson 1995; Neal *et al.* 1998).

## References

Anderson AM (1977) Parameters determining the attractiveness of stripe patterns in the honeybee. *Anim Behav* 25:80–87

Autrum HJ & Zwehl V von (1964) Die spektrale Empfindlichkeit einzelner Sehzellen des Bienenaugens. *Z vergl Physiol* 48:357–384

Backhaus W (1991) Color opponent coding in the visual system of the honeybee. *Vision Res* 31:1381–1397

Chittka L & Menzel R (1992) The evolutionary adaptation of flower colors and the insect pollinators' color vision. *J Comp Physiol A* 171:171–181

Chittka L, Thomson JD & Waser NM (1999) Flower constancy, insect psychology, and plant evolution. *Naturwiss* 86:361–377

Daumer K (1956) Reizmetrische Untersuchung des Farbensehens der Bienen. *Z vergl Physiol* 38:413–478

Free JB (1970) Effect of flower shapes and nectar guides on the behaviour of foraging bees. *Behaviour* 37:269–285

Frisch K von (1914) Der Farbensinn und Formensinn der Bienen. *Zool Jb Abt allg Zool Physiol* 35:1–182

Frisch K von (1967) *The dance language and orientation of bees*. Belknap Press, Cambridge, MA

Giger AD & Srinivasan MV (1995) Pattern recognition in honeybees: eidetic imagery and orientation discrimination. *J Comp Physiol A* 176:791–795

Giurfa M, Backhaus W & Menzel R (1995a) Color and angular orientation in the discrimination of bilateral symmetric patterns in the honeybee. *Naturwiss* 82:198–201

Giurfa M, Núñez J, Chittka L & Menzel R (1995b) Color preferences of flower-naive honeybees. *J Comp Physiol A* 177:247–259

Giurfa M, Eichmann B & Menzel R (1996*a*) Symmetry perception in an insect. *Nature* 382:458–461

Giurfa M, Vorobyev M, Kevan P & Menzel R (1996*b*) Detection of colored stimuli by honeybees: minimum visual angles and receptor specific contrasts. *J Comp Physiol A* 178:699–709

Giurfa M, Vorobyev M, Brandt R, Posner B & Menzel R (1997) Discrimination of colored stimuli by honeybees: alternative use of achromatic and chromatic signals. *J Comp Physiol A* 180:235–244

Goldstone RL (1998) Perceptual learning. *Ann Rev Psychol* 49:585–612

Gould JL & Gould CG (1988) *The honeybee*. Scientific American Library, New York

Harlow HF (1949) The formation of learning sets. *Psychol Rev* 56:51–65

Hateren JH von, Srinivasan MV & Wait PB (1990) Pattern recognition in bees: orientation discrimination. *J Comp Physiol A* 167:649–654

Heisenberg M (1995) Pattern recognition in insects. *Curr Opin Neurobiol* 5:475–481

Helversen O von (1972) Zur spektralen Unterschiedsempfindlichkeit der Honigbiene. *J Comp Physiol* 80:439–472

Hertz M (1929) Die Organisation des optischen Feldes bei der Biene. II. *Z vergl Physiol* 11:107–145

Hertz M (1933) Über figurale Intensitäten und Qualitäten in der optischen Wahrnehmung der Biene. *Biol Zentbl* 53:10–40

Horridge GA (1996) The honeybee (*Apis mellifera*) detects bilateral symmetry and discriminates its axis. *J Insect Physiol* 42:755–764

Horridge GA (1997) Pattern discrimination by the honeybee: disruption as a cue. *J Comp Physiol A* 181:267–277

Horridge GA (1999) Pattern vision of the honeybee (*Apis mellifera*): the effect of pattern on the discrimination of location. *J Comp Physiol A* 185:105–113

Kevan PG & Baker HG (1983) Insects as flower visitors and pollinators. *Ann Rev Entomol* 28:407–453

Lehrer M (1997) Honeybee's visual spatial orientation at the feeding place. In: Lehrer M (ed) *Orientation and communication in arthropods*, pp 115–134. Birkhäuser, Basel

Lehrer M (1999) Dorsoventral asymmetry of color discrimination in bees. *J Comp Physiol A* 184:195–206

Lehrer M & Bischof S (1995) Detection of model flowers by honeybees: the role of chromatic and achromatic contrast. *Naturwiss* 82:145–147

Lehrer M, Wehner R & Srinivasan MV (1985) Visual scanning behavior in honeybees. *J Comp Physiol A* 157:405–415

Lehrer M, Srinivasan MV, Zhang SW & Horridge GA (1988) Motion cues provide the bee's visual world with a third dimension. *Nature* 332:356–357

Lehrer M, Horridge GA, Zhang SW & Gadagkar R (1995) Shape vision in bees: innate preference for flower-like patterns. *Phil Trans R Soc Lond B* 347:123–137

Menzel R (1967) Untersuchungen zum Erlernen von Spektralfarben durch die Honigbiene (*Apis mellifera*). *Z vergl Physiol* 56:22–62

Menzel R (1985) Learning in honeybees in an ecological and behavioral context. In: Hölldobler B & Lindauer M (eds) *Experimental behavioral ecology*, pp 55–74. Fischer, Stuttgart

Menzel R (1999) Memory dynamics in the honeybee. *J Comp Physiol A* 185:323–340

Menzel R & Backhaus W (1991) Color vision in insects. In: Gouras P (ed) *Vision and visual dysfunction. the perception of color*, pp 262–288. Macmillan, London

Menzel R & Blakers M (1976) Color receptors in the bee eye – morphology and spectral sensitivity. *J Comp Physiol* 108:11–33

Menzel R & Müller U (1996) Learning and memory in honeybees: from behavior to neural substrates. *Ann Rev Neurosci* 19:379–404

Menzel R, Greggers U & Hammer M (1993) Functional organization of appetitive learning and memory in a generalist pollinator, the honeybee. In: Papaj D & Lewis AC (eds) *Insect learning: ecological and evolutionary perspectives*, pp 79–125. Chapman & Hall, New York

Møller AP (1995) Bumblebee preference for symmetrical flowers. *Proc Natl Acad Sci USA* 92:2288–2292

Møller AP & Eriksson M (1995) Pollinator preference for symmetrical flowers and sexual selection in plants. *Oikos* 73:15–22

Neal P, Dafni A & Giurfa M (1998) Floral symmetry and its role in plant pollinator systems: terminology, distribution, and hypotheses. *Ann Rev Ecol Syst* 29:345–373

Pearce JM (1994) Discrimination and categorization. In: Mackintosh NJ (ed) *Animal learning and cognition: handbook of perception and cognition*, pp 109–134. Academic Press, San Diego, CA

Peitsch D, Fietz A, Hertel H, de Souza J, Ventura DF & Menzel R (1992) The spectral input systems of hymenopteran insects and their receptor-based color vision. *J Comp Physiol A* 170:23–40

Srinivasan MV (1994) Pattern recognition in the honeybee: recent progress. *J Insect Physiol* 40:183–194

Vorobyev M & Brandt R (1997) How do insect pollinators discriminate colors? *Israel J Plant Sci* 45:103–113

Wehner R (1971) The generalization of directional visual stimuli in the honeybee, *Apis mellifera. J Insect Physiol* 17:1579–1591.

Wehner R (1972) Dorsoventral asymmetry in the visual field of the bee *Apis mellifera*. *J Comp Physiol* 77:256–277

Wehner R (1981) Spatial vision in arthropods. In: Autrum HJ (ed) *Handbook of sensory physiology*, vol. VII 6C, pp 287–616. Springer-Verlag, Berlin

Wehner R & Flatt I (1977) Visual fixation in freely flying bees. *Z Naturforsch* 32c:469–471

Wolf E & Zerrahn-Wolf G (1935) The effect of light intensity, area, and flicker frequency on the visual reactions of the honeybee. *J Gen Physiol* 18:853–864

Zerrahn G (1933) Formdressur und Formunterscheidung bei der Honigbiene. *Z vergl Physiol* 20:117–150

Zhang SW, Srinivasan MV, Horridge GA (1992) Pattern recognition in honeybees: Local and global analysis. *Proc R Soc Lond B* 248:55–61

**5**

# Floral scent, olfaction, and scent-driven foraging behavior

Fragrance is an ancient medium of chemical communication between flowering plants and animal pollinators (Pellmyr & Thien 1986). Pollinators use fragrance for distance orientation, approach, landing, feeding, and associative learning (Williams 1983; Metcalf 1987; Dobson 1994). In turn, scent-driven pollinator preference and constancy has been invoked as an isolating mechanism for diverse angiosperm taxa (Dodson *et al.* 1969), particularly among sympatric, synchronously blooming species with similar floral form, coloration, and pollination mechanisms (Knudsen 1999). In this chapter, I explore the potential for odor-driven floral evolution by reviewing the physiological and behavioral responses of pollinators to floral scent.

## What is fragrance? A floral scent primer

### Chemical diversity and biosynthesis

Floral scents are mixtures of small, volatile organic compounds that vary in molecular weight, vapor pressure, polarity, and oxidation state (Knudsen *et al.* 1993). Diverse chemical classes of floral volatiles are surveyed comprehensively by Croteau & Karp (1991). The analytical methods used to collect and identify floral scent have improved dramatically over the past decade, and are discussed by Raguso & Pellmyr (1998) and Agelopoulos & Pickett (1998). Floral volatiles are produced by biosynthetic pathways, through anabolic and catabolic processes. Figure 5.1 summarizes the major biosynthetic routes to fragrance production, illustrating representative products for each pathway (Azuma *et al.* 1997). These multifunctional pathways also produce plant pigments, defense

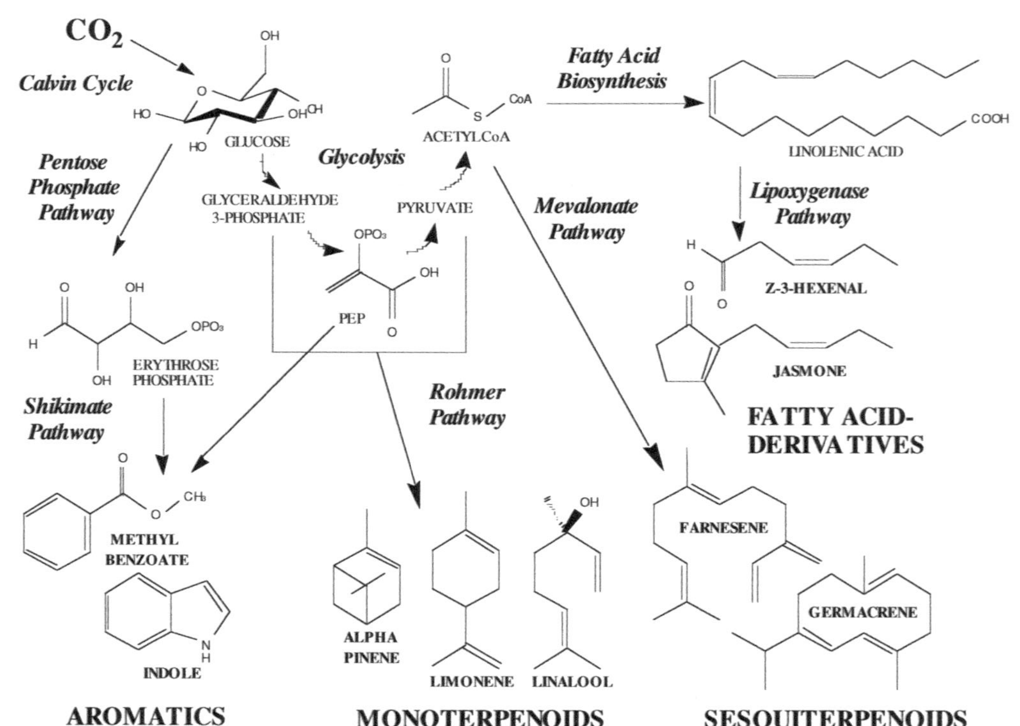


Fig. 5.1. The major biosynthetic routes to volatile production in plants. Pathway names are given in italics; representative metabolites and end products are capitalized; volatiles are bold-faced and capitalized. (Modified from Paré & Tumlinson 1997.)

compounds, structural components, growth, and signaling substances (Dixon & Paiva 1995). Recent progress in fragrance biosynthesis is reviewed by Dudareva *et al.* (1999) and Dudareva & Pichersky (2000).

### Floral scent variation: defining the phenotype

Variation in fragrance chemistry is prerequisite for scent-driven floral evolution. Geographic, altitudinal, and intrapopulation variation are frequently encountered when large samples are analyzed (Dodson *et al.* 1969; Tollsten & Bergström 1993). Fragrance varies spatially within flowers, in tissue-specific patterns and odor gradients that convey information as nectar or pollen guides (Lex 1954; von Aufsess 1960; Dobson 1987), and can be emitted from nearly any floral tissues, from surface epidermal cells to glandular trichomes or multicellular osmophores (Vogel 1963; Stern *et al.* 1987). Finally, fragrance composition and emission rates may vary temporally according to circadian rhythms and post-pollination changes (Dudareva *et al.* 1999).

The assumption of simple Mendelian inheritance of fragrance chemistry, combined with blend-specific behavioral responses by euglossine bees, prompted speculation that fragrance mutants could give rise to floral isolation and sympatric speciation in Neotropical orchids (Dressler 1968; Hills *et al.* 1972). Chemical analyses of interspecific $F_1$ hybrids of *Anthurium* (Araceae: Kuanprasert *et al.* 1998) and *Cycnoches* (Orchidaceae: Gregg 1983) yielded hybrid scent profiles that were qualitatively additive and quantitatively intermediate between parental phenotypes, suggesting polygenic inheritance or at least codominance. In the first balanced genetic analysis of fragrance, terpenoids segregated as dominant traits in interspecific *Clarkia* (Onagraceae) hybrids, but quantitative variation in emission rates was not correlated with floral morphology or environmental factors (Raguso & Pichersky 1999). This pattern may reflect the independent assortment of regulatory elements, substrate fluxes, or ultracellular factors (Curry 1987; Skubatz *et al.* 1996). Intraspecific crosses between lines of *C. breweri* that were polymorphic for methyleugenol yielded similar results, suggesting that biosynthetic enzyme activity alone does not explain phenotypic variation in fragrance (Wang & Pichersky 1999). Temperature, relative humidity, photoperiod, and edaphic conditions also contribute to plasticity in fragrance emissions (Hansted *et al.* 1994); "norm of reaction" studies with isolated genotypes would greatly contribute to defining fragrance phenotypes. Quantitative trait locus analysis of fragrance variation in a model plant system with a genetic map (e.g., *Petunia*) may be the best way to dissect phenotypic variation in floral scent (cf. Rieseberg & Noyes 1998).

### Multidimensional odor space; the natural distribution of floral scent

Distinct human-defined fragrance types traditionally have been grouped with specific pollinator classes (Knuth 1906; Vogel 1954; Faegri & van der Pijl 1979). However, no scheme relates fragrances to *pollinator* olfactory perception in the way that the color hexagon plots floral "visual space" *vis-à-vis* trichromatic visual perception (Chittka 1992; Menzel & Shmida 1993). Fragrances defy categorization along any single axis of physical properties. Various ordination techniques, such as principal components (Tollsten 1993) and multidimensional scaling (Dobson *et al.* 1997) are useful in visualizing differences in fragrance chemistry as multidimensional entities with characteristic locations in "odor space."

Does fragrance chemistry really covary with distinct pollinator guilds or foraging modes (Armbruster 1990)? Combined chemical analyses and bioassays have identified putative products of convergent evolution for pollinator attraction. For example, carvone oxide is unique to flowers pollinated by male euglossine bees, such as orchids, aroids, and *Dalechampia* euphorbs (Whitten *et al.* 1986). Knudsen & Tollsten (1993, 1995) used standard analytical methods to test the fit between fragrance chemistry and "pollinator syndromes" by analyzing fragrances from Neotropical hawkmoth- and bat-pollinated plant guilds. Hawkmoth-pollinated flowers produced "strong, sweet" scents that shared distinctive compound classes (oxygenated terpenoids, nitrogenous oximes), while the "fermented, garlicky" odors of bat-pollinated flowers shared sulfides rarely found in other angiosperms (see Winter and von Helversen, this volume). Independent surveys by Bestmann *et al.* (1997) and Miyake *et al.* (1998) confirmed these conclusions. However, interspecific fragrance variation was greater than expected, and compounds common to diurnal, generalist-pollinated flowers were present in nearly all taxa. These findings suggest that plants in pollinator guilds converge upon certain core constituents as required attractants while maintaining species-specific blends. This implies both innate and learned pollinator responses. Alternatively, fragrance may evolve in response to more than one pollinator class (Knudsen *et al.* 1999), and may be constrained by phylogenetic differences between guild members (Armbruster 1997) or selective pressures by enemies (Baldwin *et al.* 1997).

Most fragrance compounds appear to be too homoplaseous to be phylogenetically informative above the genus level (Dobson *et al.* 1997). Two recent studies (Azuma *et al.* 1999; Williams & Whitten 1999) used improved methods to re-evaluate earlier chemotaxonomic surveys of *Magnolia* trees and *Stanhopea* orchids. Fragrance data suggested multiple origins for the genus *Michelia* (Azuma *et al.* 1997), while chloroplast DNA analyses supported a monophyletic *Michelia* nested within *Magnolia*, with the loss of monoterpenoid- and shikimate-derived fragrance compounds in one species. Similarly, the addition of fragrance information to nuclear sequence data reduced the phylogenetic resolution (retention index) among subclades of *Stanhopea* (Williams & Whitten 1999). Clearly, fragrance data are most useful in identifying convergence, drift, and biogeographic patterns when mapped onto independently derived molecular phylogenies. However, rare synapomorphic compounds, when combined with other data sets, may help resolve closely related species clusters (Gerlach & Schill 1989). The application of biosynthetic pathway information to step-matrix coding of fragrance compounds as independent, multi-state characters (Barkman 2001) is the best way to combine such data.

### How do pollinators detect and perceive fragrances?

#### Odor signal detection and transduction

The structure of fragrance plumes is influenced by physical and environmental factors (Murlis *et al.* 1992). Odors are less ephemeral and less informative than visual or auditory cues, as scent trails can be followed to their sources from a distance, but convey only superficial information about species identity and patch size (Bradbury & Vehrencamp 1998). Pollinators responding to fragrance plumes face three challenges: signal detection over a range of concentrations; signal differentiation from a noisy olfactory background; and information coding and retrieval (Masson & Mustaparta 1990; Hildebrand & Shepherd 1997). Exciting progress has been made in vertebrate olfactory research (Mori *et al.* 1999), but the absence of bat, lemur, and bird studies limits my discussion to insect olfaction.

Antennae are the primary olfactory organs; their shape, size, and receptor organization reflect functional, phylogenetic, and biophysical tradeoffs (Chapman 1998). "Stereo" olfaction allows insects to detect spatial differences in scent concentration, permitting them to navigate within odor plumes (Mafra-Neto & Cardé 1994) and to use intrafloral scent guides (Lindauer & Martin 1963; Lunau 1991). Antennae are studded with microscopic sensilla (knobs, plates, or pits) that house sensory neurons (Chapman 1982). Olfactory sensilla have numerous pores that permit airborne odorants to pass to the lymph, where they contact general odorant-binding proteins (GOBPs) before reaching dendritic receptors (Steinbrecht *et al.* 1995; Vogt 1995). The binding of an odor molecule–GOBP complex to membrane-bound acceptors releases a signal-transduction cascade that opens ion channels, increases membrane calcium conductance, and generates a receptor potential (Krieger & Breer 1999). This electrochemical impulse conveys olfactory information to processing centers in the antennal lobes (Kaissling 1974). Odor-degrading enzymes inactivate odor ligands and remove them from acceptor sites (Vogt *et al.* 1990), a process that determines the kinetics of receptor recovery after stimulation (Kaissling 1986) and limits the input rate of olfactory information (Dusenbery 1992).

### Peripheral olfactory receptors

Do insect antennae preferentially amplify or filter specific fragrance compounds? Electroantennogram (EAG) recordings represent the sum of all receptor potentials evoked by an odor (White 1991). EAG differences reflect variation in receptor number or timing of response, but can also indicate uncorrected volatility differences among test stimulants (Todd & Baker 1993). Most EAG studies have measured herbivore responses to host-plant odors (e.g., Visser 1979), with a small subset of studies on pollinators and floral fragrances (Topazzini *et al.* 1990; Zhu *et al.* 1993). The fragmentary EAG literature suggests that most insects detect most plant volatiles at physiologically relevant concentrations, but the olfactory physiology of many pollinators remains unstudied.

A related technique, gas chromatographic-electroantennographic detection (GC-EAD), measures antennal responses to individual scent components as they elute from a gas chromatograph column (Thièry *et al.* 1990). When combined with behavioral assays, GC-EAD helps dissect complex fragrances. For example, Priesner (1973) surveyed the EAG responses of males from 50 species of Hymenoptera to fragrances from 18 *Ophrys* orchid taxa. These plants mimic the sex pheromones of female bees, and are pollinated through the copulatory efforts of aroused males (Kullenberg & Bergström 1976; Paulus & Gack 1990). EAG magnitudes were highly correlated with orchid–pollinator specificity, but stimulation with individual *Ophrys* volatiles rarely elicited comparable results. With GC-EAD, Schiestl *et al.* (1999) identified a series of electrophysiologically active alkanes present in comparable ratios in female bee cuticular extracts and in *Ophrys* fragrance. These compounds, not the more abundant "floral" terpenoids, elicited male attraction and copulation in field trials. The addition of conditioned proboscis extension (CPE) to GC-EAD identified the most salient components of blends to which tethered honeybees had been trained (Le Mètayer *et al.* 1997). Honda *et al.* (1998) found proboscis extension responses to be more accurate than EAG magnitude in predicting the feeding choices of naïve pierid butterflies among artificial, scented flowers. Thus, although EAGs confirm that an insect detects odors at relevant concentrations, they do not always indicate behavioral relevance (Gabel *et al.* 1992).

How broadly tuned are olfactory receptors to plant volatiles? Unlike whole-antennal studies, receptor-neuron EAGs from individual sensilla have helped define olfactory sensitivity and tuning. The high specificity of "labeled-line" pheromone receptors (Hansson 1995) is not the norm for plant odor receptors (Smith & Getz 1994), since many cell types respond to compounds sharing structural or functional similarities (Anderson *et al.* 1996; Wibe *et al.* 1997). Sensillar division of labor was observed in Kaib's (1974) study of two size-classes of sensilla basiconica in *Calliphora* blowflies, one of which responds to "floral" terpenoids and the other to aliphatic carrion odors. In other insects, diverse receptor neurons may respond to the same odor with different sensitivities, specificities, or electrophysiological properties (Dickens *et al.* 1993). Vareschi (1971) mapped the responses of honeybee sensilla placodea to hundreds of odors, revealing classes of olfactory neurons covering the entire tuning spectrum. Such variability may indicate multiple acceptor classes in the dendritic membrane, or reduced binding stringency by the acceptor (Kafka 1987).

### Odor coding, processing, and perception

In insects, olfactory information is first processed in the antennal lobe, where local interneurons connect distinct glomeruli, and projection neurons link them to the lateral protocerebrum and mushroom bodies, which are the regions of the insect brain associated with olfactory conditioning, sensory integration, and memory retrieval (Hammer 1993; Liu *et al.* 1999; Menzel, this volume). Vertebrate olfactory bulbs also include glomeruli that process inputs from receptors that share genetic identity and tuning specificity (Buck 1996). These glomeruli communicate with granule cells (the next level of odor processing) and neighboring glomeruli via dendro-dendritic synapses. Yokoi *et al.* (1995) demonstrated that synaptic inputs from flanking glomeruli modified mitral/tufted cell responses to aliphatic aldehydes in the target glomerulus, resulting in excitatory coding of 6–8 carbon compounds but inhibitory coding of longer- or shorter-chain aldehydes. Tuning specificity through lateral inhibition enhances perceptual contrast of visual signals (Bradbury & Vehrencamp 1998) and provides a mechanism for odor discrimination above the receptor level. Similar mechanisms have been invoked for pheromone processing in *Manduca* hawkmoths (Hildebrand 1995) and may be a general property of insect olfaction (Sachse *et al.* 1999).

If an odor is encoded by the activity of several glomeruli, each of which encodes several odors, then the global patterns of glomerular response to scent compounds should constitute distinct epitope maps (Buck 1996). Galizia *et al.* (1997) used calcium-sensitive dyes to visualize glomerular processing of scent compounds varying in carbon chain length and oxidation state. The resulting spatial activity patterns were characteristic for each odor tested (Joerges *et al.* 1997). Principal component analysis identified chain length as the most important variable; no glomeruli responded to functional groups independent of carbon skeleton (Sachse *et al.* 1999). This is the first depiction, albeit incomplete, of a non-pheromonal odor code in insects. Glomerular activity increases when an odor is paired with nectar but decreases when it goes unrewarded (Faber *et al.* 1999), suggesting that the same odor is now perceived as a different entity. However, spatial representations alone underestimate the richness of olfactory processing. Laurent *et al.* (1999) proposed that individual odors are coded not only by the ensemble of responding neurons and glomeruli, but also by the time course and the electrophysiological properties – spike frequencies, magnitudes, and firing patterns – of such a response. Honeybee projection neurons conditioned to a specific odor are "tuned" to show synchronized oscillations (Stopfer *et al.* 1997); exposure to neural inhibitors desynchronizes the oscillations, such that bees no longer distinguish between similar odors.

Honeybees learn odors faster than colors (Koltermann 1969), distinguish between thousands of subtly different odors (Vareschi 1971), learn faster when odors are "floral" (e.g., terpenoid or aromatic; Kriston 1971), and learn to associate floral scent, size, shape, color, and tactile cues with rewards in decreasing hierarchical rank (Menzel & Müller 1996). Most studies have measured neuroethological responses to single compounds, yet most fragrances are blends. Are blends coded as the sum of their components or as unique perceptual entities? Smith & Cobey (1994) found that pretraining with one odor diminishes the salience of a subsequent odor stimulus when both are presented together, suggesting that bees perceive individual blend components. When a visual cue is used as the pretraining stimulus, it does not block responses to the odor component of mixed visual–odor stimulus blends; this indicates a distinct processing mechanism for multi-modal signals (Gerber & Smith 1998). Finally, Hartlieb & Hansson (1999) used CPE to test whether female sex pheromone could be paired with nectar as a conditioned stimulus (CS) for male noctuid moths. Moths learned single pheromone and floral compounds equally well, but full pheromone blends impaired learning. The authors suggested that differential odor processing and learning was a consequence of pheromone blend perception within male-specific labeled-line glomeruli. While understanding pollinator perception of truly complex fragrances remains a distant goal, the tools required to dissect neuroethological responses to simple blends are now available.

## How do flower visitors respond to fragrance?

### Diversity in behavioral responses

The mechanisms of fragrance production and perception are meaningful to plant evolution only if they evoke discriminatory pollinator behavior. Such behavior can take many forms and act at different spatial scales. For example, von Knoll (1925) sandwiched *Lonicera* flowers between glass, observed *Hyles* hawkmoths probing the flattened flowers, and concluded that close orientation and feeding were entirely visual. However, the experimental arena was bathed in fragrance, which *elicits* visually guided foraging in nocturnal hawkmoths (Brantjes 1978). Additionally, studies of the same pollinator performed in different contexts can reveal unexpected behavioral flexibility. *Glossophaga* bats use fragrance and echolocation to find flowers, but learn to feed from feeders lacking these properties (Winter & von Helversen, this volume). Studies exploring how diet, physiological state, and experience modify pollinator foraging behavior are sorely needed (Simpson & Raubenheimer 1996). Recent work on scarab beetles reveals a hidden bounty of foraging modes and evolutionary patterns. Scent-driven pollination of thermogenic, nocturnal "trap flowers" by dynastine scarabs is well documented in Neotropical forests (e.g., Gottsberger & Silberbauer-Gottsberger 1991), but guilds of scentless, brightly colored, diurnal flowers are pollinated exclusively by visually foraging ruteline scarabs in South Africa (Picker & Midgley 1996; Steiner 1998) and by glaphyrid beetles in Israel (Dafni *et al.* 1990). The diversity of foraging strategies among monkey beetles alone (Donaldson *et al.* 1990) is comparable to that of the order Lepidoptera (Weiss, this volume).

Nested visual signals (display, landing, and nectar guides) are appreciated as components of complex floral phenotypes (Sprengel 1793; Waser & Price 1985), but fragrance blends also serve multiple roles during flower visitation. For example, noctuid moths visiting *Silene* flowers use fragrance as a distance attractant (Brantjes 1978), a landing cue and nectar guide (Haynes *et al.* 1991), and a conditioned stimulus (Fan *et al.* 1997). The recent discovery of "sweet" terpenoids beneath the stench of *Sauromatum guttatum* aroids and *Phallus impudicus* fungi (Borg-Karlson 1994; Skubatz *et al.* 1996) begs a behavioral and phylogenetic reassessment of sapromyophily (Kite & Hetterschieid 1997).

### Foraging strategies, pollen movement, and behavioral predictions

Bronstein (1995) provided a model for examining evolutionary conflicts of interest between animal foraging and plant reproductive strategies. Her "pollinator landscapes" define a contingency table in which specialist or generalist flower visitors interact with synchronously or asynchronously blooming plants (Fig. 5.2). The foraging strategies of strict mutualists (e.g., fig wasps), sexual dupes (mate-seeking males and ovipositing females), territorial defenders (hummingbirds), central place foragers (many bees, bats), and vagile nomads (most moths) can be condensed into three categories: (1) trapliners; (2) density-dependent visual foragers; and (3) sexual foragers.

The archetypal trapliner is a long-lived animal that learns and revisits landmarks in the absence of obvious floral cues (Feinsinger 1983). Chittka *et al.* (1995) have shown that honeybees form traplines by sequential retrieval of landmark memories without odor cues or mental maps. The heliconiine butterflies and hermit hummingbirds that trapline *Psiguria* flowers readily feed from artificial, scentless flowers (Swihart & Swihart 1970; Murawski & Gilbert 1986), but the role of pollen odors in heliconiine foraging deserves closer scrutiny. Similarly, there is no evidence for odor-dependent nectar foraging or place learning by female euglossine bees

**Pollinators**

|  | Specialized | Generalized | Migratory |
|---|---|---|---|
| **Synchronous** | *Ophrys* orchids and male bees[1]<br><br>*Yucca* and *Tegiticula* yucca moths[2]<br><br>**S** | *Calathea* and *Euglossa, Rhathymus* bees[5]<br><br>*Delphinium* and *Bombus* bumble bees[6]<br><br>**DD** | *Fouquieria* and *Selasphorus* hummingbirds[9] |
| **Asynchronous** | *Ficus* and fig wasps[3]<br><br>*Dieffenbachia* and *Cyclocephala* beetles[4]<br><br>**S** | *Lavandula* and 54 insect species[7]<br><br>*Psiguria* and *Heliconius* butterflies, *Phaethornis* hermits[8]<br><br>**T, DD** | *Agave, Pachycereus* and *Leptonycteris* bats[10]<br><br>**T, DD** |

**Plants** (row label at left)

Fig. 5.2. Bronstein's (1995) pollinator landscapes, with distributions of foraging classes and representative systems for each interaction. References: [1] Paulus & Gack (1990); [2] Pellmyr *et al.* (1996); [3] Wiebes (1979); [4] Young (1986); [5] Schemske & Horvitz (1984); [6] Waser (1982); [7] Herrera (1987); [8] Murawski & Gilbert (1986); [9] Waser (1979); [10] Fleming *et al.* (1996).

(Janzen 1971; Ackerman *et al.* 1982), despite the fact that male euglossines' attraction to fragrance is an axiom of Neotropical biology (Williams & Whitten 1983). Odor-guided distance attraction of hawkmoths and beetles between isolated rainforest plants has been likened to traplines (Schatz 1990), and features similar pollen dispersal distances (Young 1986; Nilsson & Rabakonandrianina 1988), but there is no evidence that individual hawkmoths or beetles repeat daily foraging routes or learn landmarks.

Density-dependent visual foragers are attracted from close range to aggregated floral displays (Rathcke 1983) and may use fragrance as a distance attractant, a feeding cue, or a conditioning stimulus. Flowers pollinated by such foragers produce combinations of olfactory and visual cues that may also attract herbivores or thieves (Dafni 1984). Because nectar and fragrance production incur metabolic costs (Pyke 1991) and attract unwanted plant enemies (Baldwin *et al.* 1997), autogamous species with reduced flowers should lose nectar and/or scent, while nectarless or scentless deceptive flowers are predicted to evolve under several scenarios (Knudsen & Tollsten 1993; Armstrong 1997). Depending upon its role in pollinator attraction, post-pollination fragrance emission might be (1) maintained to contribute to distance attraction (Eisikowitch & Rotem 1987), (2) eliminated to prevent futile visits (Tollsten 1993), or (3) modified to promote learned avoidance of reward-depleted flowers (Lex 1954). The chemical diversity, perceptual complexity, and salience of fragrance blends as learned cues suggest that species-specific fragrances should promote floral constancy, but direct tests of this hypothesis are needed.

Insects that visit flowers for some element of their own sexual reproduction include obligate mutualists and the victims of sexual- and brood site-deception (Faegri & van der Pijl 1979). Here, fragrances may elicit hard-wired sexual- or oviposition-related behaviors, although feeding may also be involved (Beaman *et al.* 1988). Interactions between figs and fig wasps are governed by scent from a distance and by contact chemoreception after landing (Gibernau & Hossaert-McKey 1998). Initial data support the hypothesis of ethological isolation through species-specific fragrances (Ware *et al.* 1993; Grison *et al.* 1999). In all documented cases of pseudocopulatory pollination, fragrances effectively mimic female hymenopteran sex pheromones (Borg-Karlson 1990). Peakall (1990) proposed that pollinator movement among sexually deceptive orchids should reflect optimal mate search, not optimal food foraging. Male thynnine wasps that pollinate Australian *Caladenia* and *Cryptostylis* orchids show site aversion after an attempted mating, promoting greater dispersal distances and fewer local visits than would be the case during nectar foraging (Peakall & Beattie 1996). In contrast, the post-copulatory avoidance of females and *Ophrys* flowers by male andrenid bees is based on aversive odor learning (Ayasse *et al.* 1993), resulting in bee movement toward novel odors and subtle fragrance variation among adjacent flowers (Schiestl *et al.* 1997).

**Pollinator-mediated selection on variation in floral scent**

This chapter's emphasis on pollinator attraction reflects an historical adaptive bias in fragrance research, but recent studies have explored alternative roles for floral volatiles. Pellmyr & Thien (1986), Armbruster (1997), and Schiestl *et al.* (1999) argue that plant defense and stress-response physiology are pre-adaptive for fragrance evolution. Herbivory on crop plants induces systemic vegetative emissions that are chemically "floral" and that attract parasitoids (Paré & Tumlinson 1997). Less appreciated are the deterrent components of fragrances, whether they combat microbes in standing nectar (Lawton *et al.* 1993) or repel unbidden floral visitors (Omura *et al.* 2000).

The potential for positive directional selection (by pollinators) on fragrance variation is supported by the following evidence. Galen & Kevan (1983) characterized "sweet" and "skunky" fragrance morphs of *Polemonium viscosum*, showing that bumble bees overvisit sweet scents and discriminate against skunky scents, irrespective of nectar quantity. Furthermore, sweet fragrance was correlated with floral geometry and with nectar traits favorable to bumble bee pollination (Galen & Newport 1987). Pellmyr (1986) documented scent-based assortative visitation by butterflies and bumble bees to specific *Cimicifuga simplex* chemotypes; he proposed that incipient speciation could occur through floral isolation. Finally, Pelz *et al.* (1997) discovered that increased odor concentration promotes more salient conditioning and memory consolidation in bees, at least for single compounds.

Negative directional selection (by florivores) on fragrance is supported by equally strong evidence. Galen (1983, 1999) demonstrated that ants destructively overvisit sweet-scented *Polemonium* flowers, stealing nectar and detaching styles. Ecroyd (1996) investigated the role of scented nectar in bat pollination of *Dactylanthus taylorii* in New Zealand and found that introduced opossums and rats were attracted by the fragrance and

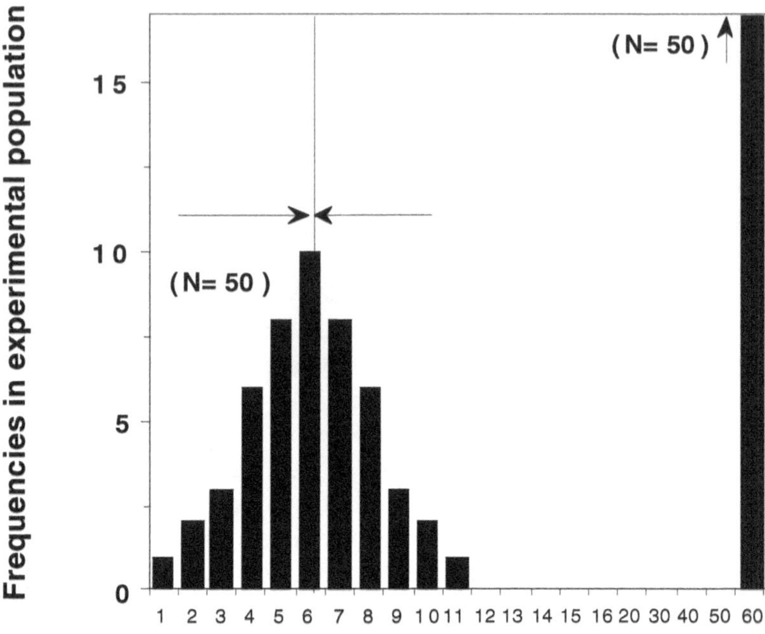



Fig. 5.3. Hypothetical stabilizing selection on fragrance emissions in *Nicotiana attenuata*. Natural distribution is assumed to be normal; note 10-fold increase in experimental treatment. Opposing arrows represent directional selection by hawkmoth pollinators and hemipteran seed predators. Mean fragrance emission (vertical line) should vary spatially and temporally among populations (after Baldwin *et al.* 1997).

destroyed the inflorescences. Baldwin *et al.* (1997) artificially enhanced benzyl acetone emissions in a wild tobacco (*Nicotiana attenuata*), anticipating increased hawkmoth attraction. Instead, they observed elevated ovule destruction by hemipteran predators (Fig. 5.3). The risks of advertising to natural enemies could counter pollinator-mediated directional selection on fragrance and other floral attractants (Brody 1997). Alternatively, fragrance variation may be non-adaptive in many cases, and may persist in populations as a consequence of drift or phenotypic plasticity (Olesen & Knudsen 1994; Ackerman *et al.* 1997).

The evidence suggests that floral scent is best defined as a mosaic product of biosynthetic pathway dynamics, phylogenetic constraints, and balancing selection due to pollinator and florivore attraction.

Continued progress in fragrance research will be enhanced by: (1) standardization of analytical methods for comparative studies; (2) the establishment of a networked "scentbank" data base; (3) integrated quantitative and molecular analysis of fragrance variation; (4) the documentation of chemical and olfactory diversity in threatened tropical mutualisms; and (5) creative graphical and theoretical approaches to representing "perceptual odor space".

## Acknowledgements

This work was supported by National Science Foundation grant DEB–9806840. Special thanks to Nina Brown, Lars Chittka, Heidi Dobson, Laurel Hester, and Rachel Levin for editorial suggestions.

## References

Ackerman JD, Mesler MR, Lu KL & Montalvo AM (1982) Food-foraging behavior of male Euglossini (Hymenoptera: Apidae): vagabonds or trapliners? *Biotropica* 14:241–248

Ackerman JD, Meléndez-Ackerman EJ & Salguero-Faría J (1997) Variation in pollinator abundance and selection on fragrance phenotypes in an epiphytic orchid. *Am J Bot* 84:1383–1390

Agelopoulos N & Pickett J (1998) Headspace analysis in chemical ecology: the effects of different sampling methods on the ratios of volatile compounds present in headspace samples. *J Chem Ecol* 24:1161–1172

Anderson P, Larsson M, Löfqvist J & Hansson BS (1996) Plant odor receptor neurons on the antennae of the two moths *Spodoptera littoralis* and *Agrotis segetum*. *Entomol Exp Appl* 80:32–34.

Armbruster WS (1990) Estimating and testing the shapes of adaptive surfaces: the morphology and pollination of *Dalechampia* blossoms. *Am Nat* 135:14–31

Armbruster WS (1997) Exaptations link evolution of plant–herbivore and plant–pollinator interactions: a phylogenetic inquiry. *Ecology* 78:1661–1672

Armstrong JE (1997) Pollination by deceit in nutmeg (*Myristica insipida*, Myristicaceae): floral displays and beetle activity at male and female trees. *Am J Bot* 84:1266–1274

Aufsess A von (1960) Geruchliche Nahorientierung der Biene bei entomophilen und ornithophilen Blüten. *Z vergl Physiol* 43:469–498

Ayasse M, Engels W, Hefetz A, Tengö J, Lübke G & Francke W (1993) Ontogenetic patterns of volatiles identified in Dufour's gland extracts from queens and workers of the primitively eusocial halictine bee, *Lasioglossum malachurum* (Hymenoptera: Halictidae). *Insectes Soc* 40:41–58

Azuma H, Thien LB & Kawano S (1999) Molecular phylogeny of *Magnolia* (Magnoliaceae) inferred from cpDNA sequences and evolutionary divergence of the floral scents. *J Plant Res* 112:291–306

Azuma H, Toyota M, Asakawa Y, Yamaoka R, Garcia-Franco JG, Dieringer G, Thien LB & Kawano S (1997) Chemical divergence in floral scents of *Magnolia* and allied genera (Magnoliaceae). *Plant Sp Biol* 12:69–83

Baldwin IT, Preston C, Euler M & Gorham D (1997) Patterns and consequences of benzyl acetone floral emissions from *Nicotiana attenuata* plants. *J Chem Ecol* 23:2327–2343

Barkman TJ (2001) Character coding of secondary chemical variation for use in phylogenetic analyses. *Biochem Syst Ecol* 29:1–20

Beaman RS, Decker PJ & Beaman JH (1988) Pollination of *Rafflesia* (Rafflesiaceae). *Am J Bot* 75:1148–1162

Bestmann HJ, Winkler L & Helversen O von (1997) Headspace analysis of volatile flower scent constituents of bat-pollinated plants. *Phytochemistry* 46:1169–1172

Borg-Karlson A-K (1990) Chemical and ethological studies of pollination in the genus *Ophrys* (Orchidaceae). *Phytochemistry* 29:1359–1387

Borg-Karlson A-K (1994) Dimethyl oligosulphides, major volatiles released from *Sauromatum guttatum* and *Phallus impudicus*. *Phytochemistry* 35:321–323

Bradbury JW & Vehrencamp SL (1998) *Principles of animal communication*. Sinauer Associates, Sunderland, MA

Brantjes NBM (1978) Sensory responses to flowers in night-flying moths. In: Richards AJ (ed) *The pollination of flowers by insects*, pp 13–19. Academic Press, London

Brody AK (1997) Effects of pollinators, herbivores and seed predators on flowering phenology. *Ecology* 78:1624–1631

Bronstein JL (1995) The plant–pollinator landscape. In: Hansson L, Fahrig L & Merriam G (eds) *Mosaic landscapes and ecological processes*, pp 256–288. Chapman & Hall, London

Buck LB (1996) Information coding in the vertebrate olfactory system. *Ann Rev Neurosci* 19:517–544.

Chapman RF (1982) Chemoreception: the significance of receptor numbers. *Adv Insect Physiol* 16:247–356

Chapman RF (1998) *The insects: structure and function*, 4th edn. Cambridge University Press, Cambridge

Chittka L (1992) The colour hexagon: a chromaticity diagram based on photoreceptor excitations as a generalized representation of colour opponency. *J Comp Physiol A* 170:533–543

Chittka L, Geiger K & Kunze J (1995) The influences of landmarks on distance estimation of honeybees. *Anim Behav* 50:23–31

Croteau R & Karp F (1991) Origin of natural odorants. In: Müller PM & Lamparsky D (eds) *Perfumes – art, science and technology*, pp 101–126. Elsevier, New York

Curry KJ (1987) Initiation of terpenoid synthesis in osmophores of *Stanhopea anfracta* (Orchidaceae): a cytochemical study. *Am J Bot* 74:1332–1338

Dafni A (1984) Mimicry and deception in pollination. *Ann Rev Ecol Syst* 15:259–278

Dafni A, Bernhardt P, Shmida A, Ivri Y, Greenbaum S, O'Toole C & Losito L (1990) Red bowl-shaped flowers: convergence for beetle pollination in the Mediterranean region. *Israel J Bot* 39:81–92

Dickens JC, Visser JH & Van der Pers JNC (1993) Detection and deactivation of pheromone and plant odor components by the beet armyworm *Spodoptera exigua* (Lepidoptera: Noctuidae). *J Insect Physiol* 39:503–516.

Dixon RA & Paiva NL (1995) Stress-induced phenylpropanoid metabolism. *Plant Cell* 7:1085–1097

Dobson HEM (1987) Role of flower and pollen aromas in host-plant recognition by solitary bees. *Oecologia* 72:618–623

Dobson HEM (1994) Floral volatiles in insect biology. In: Bernays E (ed) *Insect–plant interactions*, vol. 5, pp 47–81. CRC Press, Boca Raton, FL

Dobson HEM, Arroyo J, Bergström LG & Groth I (1997) Interspecific variation in floral fragrances within the genus *Narcissus* (Amaryllidaceae). *Biochem Syst Ecol* 25:685–706

Dodson C, Dressler R, Hills H, Adams R & Williams N (1969) Biologically active compounds in orchid fragrances. *Science* 164:1243–1249

Donaldson JMI, McGovern TP & Ladd TL Jr (1990) Floral attractants for Cetoniinae and Rutelinae (Coleoptera: Scarabaeidae). *J Econ Entomol* 83:1298–1305

Dressler RL (1968) Pollination by euglossine bees. *Evolution* 22:202–210

Dudareva N & Pichersky E (2000) Biochemical and molecular genetic aspects of floral scents. *Plant Physiol* 122:627–633

Dudareva N, Piechulla B & Pichersky E (1999) Biogenesis of floral scents. *Hort Rev* 24:31–54

Dusenbery DB (1992) *Sensory ecology: how organisms acquire and respond to information.* WH Freeman, New York

Ecroyd CE (1996) The ecology of *Dactylanthus taylorii* and threats to its survival. *NZ J Ecol* 20:81–100

Eisikowitch D & Rotem R (1987) Flower orientation and color change in *Quisqualis indica* and their possible role in pollinator partitioning. *Bot Gaz* 148:175–179

Faber T, Joerges J & Menzel R (1999) Associative learning modifies neural representations of odors in the insect brain. *Nature Neuro* 2:74–78

Faegri K & van der Pijl L (1979) *The principles of pollination ecology*, 3rd edn. Pergamon Press, Oxford

Fan R-J, Anderson P & Hansson BS (1997) Behavioural analysis of olfactory conditioning in the moth *Spodoptera littoralis* (Boisd.) (Lepidoptera: Noctuidae). *J Exp Biol* 200:2969–2976

Feinsinger P (1983) Coevolution and pollination. In: Futuyma DJ & Slatkin M (eds) *Coevolution*, pp 282–310. Sinauer Associates, Sunderland, MA

Fleming TH, Tuttle MD & Horner MA (1996) Pollination biology and the relative importance of nocturnal and diurnal pollinators in three species of Sonoran Desert cacti. *Southwest Nat* 41:257–269

Gabel B, Thièry D, Suchy V, Marion-Poll F, Hradsky P & Farkas P (1992) Floral volatiles of *Tanacetum vulgare* L. attractive to *Lobesia botrana* Den. et Schiff. females. *J Chem Ecol* 18:693–701

Galen C (1983) The effects of nectar thieving ants on seed set in floral scent morphs of *Polemonium viscosum. Oikos* 41:245–249

Galen C (1999) Flowers and enemies: predation by nectar-thieving ants in relation to variation in floral form of an alpine wildflower, *Polemonium viscosum. Oikos* 85: 426–434

Galen C & Kevan PG (1983) Bumblebee foraging and floral scent dimorphism: *Bombus kirbyellus* Curtis (Hymenoptera: Apidae) and *Polemonium viscosum. Can J Zool* 61:1207–1213

Galen C & Newport MEA (1987) Bumblebee behavior and selection on flower size in the sky pilot, *Polemonium viscosum. Oecologia* 74:20–23

Galizia CG, Joerges J, Küttner A, Faber T & Menzel R (1997) A semi-in-vivo preparation for optical recording of the insect brain. *J Neurol Meth* 76:61–69

Gerber B & Smith BH (1998) Visual modulation of olfactory learning in honeybees. *J Exp Biol* 201:2213–2217

Gerlach G & Schill R (1989) Fragrance analysis, an aid to taxonomic relationships of the genus *Coryanthes* (Orchidaceae). *Plant Syst Evol* 168:159–165

Gibernau M & Hossaert-McKey M (1998) Are olfactory signals sufficient to attract fig pollinators? *Ecoscience* 5:306–311

Gottsberger G & Silberbauer-Gottsberger I (1991) Olfactory and visual attraction of *Erioscelis emarginata* (Cyclocephalini, Dynastinae) to the inflorescences of *Philodendron selloum* (Araceae). *Biotropica* 23:23–28

Gregg KB (1983) Variation in floral fragrances and morphology: incipient speciation in *Cycnoches? Bot Gaz* 144:566–576

Grison L, Edwards AA & Hossaert-McKey M (1999) Interspecies variation in floral fragrances emitted by tropical *Ficus* species. *Phytochemistry* 52:1293–1299

Hammer M (1993) An identified neuron mediates the unconditioned stimulus in associative olfactory learning in honeybees. *Nature* 366:59–63

Hansson BS (1995) Olfaction in Lepidoptera. *Experientia* 51:1003–1027

Hansted L, Jakobsen HB & Olsen CE (1994) Influence of temperature on the rhythmic emission of volatiles from *Ribes nigrum* flowers in situ. *Plant Cell Env* 17:1069–1072

Hartlieb E & Hansson BS (1999) Sex or food? Appetitive learning of sex odors in a male moth. *Naturwiss* 86: 396–399

Haynes KF, Zhao JZ & Latif A (1991) Identification of floral compounds from *Abelia grandiflora* that stimulate upwind flight in cabbage looper moths. *J Chem Ecol* 17:637–646

Herrera CM (1987) Components of pollinator "quality": comparative analysis of a diverse insect assemblage. *Oikos* 50:79–90

Hildebrand JG (1995) Analysis of chemical signals by nervous systems. *Proc Natl Acad Sci USA* 92:67–74

Hildebrand JG & Shepherd GM (1997) Mechanisms of olfactory discrimination: converging evidence for common principles across phyla. *Ann Rev Neurosci* 20:595–631

Hills HG, Williams NH & Dodson CH (1972) Floral fragrances and isolating mechanisms in the genus *Catasetum* (Orchidaceae). *Biotropica* 4: 61–76

Honda K, Omura H & Hayashi, N (1998) Identification of floral volatiles from *Ligustrum japonicum* that stimulate flower-visiting by cabbage butterfly, *Pieris rapae. J Chem Ecol* 24:2167–2180

Janzen DH (1971) Euglossine bees as long distance pollinators of tropical plants. *Science* 171:203–205

Joerges F, Küttner A, Galizia CG & Menzel R (1997) Representation of odours and odour mixtures visualized in the honeybee brain. *Nature* 387:285–288

Kafka WA (1987) Similarity of reaction spectra and odor discrimination: single receptor cell recordings in *Antheraea polyphemus* (Saturniidae). *J Comp Physiol A* 161:867–880

Kaib M (1974) Die Fleisch- und Blumenduftrezeptoren auf der Antenne der Schmeißfliege *Calliphora vicina. J Comp Physiol* 95:105–121

Kaissling K-E (1974) Sensory transduction in insect olfactory receptors. In: Jaenicke L (ed.) *Biochemistry of sensory functions*, pp 243–273. Springer-Verlag, Heidelberg

Kaissling K-E (1986) Chemo-electrical transduction in insect olfactory receptors. *Ann Rev Neurosci* 9:121–145

Kite GC & Hetterschieid WLA (1997) Inflorescence odours of *Amorphophallus* and *Pseudodracontium* (Araceae). *Phytochemistry* 46:71–75

Knoll F von (1925) Lichtsinn und Blütenbesuch des falters von *Deilephila livornica*. *Z vergl Physiol* 2:329–380

Knudsen JT (1999) Floral scent chemistry in geonomoid palms (Palmae: Geonomeae) and its importance in maintaining reproductive isolation. *Mem NY Bot Gard* 83:141–168

Knudsen JT & Tollsten L (1993) Trends in floral scent chemistry in pollination syndromes: floral scent composition in moth-pollinated taxa. *Bot J Linn Soc* 113:263–284

Knudsen JT & Tollsten L (1995) Floral scent in bat-pollinated plants: a case of convergent evolution. *Bot J Linn Soc* 118:45–57

Knudsen JT, Tollsten L & Bergström G (1993) Floral scents – a check-list of volatile compounds isolated by head-space techniques. *Phytochemistry* 33:253–280

Knudsen JT, Andersson S & Bergman P (1999) Floral scent attraction in *Geonoma macrostachys*, an understory palm of the Amazonian rain forest. *Oikos* 85:409–418

Knuth P (1906) *Handbook of flower pollination*, vol. 1. Clarendon Press, Oxford [Based upon Hermann Müller's work "The fertilization of flowers by insects".]

Koltermann R (1969) Lern- und Vergessensprozesse bei der Honigbiene – aufgezeigt anhand von Duftdressuren. *Z vergl Physiol* 63:310–334

Krieger J & Breer H (1999) Olfactory reception in invertebrates. *Science* 286:720–723

Kriston I (1971) Zum Problem des Lernverhaltens on *Apis mellifica* L. gegenüber verschiedenen Duftstoffen. *Z vergl Physiol* 74:169–189

Kuanprasert N, Kuehnle AR & Tang CS (1998) Floral fragrance compounds of some *Anthurium* (Araceae) species and hybrids. *Phytochemistry* 49:521–524

Kullenberg B & Bergström LG (1976) The pollination of *Ophrys* orchids. *Bot Not* 129:11–19

Laurent G, MacLeod K, Stopfer M & Wehr M (1999) Dynamic representation of odours by oscillating neural assemblies. *Entomol Exp Appl* 91:7–18

Lawton RO, Alexander LD, Setzer WN & Byler KG (1993) Floral essential oil of *Guettarda poasana* inhibits yeast growth. *Biotropica* 25:483–486

Le Mètayer M, Marion-Poll F, Sandoz JC, Pham-Delègue M-H, Blight MM, Wadhams LJ, Masson C & Woodcock CM (1997) Effect of conditioning on discrimination of oilseed rape volatiles by the honeybee: use of a combined gas chromatography–proboscis extension behavioral assay. *Chem Senses* 22:391–398

Lex T (1954) Duftmale an Blüten. *Z vergl Physiol* 36:212–234

Lindauer M & Martin H (1963) Über die Orienteirung der Biene im Duftfeld. *Naturwiss* 50:509–514

Liu L, Wolf R, Ernst R & Heisenberg M (1999) Context generalization in *Drosophila* visual learning requires the mushroom bodies. *Nature* 400:753–756

Lunau K (1991) Innate recognition of flowers by bumblebees (*Bombus terrestris, B. lucorum*; Apidae): optical signals from stamens as learning reaction releasers. *Ethology* 88:203–214

Mafra-Neto A & Cardé RT (1994) Fine-scale structure of pheromone plumes modulates upwind orientation of flying moths. *Nature* 369:142–144

Masson C & Mustaparta H (1990) Chemical information processing in the olfactory system of insects. *Physiol Rev* 70:199–245

Menzel R & Müller U (1996) Learning and memory in honeybees: from behavior to neural substrates. *Ann Rev Neurosci* 19:379–404

Menzel R & Shmida A (1993) The ecology of flower colours and the natural colour vision of insect pollinators: the Israeli flora as a study case. *Biol Rev* 68:81–120

Metcalf RL (1987) Plant volatiles as insect attractants. *CRC Crit Rev Plant Sci* 5:251–301

Miyake T, Yamaoka R & Yahara,T (1998) Floral scents of hawkmoth-pollinated flowers in Japan. *J Plant Res* 111:199–205

Mori K, Nagao H & Yoshihara Y (1999) The olfactory bulb: coding and processing of odor molecule information. *Science* 286:711–715

Murawski DA & Gilbert LE (1986) Pollen flow in *Psiguria warscewiczii*: a comparison of *Heliconius* butterflies and hummingbirds. *Oecologia* 68:161–167

Murlis J, Elkinton JS & Cardé RT (1992) Odor plumes and how insects use them. *Ann Rev Entomol* 37:505–532

Nilsson LA & Rabakonandrianina E (1988) Hawkmoth scale analysis and pollination specialization in the epilithic Malagasy endemic *Aerangis ellisii* (Reichenb. fil.) Schltr. (Orchidaceae). *Bot J Linn Soc* 97:49–61

Olesen JM & Knudsen JT (1994) Scent profiles of flower colour morphs of *Corydalis cava* (Fumariaceae) in relation to foraging behaviour of bumblebee queens (*Bombus terrestris*). *Biochem Syst Ecol* 22:231–237

Omura H, Honda K & Hayashi N (2000) Floral scent of *Osmanthus fragrans* discourages foraging behavior of cabbage butterfly, *Pieris rapae. J Chem Ecol* 26:655–666

Paré P & Tumlinson JH (1997) De novo biosynthesis of volatiles induced by insect herbivory in cotton plants. *Plant Physiol* 114:1161–1167

Paulus HF & Gack C (1990) Pollinators as pre-pollinating isolation factors: evolution and speciation in *Ophrys* (Orchidaceae). *Israel J Bot* 39:43–79

Peakall R (1990) Responses of male *Zaspilothynnus trilobatus* Turner wasps to females and the sexually deceptive orchid it pollinates. *Funct Ecol* 4:159–167

Peakall R & Beattie AJ (1996) Ecological and genetic consequences of pollination by sexual deception in the orchid *Caladenia tentactulata. Evolution* 50:2207–2220

Pellmyr O (1986) Three pollination morphs in *Cimicifuga simplex*: incipient speciation due to inferiority in competition. *Oecologia* 78:304–307

Pellmyr O & Thien LB (1986) Insect reproduction and floral fragrances: keys to the evolution of the angiosperms? *Taxon* 35:76–85

Pellmyr O, Thompson JN, Brown JM & Harrison RG (1996) Evolution of pollination and mutualism in the yucca moth lineage. *Am Nat* 148:827–847

Pelz C, Gerber B & Menzel R (1997) Odorant intensity as a determinant for olfactory conditioning in honeybees: roles in discrimination, overshadowing and memory consolidation. *J Exp Biol* 200:837–847

Picker MD & Midgley JJ (1996) Pollination by monkey beetles (Coleoptera: Scarabaeidae: Hopliini): flower and colour preferences. *Afr Entomol* 4:7–14

Priesner E (1973) Reaktionen von Riechrezeptoren männlicher Solitärbienen (Hymenoptera, Apoidea) auf Inhaltsstoffe von *Ophrys* Blüten. *Zoon (Uppsala)* Suppl 1:43–54

Pyke GH (1991) What does it cost a plant to produce nectar? *Nature* 350:58–59

Raguso RA & Pellmyr O (1998) Dynamic headspace analysis of floral volatiles: a comparison of methods. *Oikos* 81:238–254

Raguso RA & Pichersky E (1999) A day in the life of a linalool molecule: chemical communication in a plant-pollinator system. Part 1: linalool biosynthesis in flowering plants. *Plant Sp Biol* 14:95–120

Rathcke BJ (1983) Competition and facilitation among plants for pollination. In: Real L (ed) *Pollination biology*, pp 305–328. Academic Press, Orlando, FL

Rieseberg LH & Noyes RD (1998) Genetic map-based studies of reticulate evolution in plants. *Trends Plant Sci* 3:254–259

Sachse S, Rappert A & Galizia CG (1999) The spatial representation of chemical structures in the antennal lobe of the honeybee: steps toward the olfactory code. *Eur J Neurosci* 11:3970–3982

Schatz GE (1990) Some aspects of pollination biology in Central American forests. In: Bawa KS & Hadley M (eds) *Reproductive ecology of tropical forest plants*, pp 69–84. UNESCO/Parthenon, Paris

Schemske DW & Horvitz CC (1984) Variation among floral visitors in pollination ability: a precondition for mutualism specialization. *Science* 225:519–521

Schiestl F, Ayasse M, Paulus H, Erdmann D & Francke W (1997) Variation of floral scent emission and post-pollination changes in individual flowers of *Ophrys sphegodes* subsp. *sphegodes*. *J Chem Ecol* 23:2881–2895

Schiestl FP, Ayasse M, Paulus HF, Löfstedt C, Hansson BS, Ibarra F & Francke W (1999) Orchid pollination by sexual swindle. *Nature* 399:421–422

Simpson SJ & Raubenheimer D (1996) Feeding behaviour, sensory physiology and nutrient feedback: a unifying model. *Entomol Exp Appl* 80:55–64

Skubatz H, Kunkel DD, Howald WN, Trenkle R & Mookherjee B (1996) The *Sauromatum guttatum* appendix as an osmophore: excretory pathways, composition of volatiles and attractiveness to insects. *New Phytol* 134:631–640

Smith BH & Cobey S (1994) The olfactory memory of the honeybee, *Apis mellifera*. II. Blocking between odorants in binary mixtures. *J Exp Biol* 195:91–108

Smith BH & Getz WM (1994) Non-pheromonal olfactory processing in insects. *Ann Rev Entomol* 39:351–375

Sprengel FC (1793) *Das entdeckte Geheimnis der Natur im Bau und in der Befruchtung der Blumen*. Englemann, Leipzig [1894 edn ed. by P. Knuth.]

Steinbrecht RA, Laue M & Ziegelberger G (1995) Immunolocalization of pheromone-binding protein and general odorant binding protein in olfactory sensilla of the silk moths *Antheraea* and *Bombyx*. *Cell Tissue Res* 282:203–217

Steiner KE (1998) Beetle pollination of peacock moraeas (Iridaceae) in South Africa. *Plant Syst Evol* 209:47–65

Stern WL, Curry KJ & Pridgeon AM (1987) Osmophores of *Stanhopea* (Orchidaceae). *Am J Bot* 74:1323–1331

Stopfer M, Bhagavan S, Smith BH & Laurent G (1997) Impaired odour discrimination on desynchronization of odour-encoding neural assemblies. *Nature* 390:70–74

Swihart CA & Swihart SL (1970) Colour selection and learned feeding preferences in the butterfly *Heliconius charitonius* Linn. *Anim Behav* 18:60–64

Thièry D, Bluet JM, Pham-Delègue M-H, Etiévant P & Masson C (1990) Sunflower aroma detection by the honeybee: study by coupling gas chromatography and electroantennography. *J Chem Ecol* 16:701–711

Todd JL & Baker TC (1993) Response of single antennal neurons of female cabbage loopers to behaviorally active attractants. *Naturwiss* 80:183–186

Tollsten L (1993) A multivariate approach to post-pollination changes in the floral scent of *Platanthera bifolia* (Orchidaceae). *Nordic J Bot* 13:495–499

Tollsten L & Bergström G (1993) Fragrance chemotypes of *Platanthera* (Orchidaceae): the result of adaptation to pollinating moths? *Nordic J Bot* 13:607–613

Topazzini A, Mazza M & Pelosi P (1990) Electroantennogram responses of five lepidopteran species to 26 general odourants. *J Insect Physiol* 36:619–624

Vareschi E (1971) Duftunterscheidung bei der Honigbiene – Einzelzell-Ableitungen und Verhaltensreaktionen. *Z vergl Physiol* 75:143–173

Visser JH (1979) Electroantennogram responses of the Colorado beetle, *Leptinotarsa decemlineata* to plant volatiles. *Entomol Exp Appl* 25:86–97

Vogel S (1954) *Blütenbiologische Typen als Elemente der Sippengliederung.* Fischer, Jena

Vogel S (1963) *The role of scent glands in pollination.* A. A. Balkema, Rotterdam

Vogt RG (1995) Molecular genetics of moth olfaction: a model for cellular identity and temporal assembly of the nervous system. In: Goldsmith MR & Wilkins AS (eds) *Molecular model systems in the Lepidoptera,* pp 341–367. Cambridge University Press, Cambridge

Vogt RG, Rybczynski R & Lerner MR (1990) The biochemistry of odorant reception and transduction. In: Schild D (ed) *Chemosensory information processing,* pp 33–76. Springer-Verlag, Berlin

Wang J & Pichersky E (1999) Identification of specific residues involved in substrate discrimination in two plant O-methyltransferases. *Arch Biochem Biophys* 368:172–180

Ware AB, Kaye PT, Compton SG & Van Noort S (1993) Fig volatiles: their role in attracting pollinators and maintaining pollinator specificity. *Plant Syst Evol* 186:147–156

Waser NM (1979) Pollinator availability as a determinant of flowering time in Ocotillo (*Fouquieria splendens*). *Oecologia* 39:107–121

Waser NM (1982) A comparison of distances flown by different visitors of the same species. *Oecologia* 36:223–236

Waser NM & Price MV (1985) The effect of nectar guides on pollinator preference: experimental studies with a montane herb. *Oecologia* 67:121–126

White PR (1991) The electroantennogram response – effects of varying sensillum numbers and recording electrode position in a clubbed antenna. *J Insect Physiol* 37:145–152

Whitten WM, Williams NH, Armbruster WS, Battiste MA, Strekowski L & Lindquist N (1986) Carvone oxide: an example of convergent evolution in euglossine-pollinated plants. *Syst Bot* 11:222–228

Wibe A, Borg-Karlson A-K, Norin T & Mustaparta H (1997) Identification of plant volatiles activating single receptor neurons in the pine weevil (*Hylobius abietis*). *J Comp Physiol A* 180:585–595

Wiebes JT (1979) Co-evolution of figs and their insect pollinators. *Ann Rev Ecol Syst* 10:1–12

Williams NH (1983) Floral fragrances as cues in animal behavior. In: Jones CE & Little RJ (eds) *Handbook of experimental pollination biology,* pp 51–69. Van Nostrand-Reinhold, New York

Williams NH & Whitten WM (1983) Orchid floral fragrances and male euglossine bees: methods and advances in the last sesquidecade. *Biol Bull* 164:355–395

Williams NH & Whitten WM (1999) Molecular phylogeny and floral fragrances of male
    euglossine bee-pollinated orchids: a study of *Stanhopea* (Orchidaceae). *Plant Sp
    Biol* 14:129–136
Yokoi M, Mori K & Nakanishi S (1995) Refinement of odor molecule tuning by
    dendrodendritic synaptic inhibition in the olfactory bulb. *Proc. Natl Acad Sci USA*
    92:3371–3375.
Young HJ (1986) Beetle pollination of *Dieffenbachia longispatha* (Araceae). *Am J Bot*
    73:931–944
Zhu Y, Keaster AJ & Gerhardt KO (1993) Field observations on attractiveness of selected
    blooming plants to noctuid moths and electroantennogram responses of black
    cutworm (Lepidoptera: Noctuidae) moths to flower volatiles. *Environ Entomol*
    22:162–166

LARS CHITTKA, JOHANNES SPAETHE, ANNETTE SCHMIDT, AND
ANJA HICKELSBERGER

6

# Adaptation, constraint, and chance in the evolution of flower color and pollinator color vision

Anciently the teaching was that nothing would have been created that did not have a definite purpose, and more recently it has been that natural selection would eliminate anything that did not serve an equally definite purpose. ... the assumed relation between the colors of flowers and the ... pollinating insects is such a classic ...

Apparently there is something about the internal mechanism ... that makes it difficult for a rose to be blue. ... therefore, the use of the idea of natural selection to explain the absence of blue roses in nature is not only not necessary but it is not justified ... It would be much better for the rose to be blue.

F. E. Lutz (1924)

We commonly think that biological signals and receivers are mutually tuned to one another. Flower colors and pollinator color vision are not exceptions. The diversity of flower colors and the differences in color vision between different classes of pollinators make speculations about their mutual adaptation tempting. Yet close inspection reveals that we know very little about evolutionary changes in flower color induced by selection pressures related to pollination, nor is there much evidence to show that color vision systems of pollinators have been tuned to flower color. We shall review cases where we think such changes have occurred, and other cases where they have not, even where a purely adaptationist scenario would predict evolutionary tuning. In such cases, we suggest alternative explanations, including phylogenetic constraint, exaptation (novel use of traits evolved for other purposes), pleiotropy (selection through correlated characters), and random evolutionary processes such as genetic drift. Because our understanding of these processes in relation to biological signals and receivers is still in its infancy, our evidence is fragmentary. We hope, however, that it will stimulate future research to add the missing pieces of the puzzle. We shall first discuss possible causes of the diversity of floral color signals and then move on to the evolution of pollinator color-vision systems.

## Pollination syndromes and flower colors

One way to explain the diversity of flower colors is to use the concept of pollination syndromes, which holds that particular classes of pollinators are specifically associated with particular floral traits, including floral color (Faegri & van der Pijl 1978). There is current debate on how tight and exclusive these associations are (Waser *et al.* 1996; Johnson & Steiner 2000; Thomson *et al.* 2000; Gegear & Laverty, this volume). One argument involves the significance of red flowers in the context of hummingbird pollination. In the classical view, red flower coloration is a strategy to kill two birds with one stone: such coloration was thought to be invisible for bees and at the same time attractive for hummingbirds (Raven 1972). Therefore, flowers that are morphologically adapted to bird pollination, and on which bees transfer pollen less efficiently than birds do, should be colored red. The premises are flawed, however. Bees do visit red hummingbird flowers, and they can be trained to distinguish red from a green, foliage-like background, as well as from yellow and orange model flowers (Chittka & Waser 1997). Researchers working on hummingbirds have not been able to find a preference for red (Lunau & Maier 1995). Thus, the association between hummingbirds and red flowers is not exclusive.

A recent study by Thomson *et al.* (2000), however, does indicate that the association exists. In seven lineages of the genera *Penstemon* and *Keckiella* (Scrophulariaceae), flowers frequented by hummingbirds are more often orange and red than their bee-visited close relatives. Also, red coloration is strongly associated with other floral traits linked to ornithophily. But what it is the significance of such coloration, if it is neither attractive for hummingbirds nor invisible for bees? In our view, there is no necessity for exclusivity: any change in floral trait may be subject to selection if it confers a change in fitness, however small. Red coloration might be an adaptation to facilitate detection by hummingbirds, or to decrease detectability by bees, or both – *even by a few percent*. For flowers that are adapted to hummingbirds, bumble bees may not only transfer pollen less efficiently than birds, thereby acting as pollen thieves

(Thomson *et al.* 2000); they may also rob nectar, further reducing plant fitness (Irwin & Brody 1999). In such circumstances selection would favor any character that diminished visitation by bees.

In many situations, hummingbirds and most bees choose nectar flowers on the basis of their net caloric rewards (Waser *et al.* 1996; Healy & Hurly, this volume; Waddington, this volume). These depend not only on the nectar content of the flowers, but also on the time taken to locate the flowers. Thus, we need to evaluate the *search times* that hummingbirds and bees take for finding red, UV-absorbing flowers, and compare these with times taken to search for flowers of other colors. Data for hummingbirds have yet to be obtained, but results for bumble bees are now available. In a flight arena, we presented *Bombus terrestris* workers with a random arrangement of three identical model flowers, all of which were rewarded. We measured the time taken from entering the arena to landing on the last flower, excluding flower-handling times. Search times strongly depended on color; the larger the color contrast of the flowers with their background, the more rapidly bees would detect the flowers. Red and white (UV-reflecting) model flowers, which make the poorest contrast with their backdrop, took more than twice as long to find than did blue or yellow flowers, for example (Fig. 6.1). Thus, red coloration may indeed be a strategy to reduce visitation by bumble bees to some degree. Another (non-exclusive) possibility is that hummingbird flowers use red color to form a mimicry ring, so that each species will be identifiable as a suitable food source by hummingbirds using experience gained on flowers of different species (Healy & Hurly, this volume).

In general, we expect sharper discontinuities between syndromes where classes of pollinators differ strongly in morphology (so that one type of pollinator transfers pollen substantially better than another) and in sensory system (so that, for example, a particular color is poorly detectable by one type of pollinator, but conspicuous for another). Red hummingbird flowers fit these prerequisites, but we stress that pollinator segregation achieved by red coloration is nowhere near exclusivity. We suspect that this observation extends beyond red flowers. The concept of "private channels" in sensory biology may apply to spectacular specializations such as ultrasonic hearing. However, in many cases, the ranges of sensory systems will overlap, sometimes heavily. In such cases, interactions between signals and signal receivers will not follow a simple crypsis vs. conspicuousness dichotomy. We may have to look for more subtle

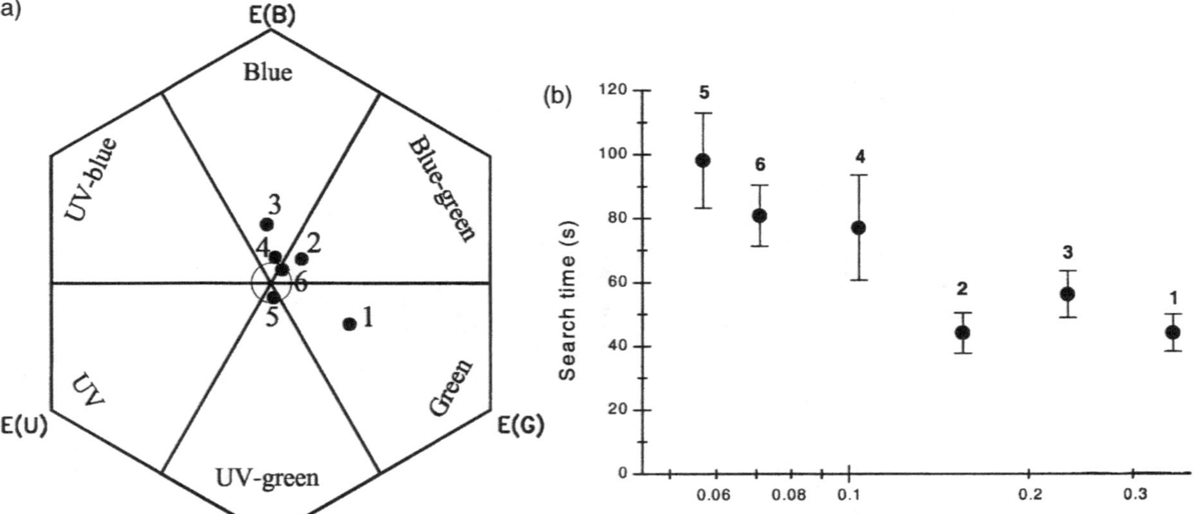



Fig. 6.1. (a) Color loci of targets for bumble bees in the color hexagon. The colors appear to humans as follows: 1–yellow; 2–white (UV absorbing); 3–blue; 4–turquoise; 5–red; 6–white (UV reflecting). The angular position of a color in the color hexagon informs us about the bee-subjective hue. We assume the photoreceptors adapt to the background against which the stimuli are presented; see Chittka (1996) for details. As a consequence of this adaptation process, the background lies in the center of color space. Thus, by definition, the color contrast of a model flower with its background is determined by the distance of its color locus from the center of color space. The distance between the center and each of the hexagon's corners is unity. (b) Correlation between color contrast (target vs. background) and search time ($r_s = -0.83$; $n = 6$; $p < 0.05$). Three colored chips of $\varnothing = 28$ mm were placed in a flight arena at random positions. We measured the time elapsed from entering the arena to landing on the third chip. Note that the correlation of detectability with color contrast is good only for large color targets; for smaller ones, an increasingly strong influence of green contrast is found (see Giurfa & Lehrer, this volume).

differences in effectiveness of different signals for different receivers, and in their actual fitness effects.

There is also the possibility of an evolutionary "arms race". If, for example, red hummingbird flowers are so profitable that bumble bees might significantly improve their fitness by exploiting them, then bees might be selected to improve their sensory skills to detect such flowers. As we discuss below, this might have happened in *Bombus occidentalis*, a bee species known for extensively robbing hummingbird flowers (Irwin & Brody 1999).

How do we explain the diversity of flower colors whose major reflectance falls within the visual range of practically all pollinators, such as UV, violet, blue, pink, white (typically UV-absorbing), or yellow (with or without UV-reflectance)? Some scientists have extended the syndrome concept to such flowers as well, but if partitioning by syndromes is the major selective pressure that drove floral color diversification, why do we not see stronger segregation? More bluntly, why are not all bumble bee flowers blue, all butterfly flowers orange, and all fly flowers white, for example? In many phylogenetic lineages, switches from one flower color to another occur without an associated morphological adaptation to a different class of pollinator (W.S. Armbruster, unpublished data). In one study on a nature reserve near Berlin, we did not find any differences among the colors of flowers visited by large and small bees, butterflies, flies, and beetles (Waser *et al.* 1996). In a phylogenetic analysis of the distribution of flower colors within two plant genera, Armbruster (unpublished data) found that all the variation occurred in association with bee pollination (see below). Thus, direct selection by pollinators in the sense of an innate affinity (as suggested by some adherents of pollination syndromes) surely cannot explain all the existing variation in floral color (Gegear & Laverty, this volume). In the following paragraphs, we highlight alternative explanations for why floral colors might diverge. Not all of these involve pollinators.

### Flower constancy and flower similarity

An alternative view to floral syndromes is that flowers differ in color as a strategy to promote flower constancy. Such fidelity by pollinators favors an efficient and directed pollen transfer between conspecifics (Chittka *et al.* 1999). Conversely, pollinators straying between flowers of different species may lose pollen during interspecific flights (Feinsinger 1987) or even reduce seed set by clogging stigmas with foreign pollen (Waser 1978).

In some closely related species, hybrids may be produced that are sometimes less viable than the parental species, thereby increasing selective pressures to diverge in floral advertising (Levin & Schaal 1970).

To understand the kind of diversity that can be expected to evolve as a strategy to promote constancy, it is critical to know the range over which pollinator-subjective color difference is correlated with flower constancy. For example, if a barely distinguishable contrast between two flower colors can produce 100% constancy, then flower constancy may drive only small-scale color differences, such as between two similar, but just distinguishable, shades of blue. However, character displacement across color categories, such as from blue to yellow, would be harder to explain by pollinator constancy if this were the case. Previous work allows us to predict how color discrimination improves with color distance (Chittka *et al.* 1992), but flower constancy and discrimination are unlikely to increase with color difference in the same way. In measuring flower constancy as a function of floral color difference, we do not ask: "How well can bees distinguish colors?" Instead, the appropriate question is: "How readily do bees retrieve memories for different flower types, depending on how similar they are to the one currently visited?" Discriminability sets the upper limit for constancy, but there is no a priori reason to assume that constancy is directly determined by discriminability.

In order to measure flower constancy as a function of color distance between flower types, we tested six species of apid bees on 15 pairs of plant species or color morphs of the same species, using a paired-flower, bee-interview protocol (Thomson 1981). We did not use the traditional Bateman's Index (Bateman 1951), because this index has a number of complications: it cannot be calculated if animals are completely constant, because the denominator in the formula becomes zero. Additionally, Bateman's Index always yields maximum constancy if the frequency of inconstant transitions from one of the flower types is 0, even if pollinators are inconstant when starting from the other flower type. Therefore, we quantified constancy using a new formula which circumvents these difficulties:

$$\mathrm{cons} = 0.5\left[(A - B)/(A + B) + (C - D)/(C + D)\right]$$

where $A$ represents the number of constant flights from $X$ to $X$, $B$ the flights from $X$ to $Y$, $C$ the flights from $Y$ to $Y$, and $D$ the flights from $Y$ to $X$. Constancy calculated in this way can range from 1 (complete constancy) through 0 (random flights between species) to $-1$ (complete inconstancy).

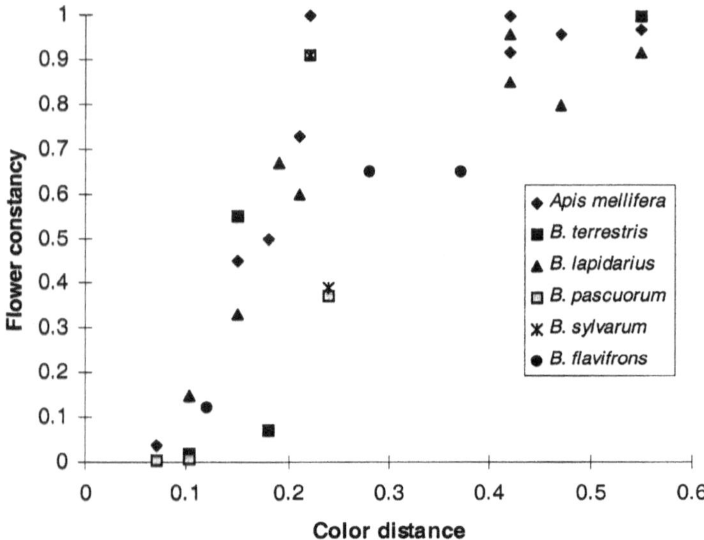


Fig. 6.2. Flower constancy in several species of bees as a function of color distance between pairs of flower types. For each pair of flower types, we recorded at least 80 choices. Flower constancy data are calculated as explained in the text.

This formula can be used only when individuals are coming to the pair of test flowers from both types of flowers.

Even though our analysis ignores differences other than color, there is a clear relationship between bee-subjective color difference and flower constancy (Fig. 6.2). Constancy does not deviate from chance at distances below 0.1 (where discrimination is already 70%; Chittka *et al.* 1992). At distances of about 0.2, constancy levels rise sharply in all pollinator species and above 0.4, constancy is generally above 80%. Thus, flower constancy is negligible at small color differences, even though bees can differentiate these colors well; it is at its maximum only in cases of pronounced differences.

Unfortunately, however, floral divergence due to benefits of constancy is not easily proven. Some authors have taken color diversity of sympatric flowers itself as evidence for character displacement (Menzel & Shmida 1993), but it is critical to test an observed distribution of phenotypes against a null model. Gumbert *et al.* (1999) examined several sets of sympatric and simultaneously flowering plants in a nature reserve near Berlin. A color distance distribution was generated for each set of flower colors by calculating bee-subjective color distances between all floral color loci in bee-color space. To test whether the flowers differ more strongly in color than would be expected by chance, we compared the real distributions with those produced by a random generator.

When common plants were examined, there were no significant differences between random sets and actual flower color distributions. The only significant deviations were detected in rare plants, and these effects varied with habitat. In one habitat, rare flowers were more similar than expected by chance, in two they were less similar, and in two others, there was no deviation from a random distribution. Thus, flower constancy may have influenced plant community structure in some habitats, but we need more research before concluding that such influences are widespread (Chittka *et al.* 1999).

Finally, pollen flow between populations at different sites may prevent local adaptations to conditions at those sites (Stanton & Galen 1997). In such a situation, plants may gain fitness if gene flow between populations is depressed. Thus, if (and only if) there is a genetic correlation between a trait favoring constancy (such as flower color) and a trait involved in, say, resource acquisition under different ecological conditions, floral signal divergence may indeed be favored (Jones, this volume).

### Pleiotropy, exaptation, constraint, and chance in the evolution of flower color

Biologists interested in the evolution of plant and animal signals tend to attribute signal diversity to selective pressures exerted by the signals' receivers (but see Newbigin 1898; Lutz 1924, for early attacks on this view). There are alternative explanations. One is pleiotropy, or indirect selection through genetically correlated characters (Armbruster *et al.* 1997). Carotenoids, responsible for yellow to orange coloration, are essential accessory pigments to chlorophyll in all plants (Scogin 1983). Many other pigment classes involved in floral coloration, or the biochemical pathways leading to the production of such pigments, may also protect against herbivores, UV radiation, and frost, or have unspecified effects on plant vigor (Onslow 1920; Levin & Brack 1995; Armbruster *et al.* 1997; Fineblum & Rausher 1997).

For example, Osche (1979) suggested that the yellow flavonoid coloration of pollen was already present in wind-pollinated ancestors of extant anthophilous plants and primarily served as protection against muta-genic UV radiation. He suggested that in the early stages of insect pollina-tion, many pollinators might have formed an innate preference for yellow floral signals, and that many plants later evolved large yellow nectar guides as supernormal stimuli to cater to this preference (Osche 1979). This hypothesis remains to be tested by phylogenetic tests, however.

To examine the possibility of pleiotropic effects in floral color evolu-tion, two plant genera with great flower-color variation, *Dalechampia* and *Acer*, were examined using phylogenetically informed analyses (Armbruster *et al.* 1997; Armbruster, unpublished data). In both, flowers shared pigments that were also found in leaves and stems. In *Dalechampia*, similar changes in flower color occurred several times independently in evolutionary history, but these changes were not associated with pollina-tion mode. Instead, in all species with pink or purple flowers, anthocya-nins were also expressed in stems and leaves, where they possibly affect plant survival in ways not related to pollination. In *Acer*, the evolution of autumn leaf color actually predates changes in flower color. Again, this suggests that evolutionary changes in flower color may have occurred without any relation to pollination: rather, selective pressures operating on vegetative traits may have first favored the expression of different chemicals (see also Newbigin 1898; Onslow 1920). Then selection to enhance floral detectability may have favored expression of the same compounds in petals. In such cases, the use of particular pigments in the flowers is an exaptation with respect to pollination (Armbruster *et al.* 1997; Armbruster, unpublished data).

Pleiotropy is not the only constraint on flower color. If the flowers of two related species (or populations of the same species) have the same colors, this may not reflect similar selective pressures, be they on floral or vegetative traits. In fact, even if optimality arguments predict different coloration of flowers blooming at two different sites (for example because of the particular competing species in each habitat), they might still have the same color. One type of constraint is ongoing gene flow between pop-ulations, which might prevent flowers from local adaptation (Stanton & Galen 1997). Positive frequency-dependent selection by pollinators might also keep floral colors from reaching a local optimum (see Smithson, this volume). In addition, there are phylogenetic constraints on flower color in several plant taxa (Chittka 1997). In many species, a change from one floral color to another may simply require an improb-able sequence of mutation events. Finally, genetic drift can act as a kind of constraint, too: evolutionary chance processes will, with some probability and depending on the size of the population, eliminate intraspecific variance, unless it is continuously added by new mutations or immigration (Adkison 1995).

Conversely, some plants show pronounced variation in flower color among populations (e.g., Beerling & Perrins 1993). These might reflect adaptations to local pollinator preferences, character displacement driven by different competing plants, or, through pleiotropy, adaptations to local selective pressures on vegetative traits. The possibility that simple genetic drift might account for these differences has been left largely unconsidered. To our knowledge, the only exception are the flowers of the *Nigella arvensis* complex, which occur not only in mainland Turkey and Greece, but also on several Aegean islands. There are strong differences in color, pattern, and floral shape among island populations; genetic drift is a likely explanation (Strid 1970). Because these island populations are small, the idea of drift is particularly palatable, but there is no a priori reason to suspect that mainland populations of plants, whose effective population sizes may be as small as those on islands, are immune to chance evolutionary processes.

### Has bee color vision adapted to flower color?

The discoveries that bees see ultraviolet and that flowers reflect it were made several decades ago (Kühn 1923; Lutz 1924, and references therein). Ever since, scientists have speculated that UV receptors in bees developed in a coevolutionary process with floral coloration (e.g., Menzel & Backhaus 1991). This notion received recent impetus from computer models showing that bee color vision is indeed the optimal system for detecting and identifying flowers (Chittka 1996). However, to prove that flower signals truly drove the evolution of bee color vision, it must be shown that the ancestors of bees possessed different sets of color receptors prior to the advent of the flowering plants. One must evaluate arthropods whose evolutionary lineages diverged from those of bees before there were flowers. If the color vision of such animals is indistinguishable from that of bees, this implies that it was present in an ancestor of bees that predated the evolution of flower color – and this is exactly what was found (Chittka 1997). The $\lambda_{max}$ values (wavelength values of maximum sensitivity) of the Crustacea and Insecta fall into three distinct clusters in the UV (around 350 nm), blue (~440 nm), and green (~520 nm). Red receptors show up irregularly both in the Crustacea and Insecta; they have evolved several times independently.

Thus, we can infer that insects were well pre-adapted for flower-color coding more than 500 million years ago, about 400 million years before the extensive radiation of the angiosperm plants that started in the middle Cretaceous (100 million years ago). Recent data on the molecular structure of photopigments support the interpretation that the basic types of arthropod visual pigments must be placed at the very roots of arthropod evolution (Chittka & Briscoe 2001). Thus, bee color vision is an exaptation with respect to flower color.

Measured peak sensitivities of receptors vary up to 30 nm across insect species, however (Chittka & Briscoe 2001). Some of this must be measurement error, but we can not exclude the possibility of actual fine-tuning of pigments to particular visuo-ecological tasks. To examine such fine tuning, it is necessary to look at closely related species with known phylogeny and distinct ecological conditions. We mapped the positions of maximum sensitivity of the color receptors of 11 species of bumble bees from five subgenera onto their phylogeny (Fig. 6.3). These species span habitats from European alpine (e.g., *Bombus monticola*) and North American temperate (e.g., *B. impatiens*) to subtropical and tropical South America (*B. morio*), but the $\lambda_{max}$ positions are very similar across species. Peitsch *et al.* (1992) claimed that bee species flying in UV-rich environments might have short-wave-shifted UV receptors, while tropical forest-dwelling bees might have long-wave-shifted UV receptors. Our analysis does not support this claim: the alpine *B. monticola* (whose altitude range is 900–2700 m; Hagen 1994) does not differ from *B. terrestris* and *B. lapidarius* (both lowland species that are not found above 1400 m; Hagen 1994). Although the tropical *B. morio* has slightly long-wave-shifted UV receptors compared to the above two, it does not differ from several temperate species.

Several types of molecular constraints, and possibly pleiotropies, that might affect the evolution of color vision have been reviewed in detail elsewhere (Chittka & Briscoe 2001). One source of inertia that is often overlooked in investigations of sensory ecology is simply *chance*. Physiologists often assume that any superior genotype will inevitably be able to spread through a population. Because this assumption is so common, we shall elaborate in some depth why this may not necessarily happen. Imagine that a bumble bee colony produces 100 new queens, one of which carries a new mutation that has a beneficial effect on foraging

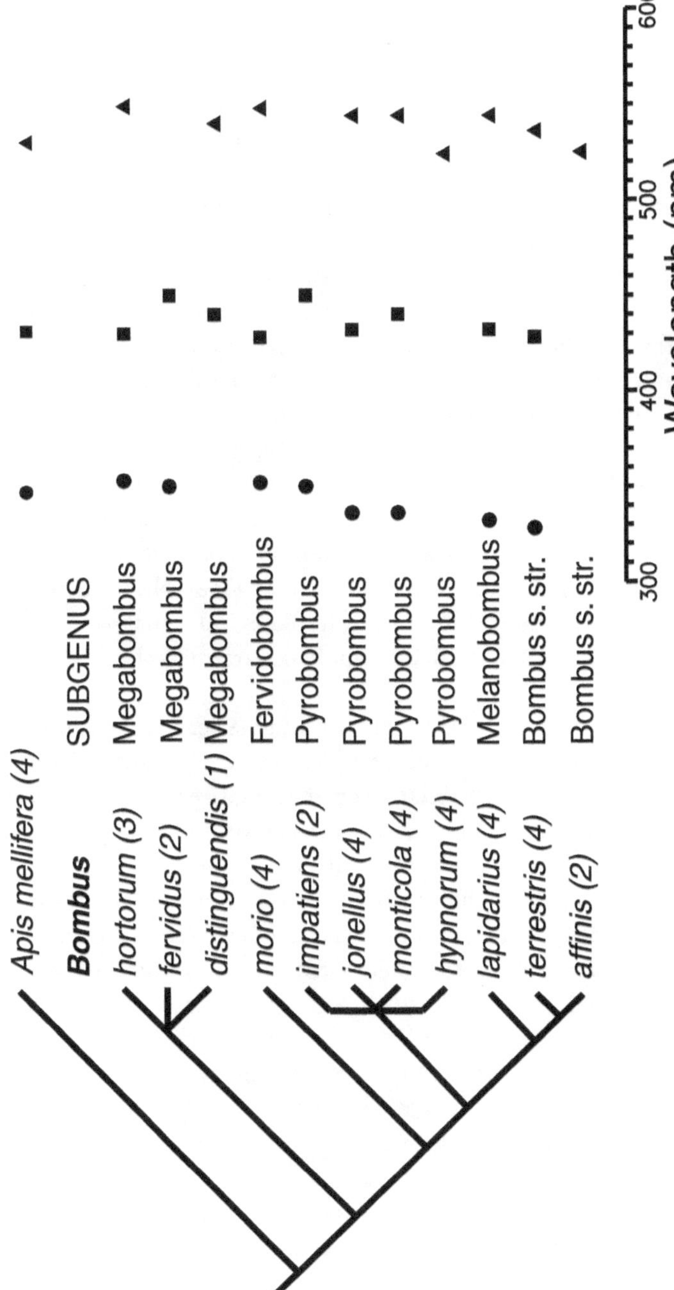


Fig. 6.3. The $\lambda_{max}$ values of photoreceptors of 11 species of bumble bees superimposed on their phylogenetic tree (according to Scholl *et al.* 1990; Williams 1994), with *Apis mellifera* for outgroup comparison. ●-UV receptors; ■-blue receptors; ▲-green receptors. Note that the absence of some receptor types in some cases does not actually mean that the species is lacking that receptor: in some cases, the authors did not seek to find all the receptors present. References: 1, Mazokhin-Porshnyakov (1969); 2, Bernard & Stavenga (1978); 3, Meyer-Rochow (1980); 4, Peitsch *et al.* (1992).

performance, such as a receptor with an altered spectral sensitivity. In a stable population, only one (or very few) of these queens will survive and again produce fertile offspring. Many will succumb to frost in the winter or bird predators, or their newly founded colonies may be attacked by cuckoo bumble bees or parasites; the question of whether a queen survives all these hardships is entirely unrelated to its foraging ability. Even if the blessed queen successfully starts a colony, and if its worker offspring forage slightly more efficiently, the next generation of queens will be subject to the same unpredictable hazards. Now imagine that the mutation in question is rare and occurs only once in several generations. It is clear, then, that its chances of spreading are very slim. Generally, the probability of the mutation spreading to fixation is correlated with the frequency of the mutation and its relative adaptive advantage and inversely correlated with population size (Ohta 1993). The influence of genetic drift will be further enhanced when reproductive success varies strongly among individuals (Adkison 1995), as is the case in bumble bees (Imhoof & Schmid-Hempel 1999), or when (repeated) bottlenecks occur, such as in Canary Island bumble bees (Widmer *et al.* 1998). The influence of these parameters on the goodness of fit in biological signal–receiver systems has generally been ignored, but should be extremely worthwhile to explore in the future.

The conservation of $\lambda_{max}$ values within the bumble bees need not necessarily reflect any kind of constraint, however. If flower-color detection and identification in all these habitats require similar receiver systems, then we would expect conservation even in a world without phylogenetic constraint. Indeed, the estimated optimal color coding systems for Brazilian, Israeli, and German flowers from several habitats were almost identical (Chittka 1996). Be that as it may, the search for sensory adaptations is predictably frustrating if several related species display the same trait. Ideally, we need to study a trait that is variable both within and between closely related species (Chittka & Briscoe 2001). The only striking variation in receptors among the Hymenoptera is the occurrence of red receptors in very few species (Peitsch *et al.* 1992). Why most bee species lack such receptors defies a simple adaptive explanation. Although pure red flowers are rare in many habitats, many flowers do present information in the red part of the spectrum. Bee color-vision systems would, in theory, gain substantially if they had red receptors in addition to UV, blue, and green receptors (Chittka & Menzel 1992).

## The evolution of flower-color preference in bumble bees

In an attempt to identify a visual trait that might reveal a more interesting pattern of adaptation to the visual environment, we evaluated the innate floral color preferences of bumble bees. We hypothesized that evolutionary changes of such preferences require changes only in the synaptic efficiency between neurons coding information from the color receptors. Therefore, color preferences might adapt more readily to environmental requirements than do the wavelength sensitivities of color receptors.

In one study, a good correlation was found between the color preferences of naïve honeybees and the nectar offerings of different flowers in a nature reserve near Berlin (Giurfa *et al.* 1995). In brief, honeybees preferred the colors violet (bee UV-blue) and blue (bee blue), which were also the colors most associated with high nectar rewards. This pattern is not unique to the German flora: a similar association of flower color with reward was found in Israel (Menzel & Shmida 1993). However, a correlation never indicates causality. To show that color preferences evolved to match floral offerings, we need to compare a set of closely related bee species (or populations of the same species) that live in habitats in which the association of floral colors with reward is different.

We tested seven species of bumble bees from three subgenera: four from central Europe (*Bombus terrestris terrestris* [229; 8; 4698], *B. lucorum* [39; 2; 547], *B. pratorum* [14; 1; 395], and *B. lapidarius* [83 ;2; 1446]); two from Japan (*B. ignitus* [89; 3; 2782] and *B. hypocrita* [54; 2; 1691]); and one from North America (*B. occidentalis* [122; 4; 3405]). Numbers in brackets give the number of individuals tested, the number of colonies, and the number of choices evaluated. All species preferred the violet–blue range, presumably a phylogenetically ancient preference (Fig. 6.4). In addition, however, *B. occidentalis* had the strongest preference for red of all mainland bumble bee populations examined. This is provocative because this species frequently robs nectar and forages heavily from red flowers apparently adapted for pollination by hummingbirds (Chittka & Waser 1997; Irwin & Brody 1999). Clearly, this preference is derived and therefore might be an adaptation unique to *B. occidentalis*.

We also tested *Bombus terrestris terrestris* from Holland [85; 3; 1670], *B. terrestris terrestris* from Germany [144; 5; 3028], *B. terrestris dalmatinus* from Israel [156; 5; 5731], *B. terrestris dalmatinus* from Rhodes; [150; 5; 5335];

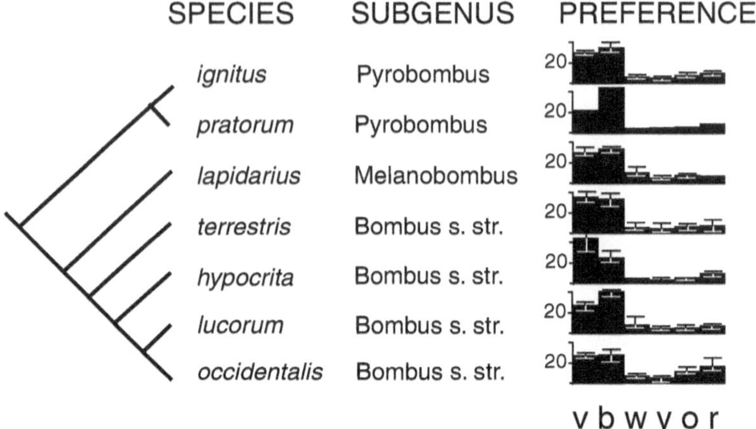



Fig. 6.4. The color preferences of seven species of bumble bees (*Bombus*) superimposed on their phylogeny (Williams 1994). Each bee was experimentally naïve at the start of the experiment, and only the first foraging bout was evaluated. Bees were individually tested in a flight arena; they were offered the colors v–violet (bee UV-blue); b–blue (bee blue); w–white (bee blue–green); y–yellow; o–orange; r–red (the latter three are all bee green). Column height denotes the percentage of cumulative choices of all bees from all colonies. Whiskers show percentages for colonies with extreme values.

*B. terrestris sassaricus* from Sardinia [133; 4; 4518], *B. terrestris xanthopus* from Corsica [58; 2; 2678], and *B. terrestris canariensis* from the Canary Islands [159; 5; 3904]. The rationale for testing island populations was that evolution often takes a different course there. Generally, the effects of chance, including those of bottleneck events, will be more manifest on islands than in large mainland populations (Adkison 1995; Barton 1998). In addition, small populations might adapt more readily to local conditions, whereas in large populations, gene flow across long distances may prevent local adaptation (Ford 1955; Stanton & Galen 1997). On the other hand, polymorphism, the raw material for evolution, is lost more easily in small populations, and deleterious mutations may spread through island populations more readily. A dramatic example known from the visual system are the totally color-blind people of Pingelap Island (Hussels & Morton 1972; Chittka & Briscoe 2001). The island populations of *B. terrestris* are particularly interesting because they are genetically diferentiated from each other and from the mainland population, whereas the entire mainland population, which stretches all through central, southern, and eastern Europe, appears to be genetically more homogenous (Estoup *et al.* 1996; Widmer *et al.* 1998).

Correspondingly, we find no strong differences in color preferences among the mainland *B. terrestris* populations: all showed the same strong preference for violet–blue shades as the other species above. But some island populations show an additional red preference (Fig. 6.5). In *B. t. sassaricus*, this preference is stronger than that for blue colors in some colonies, and is highly significant in all colonies (significance is determined both by a sign test [number of individuals per colony which prefer red over yellow] and a $\chi^2$ 2 × 2 table [colony choices for red vs. yellow]. In all colonies, both tests yielded similar results). In *B. t. canariensis*, four of five colonies showed a significant preference for red over yellow and orange. The adaptive significance of such red preference is not easy to understand. Some red, UV-absorbing, and pollen-rich flowers exist in the Mediterranean basin, particularly the eastern part, with the highest concentration in Israel (Dafni *et al.* 1990). In Israel, however, bumble bees do not show red preference, and the red flowers there appear to be predominantly visited by beetles (Dafni *et al.* 1990). Some of the red species exist in Sardinia, too, but we do not know to what extent they are exploited by bumble bees. The Canary Islands harbor several orange–red flower species (Vogel *et al.* 1984). These are probably relics of a Tertiary flora, and some seem strongly adapted to bird pollination. In fact, bird visitation has been observed at least in some of these species, but their use by bees is unknown (Vogel *et al.* 1984; Olesen 1985). Thus, we are left with an interesting observation: flower color preferences are clearly variable within *B. terrestris*, but we cannot easily correlate the color preferences in different habitats with differences in local flower colors. The possibility that genetic drift has produced the color preferences in some island populations certainly deserves consideration. To explore this possibility further, it will be necessary to sample the local floral market in more detail (as in Menzel & Shmida 1993; Giurfa *et al.* 1995) and to test whether red preference might simply evolve in some island populations because it is not selected *against*. We hope to measure the impact on foraging performance and fitness of among-colony variation in preference.

Finally, the observed patterns of floral-color preferences within bumble bees suggest that it may be worthwhile to take a closer look at the receptor level. Could some species of bumble bees (such as *B. occidentalis*) or some island populations of *B. terrestris* have actually evolved red receptors? Clearly, the observation of red preference itself cannot be taken as evidence for the existence of red receptors, because detection and identification of red flowers is possible without specific red receptors (Chittka &

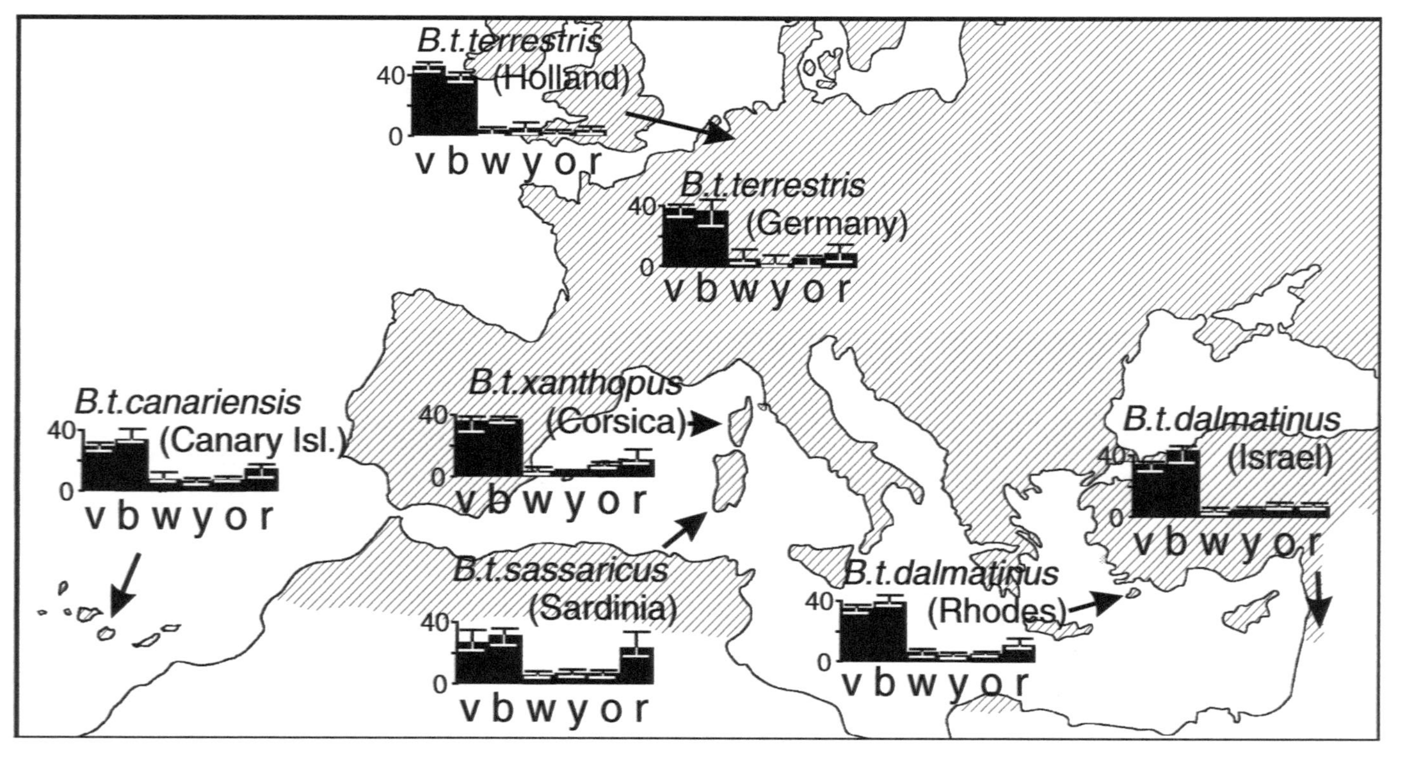


Fig. 6.5. Biogeography of floral color preferences in Bombus terrestris. Cross-hatched area: distribution of B. terrestris (this range was provided with kind permission by Prof. Pierre Rasmont; the full map will appear in Rasmont *et al.* 2001). For further explanation, see Fig. 6.4.

Waser 1997). Red flowers do take substantially longer to detect (see above), so that the evolution of red receptors might be favored in species whose ranges overlap with that of red flowers. If physiological research does reveal the existence of red receptors in bumble bees with red preference, we envision two possible evolutionary paths towards such receptors in bees. In large populations, red receptors might become fixed only in the case of a strong selective advantage, such as in bees that already exploit red flowers. Conversely, if the fitness advantage conferred by red receptors is comparatively small, new mutants that carry such receptors might be eliminated by genetic drift with very high probability. In the case of such a minor adaptive advantage, red receptors might spread only through relatively small populations, such as those on islands.

## Conclusion

We have used flower colors and bee color vision to convey the message that evolutionary matching of signals and receivers will not happen as readily and easily as physiological adaptation of, say, a receptor's sensitivity. In fact, this work contains a number of cases where behavioral, sensory, and floral traits can be better explained by the species' phylogenetic history or constraints than by the assumption that each trait in each species is individually and optimally tailored to its environment. Many paths along the way from genes to traits are intertwined, so that evolutionary changes in one trait may render another trait less efficient or nonfunctional. Finally, selection acts on individuals, and whether individuals survive and reproduce depends not only on their genetic quality, and certainly not only on the quality of *any* trait in which one happens to be interested. Chance plays an important role, and the role it plays will depend on the strength of selection (or the adaptive value of the trait in question) combined with population size and stability. We encourage readers to consider these ideas when studying the evolutionary tuning of flower signals and insect sensory systems, and to design more studies that specifically test for the above possibilities. If we take alternatives to adaptation seriously, we may ultimately understand adaptation better.

## Acknowledgements

This study was supported by DFG grants CH 147/1–2 and CH 147/2–1 to L. Chittka. For discussions or help with the experiments we thank S. Armbruster, A. Gerber-Kurz, P. F. Röseler, J. D. Thomson, and N. M. Waser.

## References


Adkison MD (1995) Population differentiation in Pacific salmon: local adaptation, genetic drift, or the environment. *Can J Fish Aquat Sci* 52:2762–2777

Armbruster WS, Howard JJ, Clausen TP, Debevec EM, Loquvam JC, Matsuki M, Cerendolo B & Andel F (1997) Do biochemical exaptations link evolution of plant defense and pollination systems? Historical hypotheses and experimental tests with *Dalechampia* vines. *Am Nat* 149:461–484

Barton NH (1998) Natural selection and random genetic drift as causes of evolution on islands. In: Grant PR (ed) *Evolution on islands*, pp 102–123. Oxford University Press, Oxford

Bateman AJ (1951) The taxonomic discrimination of bees. *Heredity* 5:271–278

Beerling DJ & Perrins JM (1993) *Impatiens glandulifera* Royle (*Impatiens roylei* Walp.). *J Ecol* 81:367–382

Bernard GD & Stavenga DG (1978) Spectral sensitivities of retinular cells measured in intact, living bumblebees by an optical method. *Naturwiss* 65:442–443

Chittka L (1996) Optimal sets of colour receptors and opponent processes for coding of natural objects in insect vision. *J Theor Biol* 181:179–196

Chittka L (1997) Bee color vision is optimal for coding flower colors, but flower colors are not optimal for being coded – why? *Israel J Plant Sci* 45:115–127

Chittka L & Briscoe A (2001) Why sensory ecology needs to become more evolutionary – insect color vision as a case in point. In: Barth FG (ed) *The ecology of sensing*. Springer-Verlag, Berlin

Chittka L & Menzel R (1992) The evolutionary adaptation of flower colors and the insect pollinators' color vision systems. *J Comp Physiol A* 171:171–181

Chittka L & Waser NM (1997) Why red flowers are not invisible for bees. *Israel J Plant Sci* 45:169–183

Chittka L, Beier W, Hertel H, Steinmann E & Menzel R (1992) Opponent colour coding is a universal strategy to evaluate the photoreceptor inputs in hymenoptera. *J Comp Physiol A* 170:545–563

Chittka L, Thomson JD & Waser NM (1999) Flower constancy, insect psychology, and plant evolution. *Naturwiss* 86:361–377

Dafni A, Bernhardt P, Shmida A, Ivri Y, Greenbaum S, O'Toole C and Losito L (1990) Beetle pollinated red bowl-shaped flowers – evolutionary convergence in the Mediterranean. *Israel J Bot* 39:81–92

Estoup A, Solignac M, Cornuet J-M, Goudet J & Scholl A (1996) Genetic differentiation of continental and island populations of *Bombus terrestris* (Hymenoptera: Apidae) in Europe. *Mol Ecol* 5:19–31

Faegri K & van der Pijl L (1978) *The principles of pollination ecology*, 3rd edn. Pergamon Press, Oxford

Feinsinger P (1987) Effects of plant species on each other's pollination: is community structure influenced? *Trends Ecol Evol* 2:123–126

Fineblum WL & Rausher MD (1997) Do floral pigmentation genes also influence resistance to enemies? The W locus in *Ipomoea purpurea*. *Ecology* 78:1646–1654

Ford EB (1955) Rapid evolution and the conditions which make it possible. *Cold Spr Harb Symp* 20:230–238

Giurfa M, Núñez J, Chittka L & Menzel R (1995) Colour preferences of flower-naive honeybees. *J Comp Physiol A* 177:247–259

Gumbert A, Kunze J & Chittka L (1999) Floral colour diversity in plant communities, bee colour space and a null model. *Proc R Soc Lond B* 266:1711–1716

Hagen E von (1994) *Hummeln*. Naturbund Verlag, Augsburg

Hussels IE & Morton NE (1972) Pingelap and Mokil atolls: achromatopsia. *Am J Hum Genet* 24:304–309

Imhoof B & Schmid-Hempel P (1999) Colony success of the bumble bee, *Bombus terrestris*, in relation to infections by two protozoan parasites, *Crithidia bombi* and *Nosema bombi*. *Insectes Soc* 46:233–238

Irwin RE & Brody AK (1999) Nectar-robbing bumble bees reduce the fitness of *Ipomopsis aggregata* (Polemoniaceae). *Ecology* 80:1703–1712

Johnson SD & Steiner KE (2000) Generalization versus specialization in plant pollination systems. *Trends Ecol Evol* 15:140–143

Kühn A (1923) Versuche über das Unterscheidungsvermögen der Bienen und Fische für Spektrallichter. *Nachr Ges Wiss, Göttingen, Math-Nat Wiss Kl l*:66–71

Levin DA & Brack ET (1995) Natural selection against white petals in phlox. *Evolution* 49:1017–1022

Levin DA Schaal BA (1970) Corolla color as an inhibitor of interspecific hybridization in *Phlox*. *Am Nat* 104:273–283

Lunau K & Maier EJ (1995) Innate colour preferences of flower visitors. *J Comp Physiol A* 177:1–19

Lutz FE (1924) Apparently non-selective characters and combinations of characters including a study of ultraviolet in relation to the flower-visiting habits of insects. *Ann NY Acad Sci* 29:181–283

Mazokhin-Porshnyakov GA (1969) *Insect vision*. Plenum Press, New York

Menzel R & Backhaus W (1991) Colour vision in insects. In: Gouras P (ed) *The perception of colour*, vol. 6, pp 262–293. Macmillan, London

Menzel R & Shmida A (1993) The ecology of flower colours and the natural colour vision of insect pollinators: the Israeli flora as a study case. *Biol Rev* 68:81–120

Meyer-Rochow VB (1980) Electrophysiologically determined spectral efficiencies of the compound eye and median ocellus in the bumblebee *Bombus hortorum tarhakimalainen* (Hymenoptera, Insecta). *J Comp Physiol A* 139:261–266

Newbigin MI (1898) *Colour in nature: a study in biology*. John Murray, London

Ohta T (1993) Interaction of selection and drift in molecular evolution. *Jap J Genet* 68:529–537

Olesen JM (1985) The Macronesian bird–flower elements and its relation to bird and bee opportunists. *Bot J Linn Soc* 91:395–414

Onslow MW (1920) *Practical plant biochemistry*. Cambridge University Press, Cambridge

Osche G (1979) Zur Evolution optischer Signale bei Blütenpflanzen. *Biol unserer Zeit* 9:161–170

Peitsch D, Fietz A, Hertel H, de Souza J, Ventura DF & Menzel R (1992) The spectral input systems of hymenopteran insects and their receptor-based colour vision. *J Comp Physiol A* 170:23–40

Rasmont P, Verhaeghe J-C, Rasmont R & Terzo M (2001) *West-Palaearctic bumblebees*. Apollo, Svenstrup (in prep.)

Raven PH (1972) Why are bird-visited flowers predominantly red? *Evolution* 26:674

Scholl A, Obrecht E & Owen RE (1990) The genetic relationship between *Bombus moderatus* Cresson and the *Bombus lucorum* auct. species complex (Hymenoptera: Apidae). *Can J Zool* 68:2264–2268

Scogin R (1983) Visible floral pigments and pollinators. In: Jones CE & Little RJ (eds) *Handbook of experimental pollination biology*, pp 160–172. Scientific & Academic Editions, New York

Stanton ML & Galen C (1997) Life on the edge: adaptation versus environmentally mediated gene flow in the snow buttercup, *Ranunculus adoneus*. *Am Nat* 150:143–178

Strid A (1970) Studies in the Aegean flora. XVI. Biosystematics of the *Nigella arvensis* complex. *Opera Bot* 28:1–169

Thomson JD (1981) Field measures of flower constancy in bumblebees. *Am Midl Nat* 105:377–380

Thomson JD, Wilson P, Valenzuela M & Malzone M (2000) Pollen presentation and pollination syndromes, with special reference to *Penstemon*. *Plant Sp Biol*, 15:11–29

Vogel S, Westerkamp C, Thiel B & Gessner K (1984) Ornithophilie auf den Canarischen Inseln. *Plant Syst Evol* 146:225–248

Waser NM (1978) Interspecific pollen transfer and competition between co-occurring plant species. *Oecologia* 36:223–236

Waser NM, Chittka L, Price MV, Williams N & Ollerton J (1996) Generalization in pollination systems, and why it matters. *Ecology* 77:1043–1060

Widmer A, Schmid-Hempel P, Estoup A & Scholl A (1998) Population genetic structure and colonization history of *Bombus terrestris* s. l. (Hymenoptera: Apidae) from the Canary Islands and Madeira. *Heredity* 81:563–572

Williams PH (1994) Phylogenetic relationships among bumble bees (*Bombus* Latr.): a reappraisal of morphological evidence. *Syst Entomol* 19:327–344

7

# Foraging and spatial learning in hummingbirds

Enthusiasm for optimal foraging theory in the 1970s and 1980s stimulated much work on foraging by bees and, to a lesser extent, hummingbirds. These animals were assumed to be energetically stressed because they were nectarivorous, small, and dependent on costly forms of flight. Researchers sought to explain foraging in terms of movement patterns that saved energy. For example, Pyke (1981) derived movement rules both within and between inflorescences. He compared observed directions and distances of movements following departure from a flower to optimal predictions, under the assumption that the animals should maximize their net rate of energy intake. Pyke (and many others) also assumed that animals would have imperfect knowledge about their environments, particularly with regard to predictions as to what and where to find food in the future. In addition to using statistical rules (Pyke 1984), such animals should always sample in order to track an ever-changing world. Despite the power and appeal of this viewpoint, we now see growing evidence that simple rules and patterns alone cannot explain foraging in hummingbirds. Here, we review how learning and memory influence hummingbird foraging and how memory might affect the ways in which hummingbirds pollinate plants.

Much of a hummingbird's diet is derived from the nectar of flowers that, in turn, rely on hummingbirds for pollination. These flowers frequently provide only a few mg of sugar daily (Kodric-Brown & Brown 1978). Hummingbirds therefore must visit many flowers on each feeding bout, transferring pollen among flowers in the process. It is claimed that replenishment of this nectar, when it occurs, requires several hours (Armstrong *et al.* 1987). If this is the case, a hummingbird that is foraging efficiently would do well to use a strategy for avoiding those flowers it has recently emptied.

Given that there is no evidence that hummingbirds detect empty flowers before visiting (see below), either by olfactory cues or by the sight of nectar, there are several types of strategy by which a bird might make decisions about which flowers (or inflorescences, plants, etc.) to visit: (1) move to new flowers following simple decision rules based on immediate past experience, the present flowers being visited, and what can be seen from the present site; (2) visit flowers based on their visual characteristics; or (3) visit flowers based on memories of the locations of individual flowers and their physical attributes (such as color or nectar). This last strategy includes combining rules and memory in a systematic fashion (e.g., traplining; Feinsinger 1978). We shall address each of these in turn.

### Simple decision rules

The application of optimal foraging theory (OFT) to investigating hummingbird foraging followed on the heels of the apparent success of applying such modeling to data on bee foraging (see Heinrich 1983). Bumble bees tend to arrive at the bottom of an inflorescence and move upwards before flying to another. Starting at the bottom might be an optimal strategy, because many vertical inflorescences display a gradient of nectar, with highest standing crops at the bottom (Pyke 1978*a*). Upward movement might not always represent an optimal response to nectar gradients, however. First, nectar gradients are not universal (see Corbet *et al.* 1981). Second, Waddington & Heinrich (1979) found it difficult to teach bees to reverse the direction of searching from a bottom-to-top to a top-to-bottom pattern. Heinrich (1983) suggests that flower morphology may also play a part in this searching strategy. Because flowers on inflorescences often hang downwards, it would be more efficient for the bee to move up the inflorescence while foraging. In addition, moving in one direction only, either up or down an inflorescence, circumvents revisiting flowers that the bee has just emptied. So, foraging by bees on inflorescences can be described using movement rules – is this true for hummingbirds also? In the flower mostly commonly used in foraging observations and experimental tests of hummingbirds foraging, scarlet gilia (*Ipomopsis aggregata*), Pyke (1978b) found no evidence that hummingbirds (broadtailed (*Selasphorus platycercus*) and rufous (*S. rufus*)) work systematically up or down an inflorescence (see also Hainsworth *et al.* 1983). At this stage, then, there is no evidence that hummingbirds forage within an inflorescence using movement rules like those of bumble bees.

Hummingbirds' movements between inflorescences have also been examined. Whereas some animals maintain a general directionality during a foraging bout (e.g., goldfish [*Carassius auratus*], Kleerekoper *et al.* 1970; European thrushes [*Turdus* spp.], Smith 1974), neither Pyke (1978*b*) nor Wolf & Hainsworth (1990) found evidence that this was how hummingbirds (broad-tailed and rufous) moved between inflorescences. Rather, a feeding hummingbird appeared to choose the next inflorescence based upon a visual impression of its proximity and/or its size. In addition, birds tended to fly further if the inflorescence on which they had just fed was of low quality. Although Wolf & Hainsworth (1990) found that the hummingbirds were foraging neither randomly nor by following simple movement rules, they concluded "that the birds have relatively little information about the rewards in an inflorescence before a visit." However, these results seem at odds with evidence (Miller & Miller 1971; Gass & Sutherland 1985) that hummingbirds can remember good feeding locations. Wolf & Hainsworth (1990) point out that foraging strategy may vary with the spatial scale at which foraging is being examined. For example, one might describe multiple visits among flowers in an inflorescence, among inflorescences within plants, or among clumps of plants.

### Use of visual cues

It is still a popularly held belief that hummingbirds have an innate preference for red flowers. Whereas considerable evidence suggests that hummingbirds exhibit color preferences, no data are available to determine whether this preference is innate (as has been demonstrated in butterflies by Weiss 1997). For example, Lyerly *et al.* (1950) found a significant avoidance of yellow feeders by a single Mexican violet-eared hummingbird (*Calibri t. thalassinus*). However, as early as 1941, Bené provided evidence that learning played a major role in flower color preference and that hummingbirds (black-chinned [*Archilochus alexandri*]) could be trained to visit specific food sources irrespective of their color (Bené 1941, 1945). Later investigations have supported this role of learning of color preferences in other species (e.g., Anna [*Archilochus anna*], Collias & Collias 1968; ruby-throated [*Archilochus colubris*], Miller & Miller 1971; rufous, Miller *et al.* 1985).

The possibility that a preference for red may arise from hummingbirds having a high visual sensitivity to red and low sensitivity to blue (Graenicher 1910) was convincingly refuted by Goldsmith & Goldsmith (1979), who demonstrated that black-chinned hummingbirds (*Archilochus alexandri*) learned to visit feeders lit by green light (546 nm) as quickly as they learned to visit feeders lit by red light (620 nm). They also showed that two different and opposing color associations could be learned simultaneously and that following experience with red feeders, the birds tended to visit red and blue (490 nm) feeders in preference to green and yellow (560 nm) feeders. Goldsmith & Goldsmith (1979) suggested that red is the least likely color to attract potential hymenopteran pollinators (although see Chittka & Waser 1997; Chittka, this volume). Thus, reduced interspecific competition between hummingbirds and the Hymenoptera may increase the relative value of red flowers and hummingbirds eventually learn this association. In addition, for hummingbirds at least, red flowers may offer a striking visual contrast against a green foliage background and may be the most conspicuous of flowers (relative to their size). This possibility has not yet been tested. Because many other floral attributes are correlated with the type of pollinator(s) plants attract, plants that offer their nectar and pollen in red flowers will also have nectar concentrations, nectar amounts, and size and shapes of flowers that differ from those of plants whose flowers are not red (Faegri & van der Pijl 1971; Thomson *et al.* 2000; but see Waser *et al.* 1996).

In hummingbirds, color preferences are learned associations between food sources and nectar amounts or concentrations (e.g., Stiles 1976; George 1980; Meléndez-Ackerman *et al.* 1997). In the field (at least in North America), this typically leads to hummingbirds feeding preferentially on red flowers. Indeed, nearly all of the Californian flora that appear to be specially adapted for pollination by hummingbirds have red flowers. These flowers are often long, tubular, and odorless, provide relatively large amounts of nectar, and have anthers and stigma positioned such that a hummingbird is that plant's most effective pollinator (e.g., Grant 1966).

Meléndez-Ackerman & Campbell (1998) attempted to dissociate morphological and color cues offered by *Ipomopsis aggregata* (red), *I. tenuituba* (whitish, longer, and more slender than *I. aggregata* ), and a hybrid of the two (intermediate in color [pink] and flower form) to foraging broadtailed *Selasphorus platycercus* and *S. rufous*. In one experiment, they painted *I. aggregata* flowers to match the colors of the three plant types; birds visited red flowers more than pink or white flowers. In another experiment, flowers of all three plant types were painted a standard red; birds showed no preference. Leaving aside possible criticisms of such painting techniques (e.g., Bennett *et al.* 1994), these manipulations appeared to show that the changing of flower color alone affected visitation rates to the three flower types, with the birds visiting red flowers in preference to pink or white flowers. In the first set of experiments, birds could not be using the relationship between flower color and nectar reward, as all the flowers were from the same plant. Therefore, they must have chosen flowers based on a previously learned association between color and reward. Despite some evidence that hummingbirds will extract more nectar from artificial flowers with wide corollas (Grant & Temeles 1992), birds in Meléndez-Ackerman & Campbell's (1998) experiment did not choose flowers based on morphology alone (the second experiment). It may be that such features are far less conspicuous than color; with flowers 0.5 m apart, as presented in these experiments, the bird might as well probe the flower once it has chosen to fly close enough to assess its morphology (*Ipomopsis aggregata*, with a wider corolla than the two alternatives, produces an average of 1.8 µl nectar per day, while *I. tenuituba* and the hybrid produce about 0.25 µl nectar per day). At least under the experimental conditions, flower color seems to explain the birds' choices. However, the experiments investigating color preferences and the role of learning in the development of preferences have shown that red flowers need to reinforce their possibly more conspicuous signal with a greater reward or easier access (e.g., via a wider corolla).

In summary, hummingbirds appear to associate color and reward, and are able to change their flower visit accordingly. However, learned preferences will persist if novel flowers tend to match the previously learned associations. Therefore, in a region in which hummingbirds have learned to associate red flowers with greater reward, novel flowers should be red in order to benefit both from high conspicuousness and from birds generalizing the learned association of red flowers and reward. This may be why the Californian flora pollinated by hummingbirds is dominated by red flowers. However, unless red flowers are also more conspicuous than other flower colors in other environments, such a relationship between flower color and hummingbird pollination need not exist. To determine whether generalization of learning has influenced floral features in this way, we need to know more about which flowers are pollinated exclusively by hummingbirds, the visual background in which they are found, and the local floral diversity.

## Memory for locations of flowers

Certain male hummingbirds defend territories of hundreds or perhaps thousands of flowers (Kodric-Brown & Brown 1978; Paton & Carpenter 1984; Armstrong *et al.* 1987). These birds may protect nectar resources by emptying the flowers on the edges of the territory early in the day, then moving toward the center of the territory as the day progresses (exploitation defense; Paton & Carpenter 1984). In order to do this the hummingbird must at least remember which flowers or patches of flowers he has emptied most recently. There are several different spatial scales at which the hummingbird might keep track of flowers he has emptied. A bird might remember: (1) patterns of movement around his territory; (2) broad areas of flowers (e.g., the southwest corner of his territory); or (3) specific flowers. The last of these has rarely been considered, possibly because such a memory capacity seems extraordinary.

The first possibility, that a bird might remember a pattern of movement around his territory, seems well within the abilities of a hummingbird, because some birds do this on a much larger scale than a territory. The lekking hummingbird species (e.g., the hermits, Trochilidae, Phaethornithinae) are thought to trapline, i.e., to visit isolated, undefended nectar sources in a regular fashion (Feinsinger & Colwell 1978; Gill 1988; Garrison & Gass 1999). To do this, the bird must remember the sequence of nectar sources (by remembering a pattern of movement and not each of the nectar sources separately) and which was the last source he visited. Despite general acceptance that some hummingbirds do forage in this manner, there are no detailed maps of routes: spatial traplining is inferred from noting regular reappearances of marked birds, not by following them (cf. Thomson, this volume). This kind of systematic foraging would be the simplest of the three memory tasks above; it seems possible, given the evidence supporting the slightly more demanding proposal, that hummingbirds can remember patches of flowers in order to avoid them for several hours (e.g., Gass & Sutherland 1985; Wolf & Hainsworth 1991). Whether or not the birds use sequences of vectors, landmark memories, or both combined is not yet known (see Chittka *et al.* 1995 for an experimental test on bees).

Given that hummingbirds appear to exhibit some sort of spatial memory, much initial research focused upon the nature of learning, rather than the capacity or function of memory in natural foraging. Laboratory tests have been usefully employed to explore the nature of spatial learning in North American species of hummingbird (black-chinned, Rivoli [now magnificent, *Eugenes fulgens*], and blue-throated [*Lampornis clemenciae*], Cole *et al.* 1982; rufous, Brown & Gass 1993; Brown 1994). Cole *et al.* (1982) found that the hummingbirds they tested (males and females) learned a "shift" task more quickly than a "stay" task. In the former, birds had to choose the new feeder of a pair; in the stay task, they had to return to the familiar feeder. Whether this difference in learning rate is based on innate biases or on experience gained from the field, the birds' responses make sense if they are accustomed to foraging on flowers that are depleted in a single visit.

Cole *et al.* (1982) suggested that stay learning might be easier to demonstrate in an experiment and easier for birds to learn, if the rewarding locations were patches of inflorescences rather than single flowers, because each visit does not produce substantial depletion. There has been, to our knowledge, no test of this suggestion. However, we have carried out two field experiments that indirectly bear on this issue. In the first, birds fed from a single artificial flower. In the first block of trials, the flower contained too much sucrose to deplete in a single visit; in the second block of trials, the flower contained 70 µl of sucrose, an amount birds take in a single visit. On their return, birds were presented with two flowers, 40 cm apart – the original flower and an alternative flower that differed from the original either in color or pattern. In some trials, the original flower stayed in its original location; in other trials, the alternative flower occupied the original position. When the alternative differed in color/pattern from the original flower, birds chose flowers apparently at random whether or not the original flower had been depleted in the first visit. Birds avoided the location of the original flower if it had been depleted, but did not show this avoidance behavior if the flower had not been depleted (see Fig. 7.1). In the second experiment, we attempted to increase the likelihood that birds would return to the original flower by allowing birds to visit a flower three times before offering an alternative. Birds were somewhat more likely to avoid the flower in the same location after three visits but were more likely to return to a flower bearing the same color/pattern as the flower they had visited three times. Therefore, it is difficult to teach male rufous to use a "stay" rule during foraging even if the "flowers" do not deplete. Of course, these birds *can* learn to stay, as is shown by their enthusiasm for using hummingbird feeders as well as the anecdotes describing birds returning from migration and hovering at the place the feeder had been hung the previous year.

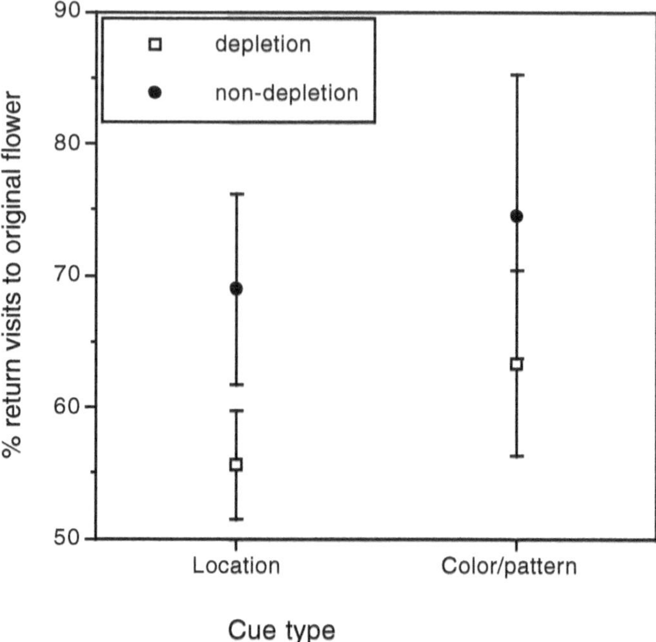



Fig. 7.1. Return visits made by hummingbirds to the original flower. In some trials the original flower was depleted after a single visit; in other trials it was not depleted. The alternative flower differed in either location or color/pattern features. Values are means and SE, n = 6 per group.

Rather than testing rules for decision-making in flower visits, recent workers have focused on the scale of spatial information remembered and used by rufous hummingbirds. Brown (1994) and Brown & Gass (1993) have shown that birds in the laboratory can use visual cues to predict which feeder of several will contain reward, even when the reward is at some distance from the relevant visual cue. Although these experiments demonstrate that hummingbirds can and do use visual cues in foraging, when birds are faced with making choices among flowers under conditions in which the spatial cues conflict with the flower's visual features (flowers 80 cm apart; Hurly & Healy 1996), the birds invariably chose to return to the flower that occupied the location previously visited, rather than the one bearing the color pattern of the rewarded flower. Several recent experiments have tried to determine which cues birds use to return to food locations by using this kind of dissociation method (e.g., Brodbeck 1994; Clayton & Krebs 1994). Typically, four possible feeders are presented in an array, all differing in their color/patterns. Only one feeder

Fig. 7.2. A schematic of a test for cue preference. The plain feeder of the four feeders on the left contains food. The animal is allowed to eat some of the food. While the animal is absent from the feeder array, the plain feeder is emptied and switched with one of the other feeders. The order in which the animal visits feeders is then observed.

contains food that the animal is unable to finish in one visit. Having eaten some part of the food, the animal is removed and the rewarded feeder is exchanged with one of the other feeders (see Fig. 7.2). When the bird returns, all the feeders are empty and the order in which the bird visits the feeders is observed. Food-storing birds, which need to remember many locations to relocate caches, are more likely to visit the feeder in the (formerly) correct location. Non-storing species are as likely to visit the feeder in the correct location as the feeder with the correct color/pattern. Although this design does not test memory capabilities for the two different types of cues, it can tell us which cue type the birds prefer to remember, or attend to, when foraging at flowers. Our hummingbirds behaved as did the food-storing birds. We interpret these data as demonstrating that, for the hummingbirds, spatial information regarding the flowers on which they feed is more important than the flowers' visual aspects. Locations and spacing of flowers and plants may, therefore, play a much greater role in hummingbird/plant pollination relationships than has been considered previously.

In order to determine whether a rufous hummingbird *can* remember specific locations of flowers it had depleted, we employed an "open-field" version of the radial-arm maze (eight flowers presented in a circle), a standard laboratory apparatus used for testing spatial memory. In Phase 1, birds in the field were allowed to visit and deplete the rewards in four "arms" of the eight-armed maze (Healy & Hurly 1995). In "free" trials, this meant birds visiting four of eight possible rewarded flowers. The birds were scared away or left after feeding from these four flowers. In "forced" trials, only four rewarded flowers were provided. On returning to the maze after being kept away for at least 5 min, the bird was presented with

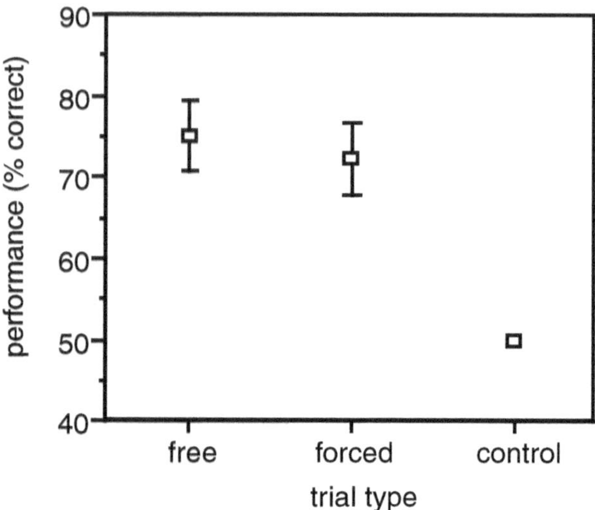



Fig. 7.3. The percentage of the first four visits to the correct flowers by hummingbirds in the single visit to the array in Phase 2 of each of the three trial types – free, forced, and control (see text for details); mean values ± SE. Chance performance is 50%. (Redrawn with permission from Healy & Hurly 1995.)

eight flowers, the four it had emptied earlier (not refilled) and four containing sucrose. If birds could remember which of the flowers they had depleted, they should avoid these when they return to the maze. Instead, they should feed from the four previously unvisited flowers. We assessed the birds' performances by comparing their ability to visit the four flowers not previously visited in Phase 1 with chance performance (50%). In both forced and free trials, birds were significantly better than expected by chance at visiting the non-depleted flowers (Fig. 7.3). In control trials, used to determine whether birds were using cues emanating from the sucrose reward to choose which flowers to visit, performance did not differ from chance. Because birds could choose which flowers to visit in Phase 1 of the free trials, performance in these trials may have been due to the birds choosing a systematic route around the maze. Although the data were not sufficient to make a thorough examination of the way in which the birds moved around the array, we did look to see whether the birds simply moved from one flower to its nearest neighbor. They did not, so some more complex behavior must explain the good performance by birds on forced trials.

Birds may have made the task easier by returning to the maze soon after Phase 1. We were not able to control a bird's return, and on some trials the retention interval was as short as a few minutes; on other trials it was longer than one hour. Performance decreased slightly with increasing delay between phases, but the effect of increasing time between the two phases of a trial was not significant. In spite of the lack of control over the timing of a bird's return to the maze, we showed that these hummingbirds are able to remember the locations of at least three flowers within such a maze for intervals of at least one hour. When the hummingbirds' performance is compared with that of other species tested under laboratory conditions, the hummingbirds do less well. However, it must be remembered that, unlike laboratory-tested animals, these birds fill the retention interval with activities that may interfere with their memories for our experimental flowers. They were away from the maze for an average of 11 min, and in that time may have fed at least once from natural flowers. Alternatively, the apparent decline in performance might not be based on a failure of memory, but on faulty assumptions on our part. For example, if flowers refill sooner than we have assumed or if there is a high probability of foraging competition from other animals, then the birds might be more likely to return to previously rewarded sites even over the time-scales observed in this experiment. None the less, our experiment was clearly not quantitatively comparable to natural foraging, and there are no data to show that these birds can remember several hundred flowers. These data do show, however, that rufous hummingbirds can and will remember point locations of flowers in the field.

Hurly (1996) carried out a one-trial learning experiment in order to determine whether rufous hummingbirds use the information about a flower's contents, in addition to its location, to make subsequent visit choices. He presented an array of four flowers, one of which contained 600 $\mu$l sucrose, too much for a bird to deplete on a single visit. The other three flowers contained equal volumes of water, which these hummingbirds prefer to avoid. The birds' performances were assessed by observing the number of flowers visited until the rewarded flower was relocated. Not only were birds very accurate at returning to the correct flower, the few errors they did make were to flowers they had not previously visited. These birds appear to remember both the contents and the locations of visited flowers, at least over time periods of less than one hour (mean time to return was 12.6 min). Given the changing nature of nectar supplies in the flowers in a bird's territory, it would be unlikely that the bird need remember all of this information for longer than a few hours.

Having demonstrated that male rufous hummingbirds are able to remember single flower locations, we then investigated how close together in space flowers can be before the birds find it difficult to discriminate between them. A bird was presented with a flower containing more sucrose than it could empty in a single visit. On its return, there were two flowers to choose from: the original, containing the remaining sucrose, and an alternative, containing water. On half of the occasions, the alternative flower was the same color as the original and on the others it was another color. Birds returned to the original flower significantly more often than expected by chance. Furthermore, birds most often chose the original flower when the distracter was of a different color, whereas performance was poor when the distracter resembled the original flower (see Fig. 7.4). However, there was a significant interaction between distracter color and distance: when the flowers were of different colors, birds chose the original flower more accurately the further it was from the distracter. When the flowers were the same color, the birds were, if anything, more likely to visit the distracter the further it was from the original flower. This result suggests that the hummingbirds were able to discriminate between the original and distracter flowers under both conditions, but that they employed different foraging tactics according to the color of the distracter in the choice phase. It seems sensible for a bird to sample new flowers that look like flowers that have been rewarding previously, but to learn very quickly to avoid flowers of a color that has never been rewarding. It also seems that these birds may be able to remember and to discriminate between flowers separated by only 3 cm. This may mean that plants could reduce or increase revisiting "errors" by producing flowers spaced apart or by clumping them very close together. Although the results from the cue-dissociation experiment show that rufous hummingbirds seem to prefer to attend to, or to remember, the global position of flowers rather than their color/patterns, it seems as if the birds also discriminate and generalize the local visual features of flowers and alter their foraging tactics accordingly. Two different learning rules, then, may be used depending on the flower cue: the color/pattern can be used to generalize across flowers or to discriminate among them (within and between species) and the likely average profitability can be assessed. The location of a flower, on the other hand, is quite specific and its expected profitability has a temporal aspect (Hurly & Healy 1996).

We have demonstrated that hummingbirds attend more to spatial

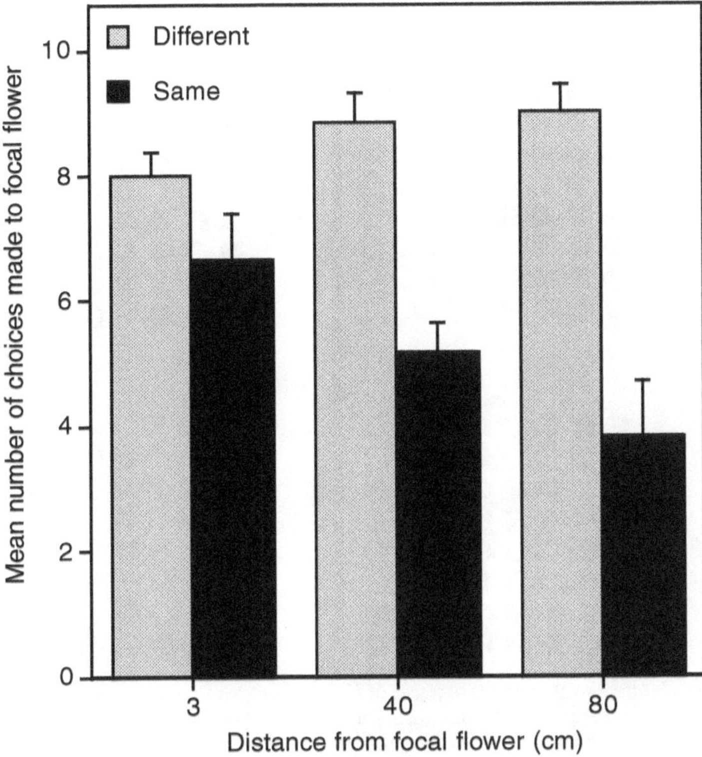

Fig. 7.4. The number of choices (out of 10) made by six rufous hummingbirds in favor of the focal flower when the distracter flower was either the same or a different color and was placed 3, 40, or 80 cm from the focal flower. Values are means and SE. (Figure reproduced with permission from Hurly & Healy 1996.)

cues than to color/pattern cues when making choices about revisiting flowers. However, the visual cues appear to enhance the rate of learning which locations are rewarded. In a laboratory task, Brown & Gass (1993) found that rufous hummingbirds learned the location of a rewarded feeder in an array faster when prominent visual cues were added. Visual conspicuousness, then, seems to work both to attract hummingbirds to flowers in the first place and, additionally, to increase the speed of learning a location (see also Healy & Hurly 1998, experiment 2). Enhancing the visual conspicuousness of its flowers may then have a dual benefit to a plant as the birds typically make very short visits to flowers (on the order of a few seconds).

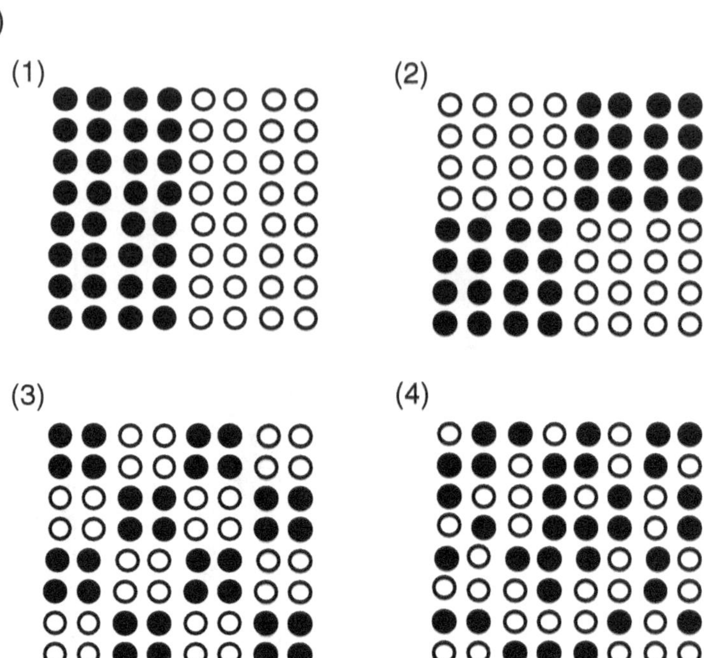

Fig. 7.5. (a) The four patterns of feeder quality: (1) halves, (2) quarters, (3) checkerboard, (4) Random. Solid circles represent feeders containing sucrose; the remaining circles represent feeders that were empty.

### Spatial patterns or point locations?

Wolf & Hainsworth (1990, 1991) could find little evidence for memory for reward locations in the five male hummingbirds (rufous and broad-tailed) they observed foraging at clumps of *Ipomopsis aggregata* inflorescences. However, rufous hummingbirds do display memory for both spatial pattern and point locations in experiments. Sutherland & Gass (1995) used an array of 64 feeders (11 cm apart), 32 of which contained reward, to test learning and memory of a range of spatial patterns in female rufous hummingbirds. The array was presented in one of four patterns: (1) halves; (2) quarters; (3) checkerboard; or, (4) random (see Fig. 7.5a) on four consecutive days, with birds having 40 trials at each pattern.

Performance on all arrays began at 50% (chance), but performance improved much more quickly and reached higher levels after 40 trials in the halves and quarters arrays than in the more complex checkerboard and random arrays. Birds' performances were, however, still improving

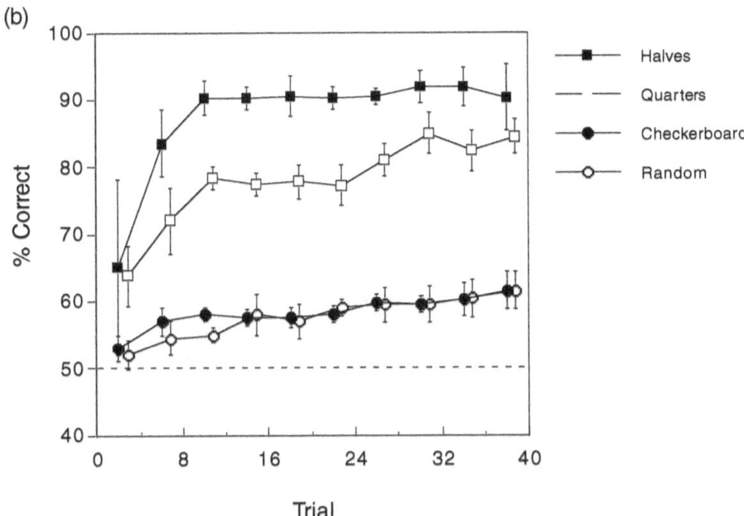


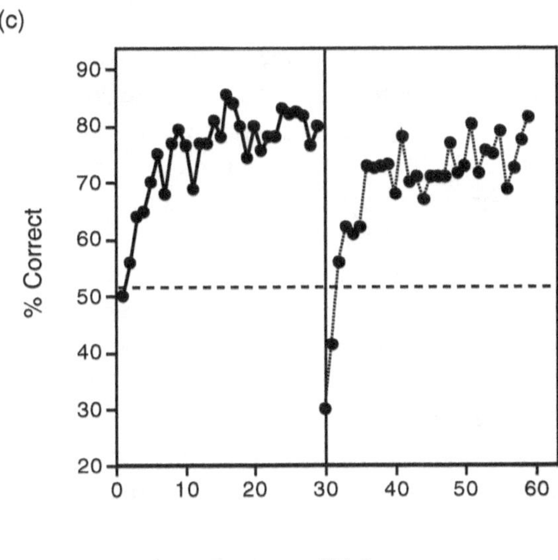


Fig. 7.5 (*cont.*). (b) Performance by four female rufous hummingbirds on the four feeder patterns. Vertical bars represent 95% confidence intervals and the broken horizontal line indicates chance performance. (c) Performance by six rufous hummingbirds on the quarters pattern. The pattern was switched to its mirror image after trial 30 (indicated by the solid vertical line). The dashed line represents chance performance. (Redrawn with permission from Sutherland & Gass 1995.)

(a)

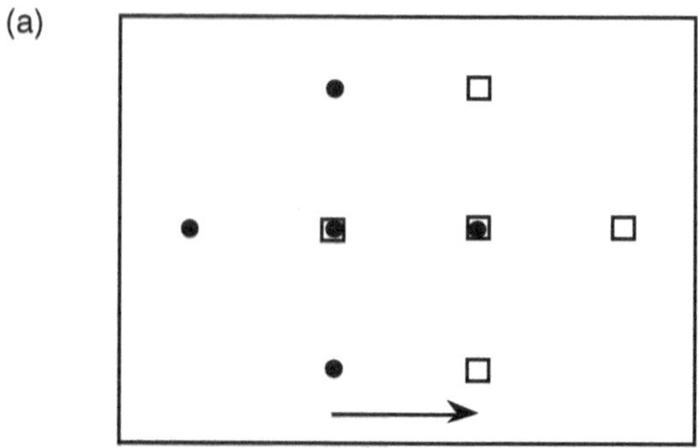

Fig. 7.6.(a) Experimental array of flowers in which the solid circles represent the array during training trials in which only the center flower was rewarded, and all others contained water. Spacing between flowers was 5, 10, 20, 40, 80, 160, and 320 cm. Open squares represent the locations of flowers during a test trial in which the array was shifted one spacing unit to the east.

on the more complex arrays after 40 trials. Therefore, they were still learning reward locations, but at a much slower rate (see Fig. 7.5b). In a second experiment, birds were trained on a quarters pattern for 30 trials, then presented with 30 trials of the mirror-image of the same pattern. All birds had been performing at 75% or more and all dropped to below 50% for at least two trials after the switch (see Fig. 7.5c). Sutherland & Gass (1995) interpreted these data as evidence that the birds knew which were the profitable locations rather than that they were using movement rules to decide which were good or bad feeders.

We carried out several field experiments similar to this latter one of Sutherland & Gass (Healy & Hurly 1998). In the first, we presented the birds (male rufous hummingbirds) with five feeders arranged in a cross with the middle feeder containing the reward (see Fig. 7.6a). The flowers were equally spaced at 5, 10, 20, 40, 80, 160 or 320 cm. Once the bird had learned to visit only the central flower, the array was shifted one spacing unit in one of the four major compass directions. When the distances between flowers were 40 cm or less, the birds returned to the central flower, but when they were greater than 40 cm, the birds visited the flower in the previously correct location, as specified by larger landmarks surrounding the array (Fig. 7.6b). In a second experiment, we used an array of 16 flowers in a quarters design with flowers either 10 or 80 cm apart, all the same color or all bearing unique color patterns. Once the birds had learned the location of the rewarded flowers, we shifted the array as in the previous experiment. Color pattern affected how quickly

(b)

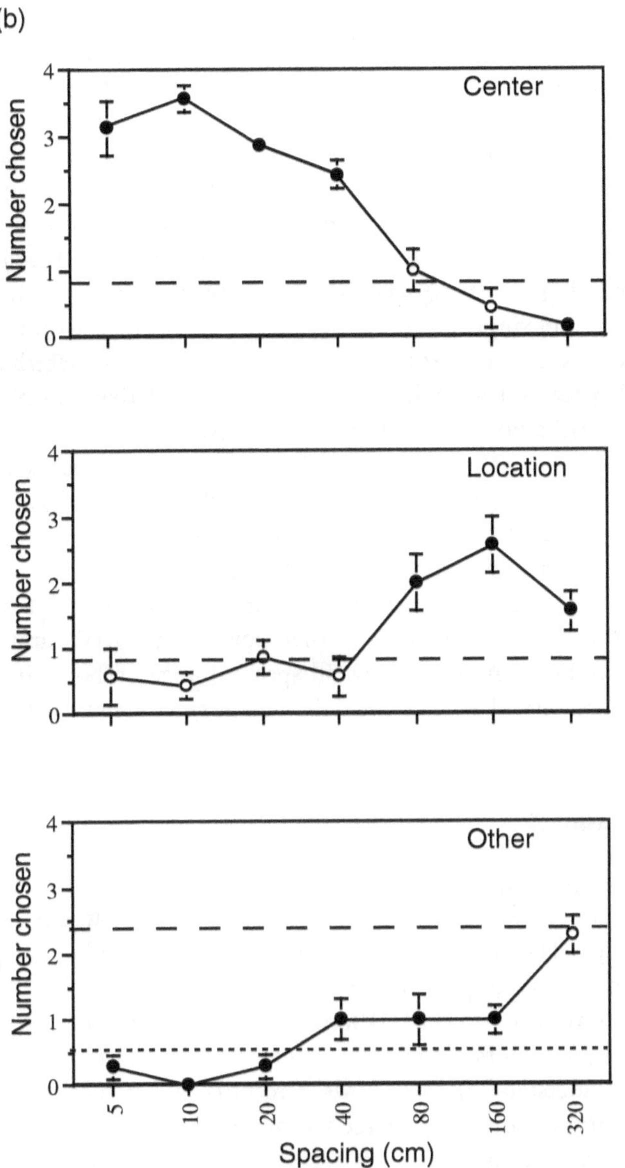


Fig. 7.6 (*cont.*). (b) For seven male rufous hummingbirds, mean (± SE) number of choices made (out of four) in favor of center, location, and other flowers during test trials. Dashed lines represent random performance and dotted line represents background sampling rate estimated from last 20 training trials for each distance for each individual. (Redrawn with permission from Healy & Hurly 1998.)

the birds learned the positions of the rewards in the array, irrespective of distance. Following the shift, all birds were worse at visiting the previously rewarded flowers, but this effect was much greater following the shift of 160 cm. Flower color pattern had no effect on post-shift performance. It appears, then, that the birds use different types of spatial information depending on the spatial scale – when the flowers are close together (<40 cm), the birds use within-array cues to relocate rewarded flowers, and when farther apart they use cues which are outside the array. When foraging on real flowers, then, it may be that birds are remembering locations of single flowers or flower clumps but that within a clump or an inflorescence the birds are more likely to remember which flowers to avoid by remembering the positions of flowers relative to each other. It is not yet clear how the performance observed in these experimental manipulations is related to that which would be demonstrated in real foraging. For example, we do not know whether there is any ecological validity to the spatial scale we used in this experiment. Different populations or species of hummingbirds might use spatial cues differently, but to understand such differences, one would need to know on what plants the birds forage (and how they and their flowers are spatially and temporally distributed). If, as is likely, all species forage on both single and clumped flowers, they may then all show a similar switch in scale of spatial cue use.

### Summary

Recent evidence supports the notion that hummingbirds, at least territorial hummingbirds, use memory for avoiding flowers on which they have recently fed. They use location cues to do this, seemingly paying little attention to the color/pattern cues of the flowers themselves. And yet, the role for flower color in the relationship between plants and their hummingbird pollinators seems to be a major one, made more explicable when an understanding of the birds' perceptual and memory capabilities is taken into account. To persuade a hummingbird to make a first visit to a flower, visual conspicuousness seems to outweigh spatial proximity. With little evidence that birds use movement rules between making flower choices, the spatial distribution pattern of flowers also seems of little consequence.

On the other hand, for plant pollination, the role of a hummingbird's spatial memory – either in accuracy, capacity, or duration – is much less clear in spite of its psychological dominance. Rufous hummingbirds, certainly, can remember after a single, very short visit the locations of small numbers (at least) of flowers, and they avoid these for short periods of time. On the other hand, locations of food sources that do not deplete require multiple visits before the birds reliably return to them. They can also discriminate between flowers that are close together (a few cm) using either other nearby flowers or more distant, larger landmarks as cues. The capacity and duration of hummingbirds' spatial memory have received little attention, not least because assessing these capabilities is logistically difficult. Whether or not plants have managed to respond to or manipulate these cognitive capacities is unclear.

In order to understand just how well plants have managed both to exploit and to be manipulated themselves by the perceptual and cognitive abilities of their hummingbird pollinators will require an interdisciplinary, integrative approach. Avenues for investigation include: field tests of learning and memory in hummingbirds using real plants, growing naturally, to assess the accuracy, capacity, and duration of spatial memory; collection of comparative data (there are almost no data of this kind to date) to determine whether hummingbird species show differences in cognitive abilities that correlate with differences amongst plants in visual, spatial, and reward features; and collection of quantitative data on the numbers of flowers, nectar refilling rates, and variation in quantity and concentration of nectar within and between plants.

## References

Armstrong DP, Gass CL & Sutherland GD (1987) Should foragers remember where they've been? Explorations of a simulation model based on the behavior and energetics of territorial hummingbirds. In: Kamil AC, Krebs JR & Pulliam HR (eds) *Foraging behavior*, pp 563–586. Plenum Press, New York

Bené, F (1941) Experiments on the color preference of the black-chinned hummingbird. *Condor* 43:237–242

Bené, F (1945) The role of learning in the feeding behavior of black-chinned hummingbirds. *Condor* 47:3–21

Bennett ATD, Cuthill IC & Norris KJ (1994) Sexual selection and the mismeasure of color. Am Nat 144:848–860

Brodbeck DR (1994) Memory for spatial and local cues: a comparison of a storing and a nonstoring species. *Anim Learn Behav* 22:119–133

Brown GS (1994) Spatial association learning by rufous hummingbirds (*Selasphorus rufus*): effects of relative spacing among stimuli. *J Comp Psychol* 108:29–35

Brown GS & Gass CL (1993) Spatial association learning by hummingbirds. *Anim Behav* 46:487–497

Chittka L & Waser NM (1997) Why red flowers are not invisible to bees. *Israel J Plant Sci* 45:169–183

Chittka L, Kunze J & Geiger K (1995) The influences of landmarks on distance estimation of honeybees. *Anim Behav* 50:23–31

Clayton NS & Krebs JR (1994) Memory for spatial and object-specific cues in food-storing and nonstoring birds. *J Comp Physiol A* 174:371–379

Cole S, Hainsworth FR, Kamil AC, Mercier T & Wolf LL (1982) Spatial learning as an adaptation in hummingbirds. *Science* 217:655–657

Collias NE & Collias EC (1968) Anna's hummingbirds trained to select different colors in feeding. *Condor* 70:273–274

Corbet SA, Cuthill I, Fallows M, Harrison T & Hartley G (1981) Why do nectar-foraging bees and wasps work upward on inflorescences? *Oecologia* 51:79–83

Faegri K & van der Pijl L (1971) *The principles of pollination ecology*. Pergamon Press, New York

Feinsinger P (1978) Ecological interactions between plants and hummingbirds in a successional tropical community. *Ecol Monogr* 48:269–287

Feinsinger P & Colwell RK (1978) Community organization among neotropical nectar-feeding birds. *Am Zool* 18:779–795

Garrison JSE & Gass CL (1999) Response of a traplining hummingbird to changes in nectar availability. *Behav Ecol* 10:714–725

Gass CE & Sutherland GD (1985) Specialization by territorial hummingbirds on experimentally enriched patches of flowers: energetic profitability and learning. *Can J Zool* 63:2125–2133

George MW (1980) Hummingbird foraging behavior at *Malvaviscus arboreus* var. *Drummondii*. *Auk* 97:790–794

Gill FB (1988) Trapline foraging by hermit hummingbirds: competition for an undefended renewable resource. *Ecology* 69:1933–1942

Goldsmith TH & Goldsmith KM (1979) Discrimination of colors by the black-chinned hummingbird, *Archilochus alexandri*. *J Comp Physiol* 130:209–220

Graenicher S (1910) On humming-bird flowers. *Bull Wisc Nat Hist Soc* 8:183–186

Grant KA (1966) A hypothesis concerning the prevalence of red coloration in California hummingbird flowers. *Am Nat* 100:85–97

Grant V & Temeles EJ (1992) Foraging ability of rufous hummingbirds on hummingbird flowers and hawkmoth flowers. *Proc Natl Acad Sci USA* 89:9400–9404

Hainsworth FR, Mercier T & Wolf LL (1983) Floral arrangements and hummingbird feeding. *Oecologia* 58:225–229

Healy SD & Hurly TA (1995) Spatial memory in rufous hummingbirds (*Selasphorus rufus*): a field test. *Anim Learn Behav* 23:63–68

Healy SD & Hurly TA (1998) Hummingbirds' memory for flowers: patterns or actual spatial locations? *J Exp Psychol: Anim Behav Proc* 24:1–9

Heinrich B (1983) Do bumblebees forage optimally, and does it matter? *Am Zool* 23:273–281

Hurly TA (1996) Spatial memory in rufous hummingbirds: memory for rewarded and non-rewarded sites. *Anim Behav* 51:177–183

Hurly TA & Healy SD (1996) Location or local visual cues? Memory for flowers in rufous hummingbirds. *Anim Behav* 51:1149–1157

Kleerekoper H, Timms AM, Westlake GF, Davy FB, Malar J & Anderson VM (1970) An analysis of locomotor behaviour of goldfish (*Carassius auratus*). *Anim Behav* 18:317–330

Kodric-Brown A & Brown JH (1978) Influence of economics, interspecific competition, and sexual dimorphism on territoriality of migrant rufous hummingbirds. *Ecology* 59:285–296

Lyerly SB, Riess BF & Ross S (1950) Color preference in the Mexican violet-eared hummingbird, *Calibri t. thalassinus* (Swainson). *Behaviour* 2:237–248

Meléndez-Ackerman E & Campbell DR (1998) Adaptive significance of flower color and inter-trait correlations in an *Ipomopsis* hybrid zone. *Evolution* 52:1293–1303

Meléndez-Ackerman E, Campbell DR & Waser NM (1997) Hummingbird behavior and mechanisms of selection on flower color in *Ipomopsis*. *Ecology* 78:2532–2541

Miller RS & Miller RE (1971) Feeding activity and color preference of ruby-throated hummingbirds. *Condor* 73:309–313

Miller RS, Tamm S, Sutherland GD & Gass CL (1985) Cues for orientation in hummingbird foraging: color and position. *Can J Zool* 63:18–21

Paton DC & Carpenter FL (1984) Peripheral foraging by territorial rufous hummingbirds: defense by exploitation. *Ecology* 65:1808–1819

Pyke GH (1978*a*) Optimal foraging in bumblebees and coevolution with their plants. *Oecologia* 36:281–292

Pyke GH (1978*b*) Optimal foraging in hummingbirds: testing the marginal value theorem. *Am Zool* 18:739–752

Pyke GH (1981) Optimal foraging in hummingbirds: rule of movement between inflorescences. *Anim Behav* 29:889–896

Pyke GH (1984) Optimal foraging theory: a critical review. *Ann Rev Ecol Syst* 15:523–575

Smith JNM (1974) The food searching behaviour of two European thrushes. I. Description and analysis of search paths. *Behaviour* 48:276–302

Stiles FG (1976) Taste preferences, color preferences and flower choice in hummingbirds. *Condor* 78:10–26

Sutherland GD & Gass CL (1995) Learning and remembering of spatial patterns by hummingbirds. *Anim Behav* 50:1273–1286

Thomson JD, Slatkin M & Thomson BA (1997) Trapline foraging by bumble bees. II. Definition and detection from sequence data. *Behav Ecol* 8:199–210

Thomson JD, Wilson P, Valenzuela M & Malzone M (2001) Pollen presentation and pollination syndromes, with special reference to *Penstemon*. *Plant Sp Biol*, 15:11–29

Waddington KD & Heinrich B (1979) The foraging movements of bumblebees on vertical "inflorescences": an experimental analysis. *J Comp Physiol* 134:113–117

Waser NM, Chittka L, Price MV, Williams N & Ollerton J (1996) Generalization in pollination systems, and why it matters. *Ecology* 77:1043–1060

Weiss MR (1997) Innate color preferences and flexible color learning in the pipevine swallowtail. *Anim Behav* 53:1043–1052

Wolf LL & Hainsworth FR (1990) Non-random foraging by hummingbirds: patterns of movement between *Ipomopsis aggregata* (Pursch). V. Grant inflorescences. *Funct Ecol* 4:149–157

Wolf LL & Hainsworth FR (1991) Hummingbird foraging patterns: visits to clumps of *Ipomopsis aggregata* inflorescences. *Anim Behav* 41:803–812

**8**

# Bats as pollinators: foraging energetics and floral adaptations

Bat pollination is a pan-tropical phenomenon, performed in the Old World by small megachiropterans (Pteropodidae) and in the New World by microchiropterans of the leaf-nosed bat family Phyllostomidae (Dobat 1985). Flower-visiting bat species total about 50 worldwide, while Dobat (1985) listed about 750 bat-pollinated plant species in 270 genera (590 for the Neo- and 160 for the Palaeotropics). Since then, many more cases have been found.

Although plants independently enlisted "megabats" and "microbats" as pollinators, it is likely that both systems have links to one common root: pollination by ancient, nocturnal, non-flying mammals dating to the late Cretaceous (Sussman & Raven 1978). The extinction of most of these early mammalian flower visitors coincided with the radiation of bats from the Eocene onward in the Old World, and during the Miocene in South America (Sussman & Raven 1978). Genera like *Parkia* may have developed mammal pollination before the separation of African and South American plates; they still retain this trait (Vogel 1969, 1980). Of the plant species found to attract bats today, however, the vast majority evolved their adaptative traits more recently (Vogel 1990).

In the Neotropics, it is useful to consider a continuum ranging from less specialized "fruit-bat" flowers to true "glossophagine" flowers (von Helversen 1993; cf. Johnson & Steiner 2000). For Costa Rica, we estimate that two-thirds of the bat-pollinated plant species are glossophagine specialists. Among the leaf-nosed bats, the subfamily Glossophaginae, with about 35 species and body masses ranging from 6 to 35 g, has evolved the highest degree of specialization for feeding from flowers. This includes the ability to feed during hovering flight, using a protrusile, brush-tipped tongue nearly as long as the body (von Helversen & von Helversen 1975). This group of bats probably accounts for the greater expansion of chiropterophily in the Neotropics than in the Old World (see above). In the Neotropics about 0.7% to 1% of the angiosperm flora is bat-pollinated, including rather delicate or herbaceous plants that can only be exploited by highly maneuverable, hover-feeding visitors.

Bats are "expensive" pollinators in that they need large amounts of nectar (even glossophagine flowers produce at least 100 $\mu$l of nectar per night, and flowers visited by large bats produce at least a few milliliters). These requirements should severely restrict the circumstances under which bats become the preferred pollinator for a plant species. On the other hand, bats offer pollen transfer over potentially large distances, which can be important for self-incompatible plants that grow sparsely. A bat may fly 60 km (von Helversen & Reyer 1984) to 100 km (Horner *et al.* 1998) in a single night's foraging and commuting, and may fly from several hundred meters up to several kilometers between successive plants.

This chapter concerns the energetic aspects of the consumer–resource relationship; we consider these to have played a pivotal role in the evolutionary interaction between glossophagines and their flowers (cf. Heinrich & Raven 1972). We first present a quantitative model of the foraging energetics by a glossophagine bat. This model calculates the minimal caloric reward required by a bat, from average flower visits, in order to balance its energy budget. One result is that selection for increased search efficiency is likely to be intense under conditions of food limitation. Second, we discuss how plant traits that enhance flower detectability may increase the energetic efficiency of pollinator foraging. Pollinators may compete for flower nectar, but flowers may also compete for pollinators, and pollinators will typically prefer the most profitable nectar sources. Profitability involves more than nectar sugar, however; if adaptations enhance a flower's detectability, a pollinator may be able to save energy in several aspects of foraging, including search, approach after detection (i.e., "pursuit" in standard foraging models), and locating and extracting nectar ("handling"; Chittka *et al.*, this volume; Gegear & Laverty, this volume; Menzel, this volume). Our energy-balance model of bat foraging demonstrates how increased detectability and any other facilitation of nectar exploitation can be converted into calories of saved foraging cost by the pollinator. *Detectability* can thus, in principle, affect profitability, and be expressed in the same "currency" as nectar sugar, even if we do not yet know the "conversion factor" quantitatively. All else being equal, plant species with higher profitabilities will be more likely to be chosen for visitation. Within species populations, more detectable individuals will be more likely to receive visits. Therefore, we expect that natural selection will tend to increase detectability, especially if adaptations for detectability are cheap compared to nectar secretion. Because detectability and nectar sugar are linked through their effects on profitability, plants that are highly detectable may be able to attract visits even if they skimp on nectar. We expect natural selection to arrive at a balance based on the functional forms of costs, benefits, and tradeoffs involving detection versus reward.

### Glossophagine bat energetics

Here we derive a quantitative estimate of the threshold requirements for nectar energy during a single flower visit by a small glossophagine bat. A pollinator can only survive in a habitat in which the available food resources allow it to balance its daily energy budget. Especially for a small homeotherm vertebrate with limited capabilities to store fat, this is a stringent requirement. This raises the question: how much energy does an individual bat need to have available in its habitat? As the energy of an average food portion must exceed the cost of acquiring it, the spatial distribution of nectar must allow for economic harvesting. The minimum energy reward that a bat needs to obtain from the average flower visit will therefore depend on the spatial distribution of flowers within the habitat. To predict resource availability for glossophagine bats, the energetic expenditures must be known.

### Daily energy requirements

We determined daily energy expenditures (DEE) of glossophagine bats both from field and laboratory measurements by using methods that included doubly-labeled water, feeding trials, and energy budget estimates derived from time budgets obtained by radio telemetry of free-ranging bats (von Helversen & Reyer 1984; Winter 1998a; unpublished data). Larger glossophagine species have higher DEEs than smaller species (Fig. 8.1, Table 8.1), and the slope of this log-linear relationship coincides with values derived for other vertebrate endotherms (Nagy

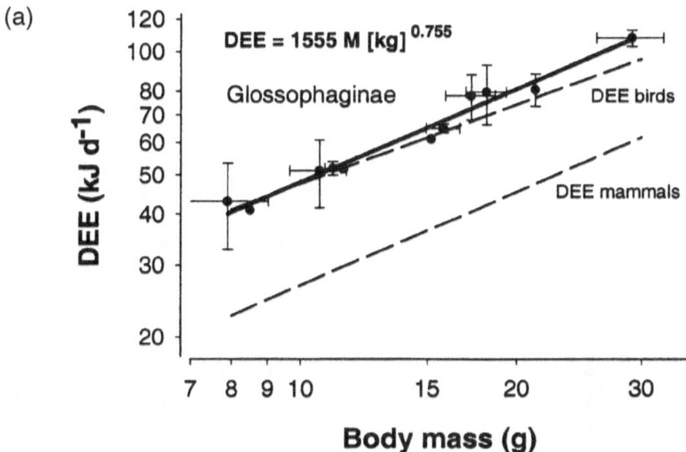



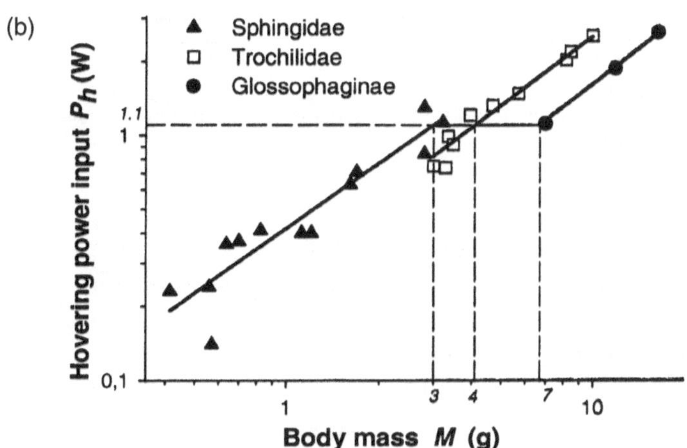



Fig. 8.1 (a) Daily energy expenditure (DEE) of nectar-feeding glossophagine bats (Phyllostomidae) as a function of body mass (data based on over 450 24-h measurements in 58 individuals from 11 species). Least-squares regression of DEE on body mass M yields DEE [kJ d$^{-1}$] = 1555 M [kg] $^{0.755}$ (Y. Winter & O. von Helversen, unpublished data). Dashed lines are regressions of DEE based on doubly-labeled water estimates for birds (Nagy 1987) and eutherian mammals (Nagy 1994). (b) The energy cost of hovering flight as a function of body mass in sphingid moths, hummingbirds (Trochilidae), and glossophagine bats (Phyllostomidae). Solid lines are regressions relating hovering power input $P_h$ to body mass M (see Table 8.1). At a power input of 1.1 W, the three groups of flower specialists overlap in energy expenditure for hovering but support very different body weights (from Voigt & Winter 1999). Axes are plotted on a logarithmic scale.

Table 8.1. *Energy relations and foraging parameters in glossophagine bats and their mass dependence*

| Parameter | Units | Equation[a] | Source |
| --- | --- | --- | --- |
| Basal metabolic rate[b] | $W^c$ | $1.15\,(2.59\,M^{0.71})$ | McNab 1988; Arends *et al.* 1995 |
| Daily energy expenditure | kJ d$^{-1}$ | $1555\,M^{0.76}$ | Y. Winter & O. von Helversen unpublished data |
| Forward flight cost | W | $50.2\,M^{0.77}$ | Winter & von Helversen 1998 |
| Hovering flight cost | W | $128\,M^{0.95}$ | Winter 1998*b*; Voigt & Winter 1999 |
| Foraging flight speed | m s$^{-1}$ | $20\,M^{0.23}$ | Winter 1999 |
| Fight acceleration | m s$^{-2}$ | $4.4$ (for *G. c.*)[d] | Winter 1999 |
| Mean hovering duration | s | 0.4 | unpublished data |
| Flower search efficiency | | ? | unknown parameter |
| Sugar content of 17% sugar (wt/wt) nectar | g l$^{-1}$ | 181.5 | Wolf *et al.* 1983 |
| Nectar sugar (26% di- and 74% monosaccharides) | kJ g$^{-1}$ | 15.92 | see Winter & von Helversen 1998 |
| Sugar assimilation efficiency | | 0.99 | Winter 1998*a* |

*Notes:*
[a] Equations based on body mass $M$ in kg.
[b] Glossophagine basal metabolic rate is 1.15 times the average basal metabolic rate for bats (Arends *et al.* 1995).
[c] $W = J\,s^{-1}$.
[d] G. c. – *Glossophaga commissarisi.*

1987, 1994). However, the average DEE of glossophagines is 60% to 70% higher than the average for other eutherian mammals of their size class, as determined on the basis of DEE estimates with the doubly-labeled water method (Nagy 1994). The DEEs for glossophagines are among the highest yet measured for mammals, and they coincide with the range typical for birds. Put differently, daily energy turnover in a glossophagine is roughly two-thirds of the total caloric content of its body – twice the value typical for a similar-sized terrestrial mammal (assuming 22 kJ g$^{-1}$ dry body mass; Masman *et al.* 1986).

Considering vertebrate nectarivores in general, nectar-feeding pteropodid bats (*Syconycteris australis;* Geiser & Coburn 1999) and 6–15 g sunbirds (Nectarinidae; Peaker 1990) have DEEs similar to glossophagines. Only hummingbird DEEs appear on average, to be about 30% higher (cf. Tiebout & Nagy 1991; Lopez Calleja *et al.* 1997).

Our DEE measurements based on 11 glossophagine species provide a basis for inferring ecological requirements. A later section addresses how this food energy must be distributed spatially within the habitat in order to meet the energy costs of foraging.

### The energy cost of flight

Horizontal forward flight

Flight is the major energetic cost for a foraging bat. The net energy gained from a flower visit by a glossophagine bat is therefore the difference between the energy content of the imbibed nectar sugar and the flight cost of commuting from the previous flower, plus the hovering expenditures during feeding. The cost of horizontal forward flight in small bats has so far been measured by indirect metabolic energy balance methods: the total energy turnover of an animal is ascertained over periods of both flight and rest. Flight cost is determined by subtracting the estimated cost entailed during the non-flight period from the total energy turnover (Speakman & Racey 1991; Winter & von Helversen 1998). Flight costs for small bats, at their lower speeds, are approximately 20%–25% lower than most estimates for birds (Table 8.1; cf. Winter & von Helversen 1998).

Through the course of a night, bats gain mass by feeding. Mass increases by about 5% during the first hour of foraging and by another 3% to 5% during the rest of the night (Winter 1998*a*), leading to a roughly proportional increase in flight cost. Forward flight cost as given by the equation in Table 8.1 is based on mean night-time body masses.

Flight speeds

The cost of flight for a given distance is a function of flight power (W or J $s^{-1}$) and flight speed (m $s^{-1}$). Commuting flight speeds have been measured in the field for several species of glossophagine bats (Sahley et al. 1993; Winter 1999). In addition, general scaling factors predicting flight speed from body mass can be derived from aerodynamic models (Norberg & Rayner 1987). According to presently available data, commuting flight speeds of glossophagines scale with body mass $M$ as $V$ $[ms^{-1}] = 20\ M[kg]^{0.23}$ (Winter 1999). When neighboring flowers are close to each other, acceleration and deceleration phases with reduced speeds constitute a significant portion of a flight interval. To determine acceleration and deceleration, we studied *Glossophaga commissarisi* (8.5 g) feeding at the bromeliad *Vriesea gladioliflora* in Costa Rica. The presence of a bat hovering and feeding at flowers was detected with photoelectric sensors at the flowers and timed by a computer (Winter 1999). The equation for flight acceleration derived from these data (see Table 8.1) can be used to estimate flight times between close flowers, and multiplication of these estimates by forward flight cost (Table 8.1) approximates the flight cost over short distances.

Fig. 8.2. Movements of the wings during hovering flight of *Glossophaga soricina* (at a flower of *Vriesea gladioliflora*). The wings need room to move in front of the body (c) and above it (a). The sequence (d)–(f) shows the supination of the "distal wing triangle" (wing tip reversal) during the backstroke of the wing, which generates lift. (From von Helversen 1993.)

Hovering flight

By hovering, glossophagine bats remain airborne while imbibing the nectar solution (Fig. 8.2). This hovering involves a kinematic feature of wing movement – the "tip reversal" – that may be unique among bats. During the wing's backstroke, the morphological underside of the hand-wing is turned upwards (supinated), thus forming a distal wing triangle that generates lift during a short phase of the wing's backstroke (von Helversen 1986).

To determine the energetic cost of hovering flight, we trained glossophagine bats to visit an artificial feeder that also served as a respirometry mask. When a bat inserted its head into the mask to feed, respiratory gases were withdrawn for oxygen and carbon dioxide analysis to estimate metabolic rates. Contrary to the expectations (from a previous quasi-steady aerodynamic analysis of hovering flight in glossophagines; Norberg *et al.* 1993), the metabolic cost of hovering flight turned out to be only slightly higher than the cost of horizontal forward flight (between 10% and 30%, depending on body mass, see Table 8.1; b; Winter 1998*b*; Voigt & Winter 1999). Although the small difference seems surprising, hummingbirds show a similar pattern (cf. Ellington 1991). It is interesting to compare the glossophagines' solution to hovering flight with that of other hovering nectar-feeders, sphingid moths and hummingbirds. Among these, glossophagines have the lowest mass-specific cost of hovering flight. At a power of 1.1 W, a glossophagine bat can support a body mass of 7 g, a hummingbird one of 4 g, and a sphingid moth a mass of 3 g (Fig. 8.1). These differences in hovering energetics are partially explained by the effect of relative wing area (Voigt & Winter 1999).

Hovering duration

The cost–benefit analysis of foraging requires knowledge of hovering durations during feeding. Infrared photoelectric devices installed at flowers both in the field and in the laboratory have shown that glossophagines normally hover for less than one second, more typically 0.3–0.6 s (Fleming *et al.* 1996; M. Tschapka & O. von Helversen, unpublished data). Hovering thus constitutes only a small fraction of the foraging time budget. The dominant factor in the flight energy budget of a glossophagine bat is the expenditure for commuting and moving between flowers in horizontal forward flight. Selection pressure to reduce the cost of forward flight may therefore account for the relatively high wing loading of glossophagines as compared to other bats and for their relatively short wings, which are shorter than predicted for optimal hovering (Norberg & Rayner 1987).

## Minimum nectar energy densities

A nectar-feeding specialist can subsist in a habitat only if the spatial density of nectar energy (kJ per foraging distance between neighboring flowers) is sufficient for balancing its energy budget. The energy relations summarized in Table 8.1 provide a basis from which minimum food energy levels can be estimated. Estimating resource densities is always difficult, but is easier for nectar-feeding bats than for most animals: (1) the caloric content of nectar can be quantified, so that energy gains during foraging are accessible to measurement; (2) the time and energy costs during foraging can now be calculated with some precision; and (3) particular plants are visited repeatedly over weeks to months, so spatial memory for flower location should enable bats to efficiently relocate flowers.

Individual *Glossophaga commissarisi* visiting *Vriesea gladioliflora* (Bromeliaceae) were tagged with transponders (PIT-tags) so that they could be identified automatically while hovering at flowers that had transponder readers installed in front of them (O. von Helversen & M. Tschapka, unpublished data). This study revealed that: (1) single flowers were visited by several bats (often two to three); (2) the feeding range of individual bats included about 40–50 *Vriesea* bromeliads; (3) individual bats visited single flowers from a few times up to about 30 times during a night; and (4) each time an individual bat consumed roughly 30 μl of 16% sugar (wt/wt) nectar (Tschapka 1998; Y. Winter & O. von Helversen, unpublished data).

Individual plants of many bat-pollinated species flower over several months and, in addition, secrete nectar throughout much of each night. The "food-resource space" for a glossophagine bat in a rainforest will therefore often be an area with predictable locations of renewable food sources. By repeatedly visiting known locations, a bat can spend most of its foraging time commuting, with only a little time spent searching. The food-energy density of the habitat from a bat's point of view is thus a function of (1) the mean distance between neighboring flowers along the foraging route and (2) the mean caloric value of nectar obtainable during a flower visit.

The resource space that can be profitably exploited by a glossophagine bat is delimited by several boundaries. First, the overall sum of energy gains from foraging must be sufficient to meet the energetic requirements during the non-foraging time period $(E_x)$.

$$E_x = \sum_{i=1}^{n} [E_{fl_i} - (t_{ff_i} P_f + t_{h_i} P_h)] \tag{8.1}$$

where $E_{fl}$ is the nectar energy obtained from flower visit $i$, $P_f$ and $P_h$ are the energy costs of horizontal forward and hovering flight per unit time, $t_{ff}$ is the commuting time between two flowers, $t_h$ is hovering duration, and $E_x$ is the energy cost of living for the rest of a day when not foraging, such as resting metabolic rate (including thermoregulation during daytime and foraging pauses during the night), food search, and social interactions. The sum of these energy expenditures constitutes the DEE.

$$DEE = \sum_{i=1}^{n} (t_{ff_i} P_f + t_{h_i} P_h)] + E_x \tag{8.2}$$

DEE is constrained by the capacity for food processing, by renal clearance capacity, and by further internal factors that limit energy turnover (Weiner 1992; Winter 1998a). Total foraging time $T_f$ is calculated according to

$$T_f = \sum_{i=1}^{n} (t_{ff_i} + t_{h_i}) \tag{8.3}$$

Foraging time is constrained by the length of the night minus the cumulative duration of digestive periods and times needed for other activities.

With these equations, for a given DEE and flight time budget we can compute the maximum number of flowers that can potentially be visited by a bat and their minimum nectar energy content requirement. The number of flower visits during a night $N_{fl}$ multiplied by the mean nectar energy $E_{fl}$ available from a flower visit must at least equal the daily energy expenditure DEE:

$$DEE \leq N_{fl} E_{fl} \tag{8.4}$$

The number of flowers a bat can visit during a night depends on the mean distance between flowers $S_{ff}$, the flight speed of the bat, the mean duration of a hover-feeding visit $t_h$, and the total duration of nightly flight $T_f$. The maximum number of flower visits thus equals the total duration of nightly flight divided by the mean time needed to approach and exploit a single flower:

$$N_{fl} = T_f/(t_{ff} + t_h) \tag{8.5}$$

Flight duration between neighboring flowers $t_{ff}$ is a function of the flight speed $V$ and the mean distance between flowers $S_{ff}$. If this distance is large, then the mean flight speed $V$ approaches commuting flight speed during foraging, which for glossophagines scales with body mass $M$ (in kg) as $V_f = 20\,M^{0.23}$ (Table 8.1; Winter 1999). Flight duration $t_{ff}$ for the distance $S_{ff}$ then simply becomes $t_{ff} = S_{ff}/V_f$. Often, however, the distance between neighboring flowers will be shorter so that acceleration and deceleration must be taken into account. Here, we approximate this by assuming a constant delay of one second for a flower visit at a cost equal to forward flight. Thus $t_{ff}$ (in seconds) is taken to equal $(S_{ff}/V_f) + 1$.

Bats will not detect every flower available in their foraging areas, and they cannot be expected to establish the minimum-length route (the traveling-salesman problem) to visit their flowers. Consequently, the mean distance between plants along a bat's foraging route will be larger than the minimum theoretically obtainable within the habitat $S_{ff}$, which we model by a factor $\text{eff}_s$ ($\leq 1$) for the search efficiency of a bat. The minimum nectar energy to be obtained from an average flower visit $E_{fl}$ then becomes

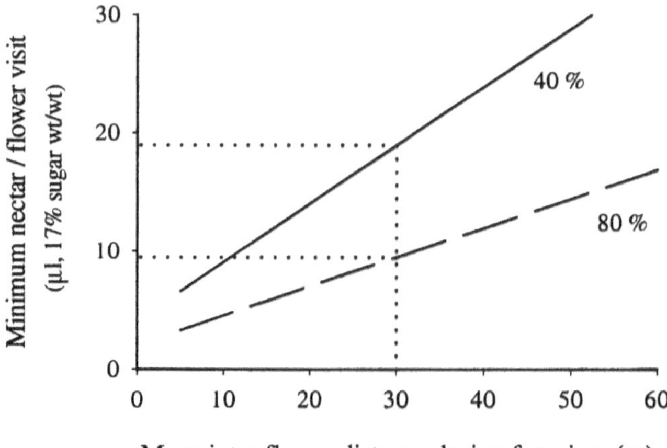


Fig. 8.3. Boundary values of minimum nectar rewards per flower visit for 7.5-g *Hylonycteris underwoodi* (Phyllostomidae, Glossophaginae), a Central American nectar and pollen specialist. Data were calculated from Eq. 8.6 assuming a nightly flight activity of 4.5 hours for foraging. Inter-flower distances are mean distances between flowers in the habitat. As bats will not forage with 100% search efficiency, the actual flight distances between flowers will surpass the theoretical minimum. For the continuous line, it was assumed that successive flowers were visited with a search efficiency eff$_s$ equal to 40% of the theoretical minimum, whereas for the broken line, an efficiency of 80% was assumed (from Y. Winter & O. von Helversen, unpublished data). The energy content of nectar with 17% sugar (wt/wt) is 2.9 J μl$^{-1}$ (see Table 8.1).

$$E_{fl} \geq DEE \frac{eff_s S_{ff}/V_f + 1 + t_h)}{T_f} \tag{8.6}$$

This equation allows an approximation of the minimum nectar energy that must be available during a flower visit for given values of daily energy expenditure (DEE), flight time budget $T_f$, and search efficiency eff$_s$. Values for DEEs to be used in this linear model can be computed from the equation given in Table 8.1. Daily flight-time budgets ($T_f$) are typically 4–5 h per night for glossophagine bats (Horner *et al.* 1998; Y. Winter & O. von Helversen, unpublished data).

For the minimum nectar energy densities shown in Fig. 8.3, we used Eq. 8.6 with two different values for search efficiency eff$_s$, 40% and 80%. The 7.5-g *Hylonycteris* rainforest bat needs 13.4 ml of 17% sugar nectar (wt/wt or 181.5 g l$^{-1}$) to balance a daily energy budget of 38.7 kJ d$^{-1}$. Let us assume a mean distance between flowers in the habitat of 30 m and a total foraging flight time of 4.5 h. Then a bat with a search efficiency of 40% would have an effective mean flight distance between flowers of 75 m. It could make approximately 700 flower visits, each of which would need to yield a minimum average of 19 µl of nectar (with 17% wt/wt sugar). At a higher search efficiency of 80%, 1400 flowers could potentially be visited within 4.5 h of flying time, and they need to provide only 9.5 µl per visit.

We consider this result to be of central importance for the relationship between plant and forager: reducing the cost of foraging (by reducing the effort per individual flower) reduces the minimum amount of nectar energy required by a bat during a flower visit in order to balance its energy budget. It is this relationship that offers plants the possibility of enhancing their attractiveness without having to invest in nectar sugar. Furthermore enhanced detectability and locatability can allow a plant to reduce the amount of energy offered to a bat without falling below the profitability threshold of the forager. The offering of smaller nectar portions may increase the number of flowers that a forager will visit per unit of time (cf. Heinrich & Raven 1972). This, in turn, may increase rates of pollen transfer.

## The "syndrome" of chiropterophily: adaptations of glossophagine flowers to their visitors' sensory physiology

Our discussion of the floral adaptations for glossophagine pollination will focus on two hypotheses. (1) Plants increase their ability to compete for pollinators by improving the cost–benefit ratio of pollinator foraging. As mentioned above, a reduction in costs eventually pays off for a bat in the same currency as an increase in rewards. Therefore, easy locatability is an important factor in reducing pollinator costs. Detectability involves a whole suite of sensory (olfaction, vision, echolocation) and cognitive (spatial memory) abilities. (2) The nectar and pollen rewards are often protected against unwanted visitors to preserve them for the energy-demanding glossophagine bats. Thus, detectability for unwanted visitors should be reduced.

### The cutting of foraging costs: addressing the senses and cognition

#### Olfaction

Olfaction is probably the primary sense for the long-distance detection of many bat flowers. Nearly all bat flowers have a strong, characteristic smell, at least to the human nose (Porsch 1931; van der Pijl 1936; Vogel 1958). To date, the scent spectra of 22 different bat flowers from at least 10 plant families have been analyzed (Knudsen & Tollsten 1995; Bestmann *et al.* 1997). Four types of scent components predominate: aliphatic, aromatic, terpenoid, and sulfur-containing compounds. Sulfur-containing compounds are important in most of the scent-bouquets. These compounds (particularly dimethyl-disulfide, dimethyl-trisulfide and dimethyl-tetrasulfide) are rare in non-bat flowers but are produced by many bat-pollinated plants that are not related to each other (Knudsen & Tollsten 1995); they seem to be the result of true convergent evolution.

Dimethylsulfides are strong attractants for glossophagines. In field experiments, free-ranging *Glossophaga commissarisi* were attracted to mock flowers when these contained either dimethyl-disulfide or 2,4-dithiapentane. Several other scent components were considerably less effective (von Helversen *et al.* 2000). This scent preference seems to be innate; laboratory-reared animals without any experience with flowers also significantly preferred dimethyl-disulfide (Fig. 8.4).

### Vision

Bats are generally believed to be color-blind (Jacobs 1993), and *Glossophaga* bats are unable to discriminate between wavelengths in color discrimination tests (J. Lopez, Y. Winter, & O. von Helversen, unpublished data). The color spectrum of bat flowers extends from greenish through whitish to brownish and brown–red; the colors are never glowing, but rather are unsaturated and usually dusky (Vogel 1969). Some glossophagine flowers are white, presumably because they have evolved from hawkmoth flowers (e.g., Bombacaceae such as *Pseudobombax septenatum* and *Bombacopsis quinatum, Bauhinia* spp., *Capparis* spp., Cactaceae, etc.). Reddish or red–brown colors may indicate evolution from bird flowers (e.g., *Erythrina glauca, Calliandra,* also Old World *Musa*). Green and brown colors probably help to make the flowers inconspicuous for other visually oriented foragers such as sphingid moths and probably also birds.

In contrast to color vision, the visual pattern-recognition ability of flower bats is well developed (Suthers *et al.* 1969). Black-and-white contrasts may help in finding flowers that are white if they contrast against dark foliage, and the exposed position of many flowers probably helps approaching bats when they fly up from below the horizon and view the flower against the light sky. Most bat flowers project into the open air, so that the bats encounter no obstacles when flying up to them. This exposure is achieved by a great variety of morphological arrangements; individual flowers may be raised above the foliage on long stems, the flowers

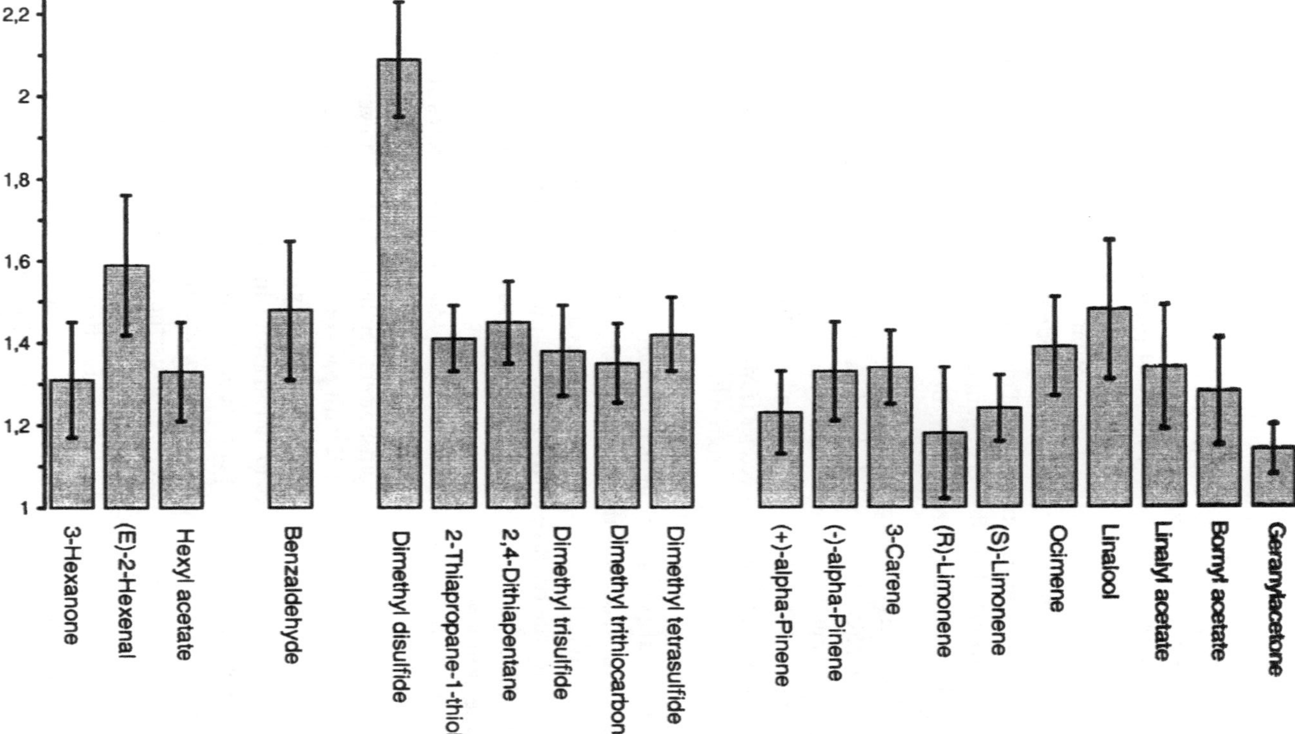


Fig. 8.4. Relative olfactory attractiveness of 20 flower scent compounds in spontaneous choice experiments conducted in the laboratory with a group of *Glossophaga soricina*. Values are means ($\pm 1$ SE) of the preference factors for the respective compounds. The preference factor describes the relative preference of the respective scent in comparison to the scentless control (without any volatile substance). A preference factor larger than 1.0 indicates that the scent compound was preferred compared to the scentless reference; a factor value of less than 1.0 would indicate a repellent effect of the scent compound. Tests were carried out on a total of 52 nights; the number of visits counted was 7551 (von Helversen *et al.* 2000).

may be situated at the ends of twigs, or flagelliflorous inflorescences may hang down for meters from the canopy (Porsch 1931; van der Pijl 1936; Vogel 1968, 1969; Tschapka *et al.* 1999). Furthermore, many chiropterophilous trees bloom when leafless (Vogel 1968, 1969). This open exposure allows unencumbered access (Fig. 8.2) and probably also increases the detectability of flowers by vision or echolocation. *Glossophaga* bats are unexpectedly sensitive to ultraviolet radiation, differing from nearly all other mammals tested except for a few rodents (Jacobs 1993; Lopez, Winter, von Helversen, unpubl. data). This sensitivity might enable the bats to detect some white flowers against a dark background, because some bat-pollinated flowers reflect UV (Burr *et al.* 1995).

Echolocation and flower shape

Glossophagine bats orient mainly by their highly developed system of echolocation. Therefore, flowers that send back conspicuous echoes should be especially well detectable for a bat. A bat-pollinated flower that attracts its pollinators with its echoes is *Mucuna holtonii* (Fabaceae). This liana grows high in the canopy, from where its many-flowered inflorescences hang down on peduncles up to several meters long. The flower's erect upper petal (vexillum), which measures about 19 by 19 mm, is formed like a small concave mirror. In field experiments, we showed (von Helversen & von Helversen 1999) that the bats detect the flowers by echolocation. Filling the concave cavity of the vexillum with a pad of cotton wool, which changes only echo reflectance, not shape or odor, drastically reduced the numbers of visits.

We examined the echoes reflected from a *Mucuna* flower exposed to artificial sound sweeps that imitated natural echolocation calls. By comparing echoes sent back by virgin flowers, by buds, and by flowers in which the vexillum had been filled with pads of cotton wool, we discovered that the echo of the entire flower was strongly dominated by the echo of the vexillum (Fig. 8.5). The echo had an astonishingly high amplitude; the spectral composition was dependent on the angle of sound incidence, but the amplitude was high within a large cone of incidence angles (about −40° to −50°, to +40 to +50°; Fig. 8.5). Thus, the vexillum of *Mucuna holtonii* acts similarly to a cat's eye or a triple mirror in the optical domain, reflecting most of the energy back into the direction of incidence. The echoes of such a concave vexillum should be acoustically conspicuous because they persist during a series of calls emitted by a passing bat. This is different from many other loud echoes, i.e., from leaves, which reflect

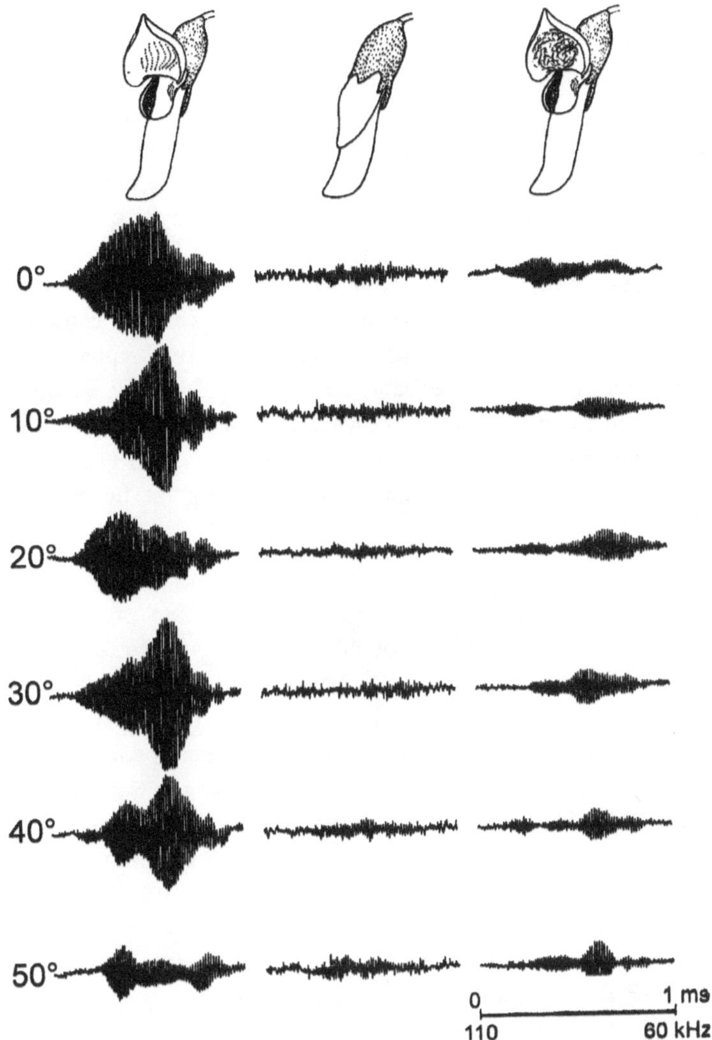


Fig. 8.5. Ultrasound echoes reflected from *Mucuna holtonii* flowers. Echoes from (left) a virgin, intact flower; (middle) a bud; (right) a flower in which the vexillum was filled with a pad of cotton wool. Degrees show the angle of sound incidence. The signal was a 1-ms sound sweep, the frequency of which was linearly modulated from 110 to 60 kHz (von Helversen & von Helversen 1999).

echoes in only one direction – and therefore for only a *single* call from a passing bat.

The peculiar concave geometry of the *Mucuna holtonii* vexillum may be a direct adaptation to the echolocation system of glossophagine pollinators: neither Palaeotropical bat-pollinated species of *Mucuna*, which are visited by small megachiropterans that do not echolocate, nor bird-pollinated species of the genus possess vexilla with the specialized shape and stiffness.

We expect similar adaptations to be found in other glossophagine-pollinated flowers. One promising case involves several columnar cacti that display their flowers within a hairy zone, the "cephalium" of the cactus. This cephalium probably absorbs sound energy, enhancing the contrast to the more reflective flowers. The surface of many bat flowers is especially smooth and waxy, and glossophagine bats examine objects with smooth surfaces when they are searching for flowers (personal observation).

### Spatial memory

In marked contrast to the single-night blooming of the individual flowers, the flowering period of chiropterophilous plants is often much longer than that of related, non-bat-pollinated species (Vogel 1958, 1968, 1969). For instance, chiropterophilous *Vriesea* species remain in bloom for up to 2 months, chiropterophilous *Cleome moritziana* for about 5 months, a single inflorescence of *Mucuna holtonii* for as long as 6 weeks – extreme cases of what has been called "steady-state flowering" (Gentry 1974). This flowering behavior may be an adaptation to the spatial memory of bats. Evidence for spatial memory in bats is fragmentary. During obstacle-avoidance experiments, bats build up a memory of exact obstacle positions that they retain for over a month (Neuweiler & Möhres 1967). We observed that after removal of an accustomed feeder in the laboratory, individuals of *Glossophaga soricina* inspected the former feeder location by hovering in mid-air at the former feeder position for several nights.

We tested for the ability to memorize feeder locations in a laboratory experiment with three *Glossophaga soricina* bats. When feeders had contained a nectar reward for a 1 hour-period during the night preceding the test, then these feeders – which remained unrewarded during the test night – were visited three times more often (38 visits per feeder, $n = 5$ feeders) than identical control feeders (11 visits per feeder, $n = 4$ feeders) that had never been rewarded (D. von Helversen, personal communication).

Spatial memory for food location is likely to be the most important mechanism enabling glossophagines to relocate flowers and minimize search costs. The experimental investigation of spatial memory and orientation will therefore be especially important.

## Securing the goods: repulsion of unwanted visitors

Secretion of nectar is much greater in bat flowers than in all other pollination syndromes (although glossophagine flowers may still have less nectar than "big bat" flowers). Typical glossophagine flowers secrete 1 to 2 ml of nectar per night, with a lower limit of about 100 μl/night; "big bat" flowers may secrete more than 20 ml/night (Dobat 1985). The sugar concentration of the nectar is only 5%–29% sugar wt/wt (often 15%–17%, thus containing about 180 mg sugar per ml of nectar; von Helversen 1993), which is much less than the concentration preferred by the bats in laboratory experiments (55% sugar; Roces *et al.* 1993). The nectar sugars are dominated by hexoses (Baker *et al.* 1998).

For bats such as *Choeronycteris* and *Leptonycteris*, ingestion of pollen seems to be an important "reward" from the flower, as these bats depend on pollen as their nitrogen source; however, less specialized glossophagines such as *Anoura* and *Glossophaga* also feed voraciously on pollen when kept on an otherwise protein-deficient diet (personal observation). It has been suggested that the pollen of bat flowers is specially adapted to the needs of the bats in its amino acid composition (Howell 1973). The pollen supply of glossophagine flowers is usually larger than that of related flowers of the same size that are not pollinated by bats (Vogel 1968, 1969). Either the number of stamens is increased or the thecae themselves are especially rich in pollen. In some species, the normal hermaphroditic flowers are accompanied by a certain number of purely male flowers (e.g., *Bauhinia* spp., Heithaus *et al.* 1974; Ramírez *et al.* 1984; *Cleome moritziana*, personal observation; and others). Flowers lose many pollen grains to glossophagines because the bats interrupt their foraging flights every 10 to 20 min, hang from a twig, and clean their fur thoroughly with their tongues, thereby ingesting the pollen (personal observation; cf. Harder *et al.*, this volume).

Because bat flowers offer unusually large amounts of nectar and pollen, they are vulnerable to parasites (Heinrich & Raven 1972). "Unwanted" visitors may deplete costly nectar or pollen (which may lead to a loss of mating opportunities), and they may damage flowers. Therefore, under certain conditions, plants should limit the spectrum of visitors. To understand how a plant may be able to "hide" its flowers from unwanted visitors or even to "repel" them, we have to know the behavior and the sensory system of the visitor to understand the plant's potential devices.

The following considerations are largely speculative, but might offer a platform for experimental investigations. All bat-pollinated plants open their flowers at night because bats are nocturnal, but, in addition, many glossophagine flowers open only after dusk and close or fade before sunrise. This is probably primarily a mechanism for excluding day-active pollinators, especially birds and bees. Only a few glossophagine flowers remain open for two or three days, and most of these are protandrous, i.e., male during the first night and female during the second (e.g., *Cobaea, Paliavana, Macrocarpaea, Agave,* etc.). In these cases, nectar secretion is often restricted to night. Exceptions include some generalist flowers that also attract birds; in those, some nectar is secreted diurnally (e.g., different species of *Macrocarpaea* and *Puya* in the paramo of Ecuador; F. Matt & H. Schmid, personal communication).

Bolten & Feinsinger (1978) suggested that the low nectar concentration of hummingbird flowers may be a characteristic to deter bees. This idea could possibly also hold for bat flowers, with their even lower nectar concentrations, but only a few species of wasps and bees are night-active. Nocturnal wasps can sometimes be observed on bat flowers (personal observation). Bees can commonly be observed gathering nectar and pollen left in bat flowers in the morning, or robbing nectar and/or pollen in the evening by forcing their way into buds.

The dark colors of many bat flowers should make them difficult to find and approach for sphingid moths, which orient visually. In *Markea neurantha*, for instance, the opening of the corolla tube – which is just the place where a hawkmoth would have to introduce its proboscis – is dark purple–brown, whereas the outer surface is greenish and hardly stands out against the foliage even in daylight.

Bat-flower nectar in some cases seems to have a higher viscosity than expected on the basis of its sugar concentration, due to the additional secretion of mucous substances (van der Pijl 1936; personal observation). As sphingids have to suck nectar through a very long capillary tube in their mouthparts, whereas bats lick nectar, a high viscosity may well present much more difficulty to the moths (see Heyneman 1983). Ants can steal bat-flower nectar (Haber *et al.* 1981; personal observation). Therefore, many mechanisms, mostly mechanical, have been developed to repel ants (Kerner von Marilaun 1876; Guerrant & Fiedler 1981; von Helversen 1993).

## Conclusion

Many neotropical glossophagine pollinated flowers, which most likely evolved from flowers visited by non-flying mammals, presently show characteristics (i.e. pendant peduncles, delicate supports) that might be adaptations both to deter visits from non-flying mammals and, in combination with other cues (scent, echo reflectance), to increase detectability and accessibility to a hovering visitor. This selects in the bat population for greater agility (smaller bodies, better hovering skills). Selection pressure for small pollinator size may also be caused by interspecific exploitation competition because energy requirements decrease with body size. Because flower detectability and nectar sugar are linked through their effects on profitability, plants that are highly detectable may be able to attract visits even if they skimp on nectar. These characteristics should interact and could reinforce each other until the system runs into some constraints which can adequately be described only with quantitative physiology.

## References

Arends A, Bonaccorso FJ & Genoud M (1995) Basal rates of metabolism of nectarivorous bats (Phyllostomidae) from a semiarid thorn forest in Venezuela. *J Mamm* 76:947–956

Baker HG, Baker I & Hodges SA (1998) Sugar composition of nectars and fruits consumed by birds and bats in the tropics and subtropics. *Biotropica* 30:559–586

Bestmann HJ, Winkler L & Helversen O von (1997) Headspace analysis of volatile flower scent constituents of bat-pollinated plants. *Phytochemistry* 46:1169–1172

Bolten AB & Feinsinger P (1978) Why do hummingbird flowers secrete dilute nectar? *Biotropica* 10:307–309

Burr B, Rosen D & Barthlott W (1995) Untersuchungen zur Ultraviolettreflexion von Angiospermenblüten III. Dillenidae und Asteridae s.l. *Trop Subtrop Pflanzenwelt* 95:1–185

Dobat K (1985) *Blüten und Fledermäuse (Chiropterophilie)*. Kramer Verlag, Frankfurt

Ellington CP (1991) Limitations on animal flight performance. *J Exp Biol* 160:71–91

Fleming TH, Tuttle MD & Horner MA (1996) Pollination biology and the relative importance of nocturnal and diurnal pollinators in three species of sonoran desert columnar cacti. *Southwest Nat* 41:257–269

Geiser F & Coburn DK (1999) Field metabolic rates and water uptake in the blossom-bat *Syconycteris australis* (Megachiroptera). *J Comp Physiol B* 169:133–138

Gentry AH (1974) Flowering phenology and diversity in tropical Bignoniaceae. *Biotropica* 6:64–68

Guerrant EO & Fiedler PL (1981) Flower defense against nectar-pilferage by ants. *Biotropica* 13:25–33

Haber WA, Frankie GW, Baker HG, Baker I & Koptur S (1981) Ants like flower nectar. *Biotropica* 13:211–214

Heinrich B & Raven PH (1972) Energetics and pollination ecology. *Science* 176:597–602

Heithaus ER, Opler PR & Baker HG (1974) Bat activity and pollination of *Bauhinia pauletia*: plant-pollinator coevolution. *Ecology* 55:412–419

Helversen D von & Helversen O von (1975) *Glossophaga soricina* (Phyllostomatidae) Nahrungsaufnahme (Lecken) [feeding, high speed film]. In: Wolf G. *Encyclopaedia cinematographica*, vol. E, 1837:3–17. Institute for the Scientific Film, Göttingen

Helversen D von & Helversen O von (1999) Acoustic guide in bat-pollinated flower. *Nature* 398:759–760

Helversen O von (1986) Blütenbesuch bei Blumenfledermäusen: Kinematik des Schwirrfluges und Energiebudget im Freiland. In: Nachtigall W (ed) *Biona-Report 5: Fledermausflug – Bat Flight*, pp 107–126. Fischer, Stuttgart

Helversen O von (1993) Adaptations of flowers to the pollination by glossophagine bats. In: Barthlott W (ed) *Animal–plant interactions in tropical environments*, pp 41–59. Museum König, Bonn

Helversen O von & Reyer HU (1984) Nectar intake and energy expenditure in a flower-visiting bat. *Oecologia* 63:178–184

Helversen O von, Winkler L & Bestmann HJ (2000) Sulphur containing "perfumes" attract flower-visiting bats. *J Comp Physiol A* 186:143–153

Heyneman A (1983) Optimal sugar concentration of floral nectars – dependence on sugar intake efficiency and foraging costs. *Oecologia* 60:198–213

Horner MA, Fleming TH & Sahley CT (1998) Foraging behaviour and energetics of a nectar-feeding bat, *Leptonycteris curasoae* (Chiroptera: Phyllostomidae). *J Zool Lond* 244:575–586

Howell DJ (1973) Bats and pollen: physiological aspects of the syndrome of chiropterophily. *Comp Biochem Physiol* 48A:263–276

Jacobs GH (1993) The distribution and nature of colour vision among the mammals. *Biol Rev* 68:413–471

Johnson SD & Steiner KE (2000) Generalization versus specialization in plant pollination systems. *Trends Ecol Evol* 15:140–143

Kerner von Marilaun A (1876) *Die Schutzmittel der Blüten gegen ungebetene Gäste*. Festschrift Zoologische-Botanische Gesellschaft Wien, Vienna

Knudsen JT & Tollsten L (1995) Floral scent in bat-pollinated plants: a case of convergent evolution. *Bot J Linn Soc* 119:45–57

Lopez Calleja MV, Bozinovic F & Martinez Del Rio C (1997) Effects of sugar concentration on hummingbird feeding and energy use. *Comp Biochem Physiol A* 118:1291–1299

Masman D, Gordijn M, Daan S & Dijkstra C (1986) Ecological energetics of the kestrel: field estimates of energy intake throughout the year. *Ardea* 74:24–39

McNab BK (1988) Complications inherent in scaling the basal rate of metabolism in mammals. *Q Rev Biol* 63:25–53

Nagy KA (1987) Field metabolic rate and food requirement scaling in mammals and birds. *Ecol Monogr* 57:111–128

Nagy KA (1994) Field bioenergetics of mammals: what determines field metabolic rates? *Austral J Zool* 42:43–53

Neuweiler G & Möhres FP (1967) Die Rolle des Ortsgedächtnisses bei der Orientierung der Grossblatt-Fledermaus. *Z vergl Physiol* 57:147–171

Norberg UM & Rayner JMV (1987) Ecological morphology and flight in bats (Mammalia: Chiroptera): wing adaptations, flight performance, foraging strategy and echolocation. *Phil Trans R Soc Lond* 316:337–419

Norberg UM, Kunz TH, Steffensen J, Winter Y &Helversen O von (1993) The cost of hovering and forward flight in a nectar-feeding bat, *Glossophaga soricina*, estimated from aerodynamic theory. *J Exp Biol* 182:207–227

Peaker M (1990). Nutritional requirements and diets for hummingbirds and sunbirds. *Int Zoo Yb* 29:109–118

Pijl L van der (1936) Fledermäuse und Blumen. *Flora* N.F. 31:1–40

Porsch O (1931) Das Problem Fledermausblume. *Anz Akad Wiss Wien, Math-Naturwiss Kl* 69:27–28

Ramírez N, Sobrevila C, De Enrech NX & Ruiz-Zapata T (1984) Floral biology and breeding system of *Bauhinia ungulata*, Leguminosae, a bat pollinated tree in Venezuelan llanos. *Am J Bot* 71:273–280

Roces F, Winter Y & Helversen O von (1993) Nectar concentration preference and water balance in a flower visiting bat, *Glossophaga soricina antillarum*. In: Barthlott W (ed) *Animal–plant interactions in tropical environments*, pp 159–165. Museum König, Bonn

Sahley CT, Horner MA & Fleming TH (1993) Flight speeds and mechanical power outputs of the nectar-feeding bat, *Leptonycteris curasoae* (Phyllostomatidae: Glossophaginae). *J Mamm* 74:594–600

Speakman JR & Racey PA (1991) No cost of echolocation for bats in flight. *Nature* 350:421–423

Sussman RW & Raven PH (1978) Pollination by lemurs and marsupials: an archaic coevolutionary system. *Science* 200:731–736

Suthers RA, Chase J & Braford B (1969) Visual form discrimination by echolocating bats. *Biol Bull* 137:535–546

Tiebout HM & Nagy KA (1991) Validation of the doubly labeled water method ($^3HH^{18}O$) for measuring water flux and $CO_2$ production in the tropical hummingbird *Amazilia saucerottei*. *Physiol Zool* 64:362–374

Tschapka M (1998) Koexistenz und Ressourcennutzung in einer Artengemeinschaft von Blumenfledermäusen (Phyllostomidae: Glossophaginae) im atlantischen Tieflandregenwald Costa Ricas. Dissertation, Erlangen University.

Tschapka M, Helversen O von & Barthlott W (1999) Bat pollination of *Weberocereus tunilla*, an epiphytic rainforest cactus with functional flagelliflory. *Plant Biol* 1:554–559

Vogel S (1958) Fledermausblumen in Südamerika. *Österr Bot Z* 104:491–530

Vogel S (1968, 1969) Chiropterophilie in der neotropischen Flora. Neue Mitteilungen I–III. *Flora* 157:562–602, 158:185–222, 158:289–323

Vogel S (1980) Florengeschichte im Spiegel blütenbiologischer Erkenntnisse. *Vortr rhein-westfäl Akad Wiss* 291:7–48

Vogel S (1990) Radiación adaptiva del sindrome floral en las familia Neotropicales. *Boln Acad Nac Cienc Cordoba* 59:5–30

Voigt CC & Winter Y (1999) The energetic cost of hovering flight in nectar-feeding bats (Phyllostomidae: Glossophaginae) and its scaling in moths, birds and bats. *J Comp Physiol B* 169:38–48

Weiner J (1992) Physiological limits to sustainable energy budgets in birds and mammals: ecological implications. *Trends Ecol Evol* 7:384–388

Winter Y (1998*a*) *In vivo* measurement of near maximal rates of nutrient absorption in a
    mammal. *Comp Biochem Physiol* 119A:853–859
Winter Y (1998b) Energetic cost of hovering flight in a nectar-feeding bat measured with
    fast-response respirometry. *J Comp Physiol B* 168:434–444
Winter Y (1999) Flight speed and body mass of nectar-feeding bats (Glossophaginae)
    during foraging. *J Exp Biol* 202:1917–1930
Winter Y & Helversen O von (1998) The energy cost of flight: do small bats fly more
    cheaply than birds? *J Comp Physiol B* 168:105–111
Wolf AV, Brown MG &Prentiss PG (1983) Concentrative properties of aqueous solutions:
    conversion tables. In: *CRC Handbook of chemistry and physics*, 64th edn., pp
    D223–D272. CRC Press, Boca Raton, FL

9

# Vision and learning in some neglected pollinators: beetles, flies, moths, and butterflies

Ask a member of the general public what kinds of insects pollinate flowers and chances are she'll say bees. Certainly hymenopterans pollinate a tremendous variety of plant taxa, and honeybees and bumble bees in particular are economically important and visible pollinators (McGregor 1976; Buchmann & Nabhan 1996; Proctor *et al.* 1996). However, studies of social bees have long dominated academic and applied pollination arenas (Lindauer 1963; von Frisch 1967; Menzel 1967), to the relative neglect of other taxa. Insects in three major orders, Coleoptera, Diptera, and Lepidoptera, are key pollinators of a broad range of angiosperm taxa (Kevan & Baker 1983; Proctor *et al.* 1996), but in comparison with bees, much less is known about their effectiveness as pollinators, or about the sensory attributes and learning abilities that guide their behaviors. This lack of study has several causes, including the lesser importance of non-hymenopteran insects as pollinators of crop plants (notwithstanding their role in pollination of mangos, cacao, papayas, parsnips, pomegranates, carrots, and onions; McGregor 1976), their relative infrequency as major pollinators in European and North American systems (Johnson & Steiner 2000), and the difficulty in raising and studying solitary rather than social insects.

Further study of these neglected pollinators will help us to understand the breadth and diversity of insect sensory systems and learning abilities. We might expect that beetles, flies, moths, and butterflies would have much in common with bees, based on the monophyly of Insecta and the common dependence of all anthophilous insects on flowers, which should subject them to similar selection pressures. We might also expect a number of differences both across and within taxa, given the independent evolutionary histories of Coleoptera, Diptera, Lepidoptera, and

Hymenoptera, and the attendant differences in each group's lifestyles.

Investigation of the sensory capacities and learning abilities of non-hymenopteran insects will also help to elucidate the pathways by which flowers have evolved. Although angiosperms appear in the fossil record at least 130 million years ago (Ren 1998; Sun *et al.* 1998), eusocial hymenopterans are relatively late arrivals evolutionarily, appearing somewhere between 40 and 80 million years ago (Michener & Grimaldi 1988; Poinar 1994). Thus other pollinating insects, including beetles, flies, butterflies, moths, and non-social bees, some of which antedate angiosperms in the fossil record (Crepet & Friis 1987; Ren 1998), have undoubtedly played an important role as agents of natural selection on floral features.

In this chapter, I focus on vision and learning in flower-visiting coleopterans, dipterans, and lepidopterans. I briefly review what is known about their performance as pollinators, the senses they use to locate flowers, their color vision and innate color preferences, and finally their learning ability. I then discuss some important lifestyle characteristics that affect flower visitation patterns and thus have implications for differences in the aforementioned traits. I conclude with some open-ended questions that I hope will suggest directions for future research on these neglected pollinators.

### Non-hymenopteran insects are important pollinators in a range of habitats

The relative importance of non-hymenopteran pollinators varies across habitats. Relatively few specialized pollination systems involving non-hymenopteran insects are known from Europe and eastern North America, where pollination is commonly carried out by opportunistic social bees (Johnson & Steiner 2000). In these areas, many flowering plants receive visits from a range of insect taxa, including bees, beetles, flies, and lepidopterans, which vary in their importance as pollinators (Herrera 1987; Fishbein & Venable 1996; Waser *et al.* 1996). However, coleopterans, dipterans, and lepidopterans also specialize on plants that are in turn specialized for them with respect to floral morphology, phenology, and type of reward, and that in some cases depend exclusively on them for pollination. These systems are particularly common in tropical and southern hemisphere temperate habitats (Johnson & Steiner 2000).

Plants pollinated primarily by beetles occur in temperate regions (Dafni *et al.* 1990; Englund 1993), but are more abundant in the tropics

(Young 1986; Momose *et al.* 1998). Flower-visiting dipterans are abundant in montane and Arctic areas, where they are probably major pollinators (Kearns 1992). Long-tongued flies in the family Nemestrinidae are important and sometimes exclusive pollinators of a range of South African taxa (Johnson & Steiner 1997; Manning & Goldblatt 1997), and bombyliids are key pollinators in semi-arid regions (Johnson & Midgley 1997; Johnson & Dafni 1998). Hawkmoths are particularly important in habitats with warm temperatures at dusk, as flight activity is limited on cold nights (Martínez del Rio & Búrquez 1986; Haber & Frankie 1989; Willmott & Búrquez 1996; Nilsson *et al.* 1997), and butterfly-pollinated species occur in both temperate and tropical habitats (Gilbert 1972; Levin & Berube 1972; Cruden & Hermann-Parker 1979). The relatively greater role of flower-visiting non-hymenopterans in tropical, alpine, and desert areas may contribute to an under-appreciation of their importance as pollinators by temperate biologists.

Quantitative data on pollinator performance, including amounts of pollen removed and deposited, extent of pollen carryover, distances flown between plants, etc., are scarce for all groups except bees (e.g., Wilson & Thomson 1991). Beetles often travel long distances between successively visited plants (e.g., mean of 18.2 m in a temperate system and 83 m in a tropical system), and so may be particularly important in effecting cross-pollination (Young 1986; Englund 1993). Flies can carry substantial amounts of pollen (Yeboah Gyan & Woodell 1987; Kearns 1992), and move relatively short (0.2–2.7 m) distances between plants (Schmitt 1983; Olesen & Warncke 1989; Widen & Widen 1990). Moths can deliver significant amounts of pollen to stigmas (Pettersson 1991; Willmott & Búrquez 1996), and may fly long distances (up to 400 m) between successively visited plants (Linhart & Mendenhall 1977). Some studies have concluded that temperate-zone butterflies serve mainly as nectar thieves, as they carry relatively small amounts of pollen (Wiklund *et al.* 1979; Venables & Barrows 1985), while others have documented relatively large pollen loads and ascribe a significant role in pollination to these insects (Levin & Berube 1972; Courtney *et al.* 1982; Murphy 1984). Pollen dispersal distances by butterflies in temperate systems are on the order of less than 1 meter up to about 12 meters (Levin & Kerster 1968; Schmitt 1980; Waser 1982). In a tropical system, mean pollen dispersal distances for five *Heliconius* species ranged between 23 and 76 m, with a maximum distance of 350 m (Murawski & Gilbert 1986). Long-distance inter-plant movements by beetles, moths, and butterflies are likely to result in more cross-pollination and lower genetic microdifferentiation within populations of plants pollinated by these insects (Schmitt 1980; Young 1986; Herrera 1987; Englund 1993).

Floral constancy, commonly thought of as a characteristic of bee pollination, has also been reported for beetles, flies, and lepidopterans. Several studies of beetle pollinators report floral constancy, though few if any actually quantify both the flowers chosen by the beetles as well as the background of available flowers from which they made their choice (Pellmyr 1985; De Los Mozos Pascual & Domingo 1991; Englund 1993; Listabarth 1996). Syrphid flies foraging in a mixed array show marked constancy to a given floral species (Goulson & Wright 1998), and many species of butterflies are constant when foraging in a mixed patch of real or artificial flowers (Murphy 1984; Lewis 1986, 1989; Goulson & Cory 1993; Kandori & Ohsaki 1996; Goulson *et al.* 1997).

### Cues used to locate flowers

Although most pollinating insects rely on visual and/or olfactory cues to locate their flowers, the relative importance of these stimuli varies within and across orders (Proctor *et al.* 1996). Many beetle taxa depend on odor, in the absence of visual cues, to reach their flowers (Young 1986; Eriksson 1994), while for others, color cues alone may suffice (Dafni *et al.* 1990; Steiner 1998). In some cases, floral odor attracts beetles towards an inflorescence, and then releases searching behavior for a particular color or visual attractant at close range (Pellmyr & Patt 1986). Flies approaching non-deceptive flowers seem to depend on visual cues from a distance and olfactory cues at close range (Knoll 1921; Kugler 1951; Dobson 1994), but their approach to deceptive flowers is generally based on scent (Dafni 1984). Crepuscular or nocturnal hawkmoths and settling moths generally rely on scent to locate their flowers from a distance (Dobson 1994; Raguso, this volume), and vision plays a role at close range. Diurnally foraging butterflies, on the other hand, tend to use long-distance visual cues to locate their nectar sources, though odor may be important for releasing color searching behavior (Tinbergen 1968; Proctor *et al.* 1996) or for direct attraction to flowers (DeVries & Stiles 1990).

These different sensory weightings affect the suites of characters (floral "syndromes") that sometimes occur in flowers predominately pollinated by particular groups (Faegri & van der Pijl 1979). In general, because moths and beetles are likely to be guided by olfactory cues, flowers that depend on them as primary pollinators will tend to be strongly scented, while flowers that rely on the more visually oriented butterflies and bees will tend to emphasize colors over odors. Within these broad categories, some groups of pollinators pay particular attention to subsets of visual and olfactory space. Raguso (this volume) identifies scents characteristic of many moth-pollinated flowers; the putrid odors of carrion-mimic flowers attract only some taxa of flies and beetles (Dafni 1984).

That many beetles, flies, lepidopterans, and bees use an overlapping set of cues to locate their flowers has implications for generalization of pollination systems as well as for opportunities for pollinator shifts. Generalist foragers in all four orders can opportunistically visit flowers of a given species because they all respond to more or less the same cues. If each insect group depended on an entirely different sensory modality to locate its flowers, specialization would be the norm. Shifts from one set of pollinators to another (e.g., Steiner 1998) also depend on overlap in cues used by different taxa.

### Color vision and innate color preferences

Both phylogenetic history and selection have influenced the development and elaboration of sensory systems in insects. Chittka (1996) has suggested that a set of UV, blue, and green photoreceptors is ancestral to Insecta, and that some lineages have since lost or added receptors, presumably as a result of selection.

Investigations into the visual capabilities of beetles have been few (Hasselmann 1962; Lall *et al.* 1982; Agee *et al.* 1990; Lin & Wu 1992). Lampyrids (fireflies) have UV, blue, and green receptors (Lall *et al.* 1982), and spectral sensitivity curves obtained from ERG recordings revealed UV and green receptors in three other beetle families (Lin & Wu 1992). An earlier electroretinogram (ERG) study also found red reception in a carabid beetle (Hasselmann 1962). Behavioral studies have also suggested that beetles have the ability to recognize and distinguish colors (Dafni *et al.* 1990).

Calliphorid, drosophilid, and muscid flies have been shown to perceive colors (Hernández de Salomon & Spatz 1983; Fukushi 1989; Pickens 1990; Troje 1993). Foraging syrphids, calliphorids, tephritids, and anthomyiids innately prefer the color yellow (Lunau 1988; Fukushi 1989; Lunau & Maier 1995; Sutherland *et al.* 1999), while bombyliids often visit pink, blue, or violet flowers (Proctor *et al.* 1996; Johnson & Dafni 1998).

The spectral range of lepidopterans varies across taxa, but in some species is among the widest reported for any animal, covering wavelengths from 300 to 700 nm (UV to red) (Silberglied 1984; Lunau & Maier 1995). Reported numbers of visual pigment types vary from three in various moth and butterfly taxa (e.g., Shimohigashi & Tominaga 1991) to five or six in swallowtail butterflies (Arikawa *et al.* 1987; Briscoe & Chittka 2001). Innate color preferences in the context of foraging (often for yellow, blue, and sometimes orange–red) have been found in a broad taxonomic range of lepidopterans, including nymphalids, papilionids, satyrids, pierids, and sphingids (Crane 1955; Ilse & Vaidya 1956; Swihart & Swihart 1970; Silberglied 1984; Traynier 1986; Scherer & Kolb 1987*a*, *b*; Arikawa *et al.* 1987; Goulson & Cory 1993; Weiss 1995, 1997; Kelber & Pfaff 1997; Kinoshita *et al.* 1999). Reported color preferences may differ between genera in a family, between species in a genus, or between sexes of species (Ilse 1928; Ilse & Vaidya 1956; Weiss 1997; Kinoshita *et al.* 1999).

### Learning ability

It should not be surprising that beetles, flies, and lepidopterans can learn, as learning ability of one sort or another has been found in virtually all animals tested (Alloway 1973). However, the kind of learning (e.g., habituation, sensitization, associative learning), the extent to which an insect depends on learned rather than innate responses, and the behavioral context or sensory modality in which learning is expressed, vary between taxa, probably as a result of selection (Papaj & Prokopy 1989). Stephens (1993) argues that learning ability should evolve under environmental conditions of intermediate predictability – that is, in situations in which the environment is too unpredictable within one or a few generations for fixed behavior patterns to be favored, but not so unpredictable that the individual cannot behaviorally track changes. Floral resources do vary in this way, and indeed, most flower-visiting insects that have been investigated have been shown to be capable learners.

I know of no investigation of beetle learning in relation to flower-visitation behavior. It is not clear whether this truly reflects beetles' inability to learn in the context of flower visiting, or is due to a lack of experimental investigation or non-reporting of negative results. Plotkin (1979) reports that a predaceous ground beetle could learn to reduce a strong thigmotactic response in an open field and hence locate a centrally placed water source. Alloway (1973) summarized other reports of spatial maze learning in larval and adult grain beetles (Tenebrionidae).

Various fly taxa possess efficient capacities for associative learning of visual and olfactory stimuli. Sheep blowflies rapidly learn to associate color with reward (Fukushi 1989), and although hoverflies could be trained to land on colored artificial flowers (Kugler 1950), they could not be trained to extend their proboscides towards a rewarded color, responding instead to their innately preferred yellow (Lunau 1992). Olfactory conditioning has been reported for drosophilid, muscid, and tephritid flies (Fukushi 1973; Spatz *et al.* 1974; Prokopy *et al.* 1982). Additionally, some flies can, like honeybees, learn and make use of spatial landmarks. Male hoverflies hover stably in mid-air and, using visual cues, return to approximately the same position in space after chases (Collett & Land 1975).

Lepidopterans can associatively learn a number of different stimuli in a range of behavioral contexts, including oviposition, nectar foraging, and perhaps navigation. In the context of oviposition, moths and butterflies can learn to associate odors or tastes with leaf shape (Papaj 1986) or with color (Traynier 1986). Ovipositing females preferentially select hostplants with which they have had experience, and consequently make fewer landing mistakes on non-hosts in the field (Stanton 1984). Increased oviposition experience on a novel host does not, however, result in female *Euphydryas editha* (Nymphalidae) butterflies becoming more efficient at locating that novel host in the field (Parmesan *et al.* 1995).

Foraging moths and butterflies can rapidly learn to associate a sugar reward with odor (Hartlieb 1996; Fan *et al.* 1997) or color (Swihart & Swihart 1970; Swihart 1971; Goulson & Cory 1993; Weiss 1995, 1997; Kelber 1996; Kelber & Hénique 1999; Kinoshita *et al.* 1999), sometimes after only a single rewarded trial. Insects trained to one color can rapidly learn to reverse their preferences when the reward is switched to a previously unrewarding color (Goulson & Cory 1993; Kelber 1996; Weiss 1997; Kelber & Hénique 1999; Kinoshita *et al.* 1999). *Macroglossum stellatarum* (Sphingidae) moths trained in a dual-choice situation learned not only to visit the rewarding color but also to avoid the unrewarding color when given a choice of three; such avoidance of unrewarding stimuli has also been found in honeybees (Kelber 1996). Both moths and butterflies improve at finding nectar in real and artificial flowers with increased experience, and can also learn to access nectar in a new floral location after learning an initial pattern (Lewis 1986; Kandori & Ohsaki 1996; Cunningham *et al.* 1998).

Some moths and butterflies seem to be able to use landmarks to return to a given spatial location. *Heliconius* (Nymphalidae) butterflies return to nocturnal roost sites, "trapline" from flower to flower, and avoid areas where they have been captured and released, all of which may involve learned use of visual landmarks (Turner 1981; Waller & Gilbert 1982; Mallet *et al.* 1987). *Macroglossum stellatarum* moths return to a location where they were previously fed (Kelber & Pfaff 1997), again suggesting spatial learning ability.

As the research reviewed above demonstrates, non-hymenopteran insects can readily learn a range of floral parameters, including color, odor, morphology, and perhaps even location. Varying experimental designs, training, and testing protocols make it difficult to compare learning abilities across taxa. However, Fig. 9.1 shows that flies and butterflies can, like bees, associate a color with a sugar reward after only a single exposure.

## Pollinator learning ability has implications for floral evolution

Learning abilities in pollinators have important implications for the evolution of their flowers. Although a simple model of floral evolution could involve pollinator behaviors based solely on innate responses, a range of floral features, including complex morphologies, color changes, temporally variable patterns of anthesis and nectar availability, and spatial arrangement of flowers along a trapline, may be better explained by the pollinators' ability to learn. Below, I briefly discuss two such floral features – complex morphologies and color changes – responses to which have been shown to involve learning by non-hymenopteran pollinators.

### Complex morphologies

Plants take advantage of pollinator motor learning ability by producing flowers with complex morphologies, in which access to nectar or pollen rewards is not immediately obvious. Once a pollinator arrives at a flower, initial motor patterns are likely to be innate, and may be released by particular colors or patterns (Lunau 1988; Lunau & Maier 1995). Beyond the innate responses, however, learning is also involved. For bumble bees,

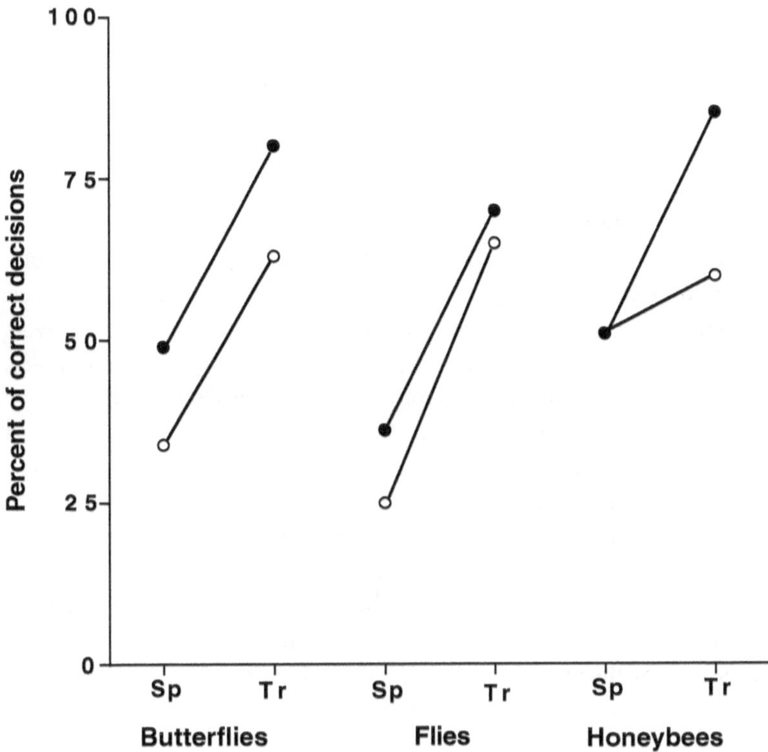


Figure 9.1. Single-trial color learning in butterflies, flies, and honeybees. For all three taxa, a single rewarded exposure to sucrose on a training color yielded a significant increase in choice of that color. Sp = spontaneous color choice; Tr = trained choice after one sucrose reward. Butterfly (*Papilio xuthus*, Papilionidae) data taken from Kinoshita *et al.* (1999); solid circle = yellow; open circle = red. Naïve and trained insects were offered four color choices. Fly (*Lucilia cuprina*, Calliphoridae) data taken from Fukushi (1989); solid circle = yellow; open circle = blue. Naïve and trained insects were offered four color choices. Honeybee (*Apis mellifera*) data taken from Menzel (1967); solid circle = violet; open circle = blue. Trained insects were offered two color choices, each of which received 50% of visits prior to training.

honeybees, and *Pieris* butterflies, floral handling time decreases as pollinators gain experience in accessing rewards from a given plant species, and often increases when the pollinator switches to a new species (Heinrich 1979; Lewis 1986, 1993; Laverty 1994; Goulson *et al.* 1997). Lewis (1993) suggested that flower morphology can be seen as shaped in part by the advantages conferred when a pollinator must make a substantial investment in learning to handle a flower. She posited that plants should evolve to attract generalist pollinators, which then become facultative specialists based on "training" by the flower (Lewis 1993).

### Floral color change

Plants take advantage of pollinators' color-learning ability by producing flowers that change color to provide accurate indications of reward status. Flowers in over 80 angiosperm families undergo localized or whole-flower color changes that are highly correlated with nectar and pollen availability; pre-change flowers are rewarding, but post-change flowers are not (Gori 1983; Weiss 1991, 1995; Weiss & Lamont 1997). Beetles, flies, moths, butterflies, and bees preferentially visit variously colored pre-change flowers even if they are in the minority in a floral display (Gori 1983; Weiss 1991, 1995; Weiss & Lamont 1997). A learned association between floral color phase and nectar availability has been explicitly demonstrated for butterflies (Weiss 1995, 1997), and is likely for bees and flies. Beetle-pollinated color-changing flowers are not common (Weiss 1995; Weiss & Lamont 1997), and it is possible that beetles could be guided to pre-change flowers by innate attraction to particular odors or colors, rather than by the more flexible learned response that seems to characterize visits by other pollinators (Weiss 1995).

## Lifestyle traits of solitary insects may impact learning abilities and pollinator performance

Fundamental differences in the lifestyles of solitary vs. social flower-visiting insects may affect the development and/or expression of their innate preferences and learning abilities, and will affect their performance as pollinators. Below, I discuss some possible implications of three such differences.

### Solitary insects may confuse stimuli associated with different tasks

A division of labor in social insects allows workers to specialize on food collection at a given time in their lives (Winston 1987), while solitary insects typically combine food-gathering with other tasks. Butterflies, beetles, and flies frequently intermix foraging, mating, and oviposition activities in a single flight or at a single location and time period (Faegri &

van der Pijl 1979; Dafni 1984; Stanton 1984; Young 1986; Eriksson 1994). Such "multi-tasking" may provide opportunities for interference or confusion of stimuli. Free-flying *Colias* butterflies, for example, showed decreased landing accuracy on oviposition host-plants after periods of nectar feeding, suggesting a tradeoff between the two searching modalities (Stanton 1984). Similarly, *Pieris rapae* butterflies that alternated nectaring with oviposition flights switched among species of nectar plant significantly more often than did males or non-ovipositing females (Lewis 1989).

Solitary insects may reduce the problems of interference behaviorally, or via innate or learned responses. For butterflies, temporal segregation of feeding and oviposition flights may minimize opportunities for confusion of stimuli (Wiklund 1977; Dukas 1998). Context-dependent innate responses may also help solitary insects differentiate between separate behaviors. For example, nectar-feeding calliphorid and sarcophagid flies, which lay their eggs in dung and carrion, prefer yellow over brown–purple colored models in the presence of sweet scents, and the reverse in the presence of excremental scents (Kugler 1956). Similarly, exposure of *Pieris* butterflies to well-defined spectral regions elicits specific behaviors; e.g., insects extend their proboscides on blue and orange–red colors, and drum their tarsi on yellow–green (Scherer & Kolb 1987*a*). Such innate recognition systems may also serve as contextual triggers for learning (Gould 1984).

### Differing energetic requirements will lead to different flower-visitation patterns

Energy requirements of pollinating insects also have important implications for their flower-visitation patterns. Because bees provision the nest for their young, they must collect more reward per unit time than do solitary insects, which generally forage only to meet their daily needs (Heinrich 1975). Metabolic energy costs of foraging and thermoregulation also vary between groups of pollinators; for example, bumble bees, which thermoregulate metabolically, will have higher energy requirements than will butterflies, which regulate their temperatures by basking (Heinrich 1975).

The high energetic requirements of social bees may select for "optimal" foraging rules that lead to efficient flower handling and predominately near-neighbor visits (Heinrich 1975; Pyke *et al.* 1977). Solitary pollinators, on the other hand, may remain on flowers for longer periods of time, fly greater distances between flowers, and visit fewer flowers in a given time period. Schmitt (1980), for example, found that bumble bees visiting *Senecio* (Asteraceae) flowers typically visit near-neighbor plants, while butterflies frequently bypass neighbors and fly significantly greater distances between plants. Such differing patterns of flower visitation may in turn affect parameters such as speed of learning, degree of interference between learned associations, duration of memory, and timing of transfer between short- and long-term memory (Greggers & Menzel 1993; see Menzel, this volume), all of which are virtually unexplored for non-hymenopteran insect pollinators.

### Multiple uses of flowers may affect innate preferences

Some solitary insects use flowers for more than one activity; for example, beetles may congregate, mate, and lay eggs within a flower or inflorescence, and some moths both pollinate flowers and oviposit on leaves of the same plant. In such cases, innate preferences can serve multiple functions, leading an insect to either a food source, a mate, or a host plant. Beetles and flies that aggregate and mate in flowers or inflorescences are attracted to dark spots on petals, which may mimic resting insects and so attract potential pollinators (Dafni *et al.* 1990; Johnson & Midgley 1997). Deceptive pollination systems, including carrion and pseudocopulation mimicry, take further advantage of innate attractions, luring insects to flowers using cues associated with other behavioral contexts (Dafni 1984).

### Open-ended questions

### What makes an insect a good pollinator?
The answer to this seemingly basic question is poorly understood. It is often assumed that bees are superior pollinators (Kevan & Baker 1983), based in part on their removal of large amounts of pollen, rapid visitation rates, efficient flower handling, constancy, and ability to learn. These characteristics allow bees to collect resources efficiently, but are not necessarily ideal for the plant, as pollen removal does not necessarily correlate with pollen deposition (Wilson & Thomson 1991), and multiple visits to flowers on a single plant or to near neighbors may result in a high level of inbreeding. And while learning ability and floral constancy may facilitate intraspecific pollen transfer, these attributes are also shared by many non-hymenopteran insects. In some cases, insects that remove less pollen from

a flower, groom less of it off their bodies, visit fewer flowers per plant, and travel further between plants, may be better pollinators (Herrera 1987; Thomson & Thomson 1992; Harder *et al.*, this volume). As some non-hymenopteran insects meet many of these criteria, a critical examination of their importance as generalist pollinators is in order. Components of pollinator "quality" are likely to vary across taxa. Herrera (1987) found, for example, that bees pollinate *Lavandula* (Labiatae) flowers frequently, but generally promote geitonogamy with short inter-flower flights; butterflies, on the other hand, pollinate flowers less often but tend to do so with cross-pollen, based on long inter-flower flight distances. Quantitative estimates of pollination parameters for diverse insect visitors on a range of plants will help to address the question of what makes a good pollinator.

### How broad-based or species-specific are sensory attributes and learning abilities in pollinators?

In considering the sensory attributes and learning abilities of pollinators as they have influenced floral evolution, it would be interesting to know which are general to all insects, which are variable across and within taxonomic levels, and how these patterns came about. If associative learning ability, for example, is found in some taxa and not in others, has it evolved independently as needed, was it ancestral to a lineage and subsequently lost in taxa that did not make use of it, or are both scenarios plausible? Asking ecologically based questions within a phylogenetic framework will shed light on evolutionary patterns. For example, do flower-visiting flies possess a more highly developed sense of color vision than coprophagous flies? Do fruit- and flower-feeding butterflies differ in their processing of and response to colors and odors? Can anthophilous beetles learn to associate colors or odors with rewards more readily than non-flower-visiting beetles?

### How important were early pollinators in shaping flower form?

If beetles indeed "stood at the cradle of the flower" (Faegri & van der Pijl 1979), they (as well as contemporaneous flies and non-social hymenopterans) would have had almost 100 million years to influence floral evolution before the arrival of social bees – ample time to establish an insect-pollinated floral *Bauplan*. Thus, the innate color and odor preferences of these early pollinators, as well as whatever learning abilities they may have possessed, are likely to have been formative in early angiosperm evolution. We really have no idea how important the early pollinators were in establishing the fundamental features of the angiosperm flower, and the extent to which later pollinators have had to modify these early designs or have been able to select on floral features from scratch. A fuller understanding of the sensory and behavioral attributes of non-hymenopteran pollinators will be important in our attempts to reconstruct patterns of floral evolution.

## Conclusion

Though beetles, flies, moths, and butterflies visit a broad range of angiosperm taxa relative to social bees, much less is known about their effectiveness as pollinators, or about the sensory attributes and learning abilities that guide their behaviors. The research reviewed above suggests that non-hymenopteran insects can be effective pollinators of a range of taxa, and that many species in these groups are adept and flexible learners. Quantitative data on pollinator performance are critical in order to evaluate the importance of non-hymenopterans as pollinators in generalized as well as specialized systems. Comparative studies of learning and memory, using identical training and testing protocols, will allow us to evaluate differences across taxa, and will help us to tease apart the relative contributions of phylogenetic history and lifestyle-mediated selection.

## Acknowledgements

Thanks to Lars Chittka, James Thomson, Almut Kelber, and an anonymous reviewer for insightful comments, and to Josh Rosenthal and Jennifer Maupin for assistance of all kinds. This work is dedicated to Isabel, whose gestation and birth coincided with that of the chapter.

## References

Agee HR, Mitchell ER & Flanders RV (1990) Spectral sensitivity of the compound eye of *Coccinella septempunctata* (Coleoptera: Coccinellidae). *Ann Entomol Soc Am* 83:817–819

Alloway TM (1973) Learning in insects except Apoidea. In: Corning WC, Dyal JA & Willows AOD (eds) *Invertebrate learning*, vol. 2, *Arthropods and gastropod mollusks*, pp 131–171. Plenum Press, New York

Arikawa K, Inokuma K & Eguchi E (1987) Pentachromatic visual system in a butterfly. *Naturwiss* 74:297–298

Briscoe A & Chittka L (2001) The evolution of color vision in insects. *Ann Rev Entomol*, 46:471–510

Buchmann SL & Nabhan GP (1996) *The forgotten pollinators.* Island Press, Washington, DC

Chittka L (1996) Does bee color vision predate the evolution of flower color? *Naturwiss* 83:136–138

Collett TS & Land MF (1975) Visual spatial memory in a hoverfly. *J Comp Physiol* 100:59–84

Courtney SP, Hill CJ & Westerman A (1982) Pollen carried for long periods by butterflies. *Oikos* 38:260–263

Crane J (1955) Imaginal behavior of a Trinidad butterfly, *Heliconius erato hydara* Hewitson, with special reference to the social use of color. *Zoologica* (NY) 40:167–196

Crepet WL & Friis EM (1987) The evolution of insect pollination in angiosperms. In: Friis EM, Chaloner WG & Crane PR (eds) *The origins of angiosperms and their biological consequences*, pp 181–201. Cambridge University Press, Cambridge

Cruden RW & Hermann-Parker SM (1979) Butterfly pollination of *Caesalpinia pulcherrima*, with observations on a psychophilous syndrome. *J Ecol* 67:155–168

Cunningham JP, West SA & Wright DJ (1998) Learning in the nectar foraging behaviour of *Helicoverpa armigera*. *Ecol Entomol* 23:363–369

Dafni A (1984) Mimicry and deception in pollination. *Ann Rev Ecol Syst* 15:259–278

Dafni A, Bernhardt P, Shmida A, Ivri Y, Greenbaum S, O'Toole C & Losito L (1990) Red bowl-shaped flowers: convergence for beetle pollination in the Mediterranean region. *Israel J Bot* 39:81–92

De Los Mozos Pascual M & Domingo LM (1991) Flower constancy in *Heliotaurus ruficollis* (Fabricius 1781), Coleoptera, Alleculidae. *Elytron (Barc)* 5:9–12

DeVries PJ & Stiles FG (1990) Attraction of pyrrolizidine-alkaloid-seeking Lepidoptera to *Epidendrum paniculatum* orchids. *Biotropica* 22:290–297

Dobson HEM (1994) Floral volatiles in insect biology. In: Bernays EA (ed) *Insect–plant interactions*, vol. 5, pp 47–81. CRC Press, Boca Raton, FL

Dukas R (1998) Constraints on information processing and their effects on behavior. In: Dukas R (ed) *Cognitive ecology*, pp 89–128. University of Chicago Press, Chicago

Englund R (1993) Movement patterns of *Cetonia* beetles (Scarabaeidae) among flowering *Viburnum opulus* (Caprifoliaceae). *Oecologia* 94:295–302

Eriksson R (1994) The remarkable weevil pollination of the neotropical *Carludovicoideae* (Cyclanthaceae). *Plant Syst Evol* 189:75–81

Faegri K & van der Pijl L (1979) *The principles of pollination biology*, 3rd edn. Pergamon Press, New York

Fan R-J, Anderson P & Hansson BS (1997) Behavioural analysis of olfactory conditioning in the moth *Spodoptera littoralis* (Boisd.) (Lepidoptera: Noctuidae). *J Exp Biol* 200:2969–2976

Fishbein M & Venable DL (1996) Diversity and temporal change in the effective pollinators of *Asclepias tuberosa*. *Ecology* 77:1061–1073

Frisch K von (1967) *The dance language and orientation of bees.* Belknap Press, Cambridge, MA

Fukushi T (1973) Classical conditioning in the housefly, *Musca domestica*. *Annot Zool Jap* 46:135

Fukushi T (1989) Learning and discrimination of coloured papers in the walking blowfly, *Lucilia cuprina*. *J Comp Physiol A* 166:57–64

Gilbert LE (1972) Pollen feeding and reproductive biology of *Heliconius* butterflies. *Proc Natl Acad Sci USA* 69:1403–1407

Gori DF (1983) Post-pollination phenomena and adaptive floral changes. In: Jones CE & Little RJ (eds) *Handbook of experimental pollination biology*, pp 31–49. Scientific & Academic Editions, New York

Gould JL (1984) Natural history of bee learning. In: Marler P & Terrace HS (eds) *The biology of learning*, pp 149–180. Springer-Verlag, Berlin

Goulson D & Cory JS (1993) Flower constancy and learning in foraging preferences of the green-veined white butterfly *Pieris napi*. *Ecol Entomol* 18:315–320

Goulson D & Wright NP (1998) Flower constancy in the hoverflies *Episyrphus balteatus* (Degeer) and *Syrphus ribesii* (L.) (Syrphidae). *Behav Ecol* 9:213–219

Goulson D, Stout JC & Hawson SA (1997) Can flower constancy in nectaring butterflies be explained by Darwin's interference hypothesis? *Oecologia* 112:225–231

Greggers U & Menzel R (1993) Memory dynamics and foraging strategies of honeybees. *Behav Ecol Sociobiol* 32:17–29

Haber WA & Frankie GW (1989) A tropical hawkmoth community: Costa Rican dry forest Sphingidae. *Biotropica* 21:155–172

Hartlieb E (1996) Olfactory conditioning in the moth *Heliothis virescens*. *Naturwiss* 83:87–88

Hasselmann EM (1962) Über die relative spektrale Empfindlichkeit von Käfer- und Schmetterlingsaugen bei verschiedenen Helligkeiten. *Zool J Physiol* 69:537–576

Heinrich B (1975) Energetics of pollination. *Ann Rev Ecol Syst* 6:139–170

Heinrich B (1979) Resource heterogeneity and patterns of movement in foraging bumblebees. *Oecologia* 40:235–245

Hernández de Salomon C & Spatz H-C (1983) Colour vision in *Drosophila melanogaster*: wavelength discrimination. *J Comp Physiol* 150:31–37

Herrera CM (1987) Components of pollinator "quality": comparative analysis of a diverse insect assemblage. *Oikos* 50:79–90

Ilse D (1928) Über den Farbensinn der Tagfalter. *Z vergl Physiol* 8:658–692

Ilse D & Vaidya VG (1956) Spontaneous feeding response to colours in *Papilio demoleus* L. *Indian Acad Sci Proc (B)* 43:23–31

Johnson SD & Dafni A (1998) Response of bee-flies to the shape and pattern of model flowers: implications for floral evolution in a Mediterranean herb. *Funct Ecol* 12:289–297

Johnson SD & Midgley JJ (1997) Fly pollination of *Gorteria diffusa* (Asteraceae), and a possible mimetic function for dark spots on the capitulum. *Am J Bot* 84:429–436

Johnson SD & Steiner KE (1997) Long-tongued fly pollination and evolution of floral spur length in the *Disa draconis* complex (Orchidaceae). *Evolution* 51:45–53

Johnson SD & Steiner KE (2000) Generalization versus specialization in plant pollination systems. *Trends Ecol Evol* 15:140–143

Kandori I & Ohsaki N (1996) The learning abilities of the white cabbage butterfly, *Pieris rapae*, foraging for flowers. *Res Popul Ecol* 38:111–117

Kearns CA (1992) Anthophilous fly distribution across an elevation gradient. *Am Midl Nat* 127:172–182

Kelber A (1996) Colour learning in the hawkmoth *Macroglossum stellatarum*. *J Exp Biol* 199:1127–1131

Kelber A & Hénique U (1999) Trichromatic colour vision in the hummingbird hawkmoth, *Macroglossum stellatarum* L. *J Comp Physiol A* 184:535–541

Kelber A & Pfaff M (1997) Spontaneous and learned preferences for visual flower features in a diurnal moth. *Israel J Plant Sci* 45:235–246

Kevan PG & Baker HG (1983) Insects as flower visitors and pollinators. *Ann Rev Entomol* 28:407–453

Kinoshita M, Shimada N & Arikawa K (1999) Colour vision of the foraging swallowtail butterfly *Papilio xuthus*. *J Exp Biol* 202:95–102

Knoll F (1921) *Bombylius fuliginosus* und die Farbe der Blumen (Insekten und Blumen I). *Abh Zool Bot Ges Wien* 12:17–119

Kugler H (1950) Der Blutenbesuch der Schlammfliege (*Eristalomyia tenax*). *Z vergl Physiol* 32:328–347

Kugler H (1951) Blutenokologische Untersuchungen mit Goldfliegen (Lucilien). *Berichte D Bot Ges* 64:327–341

Kugler H (1956) Über die optische Wirkung von Fliegenblumen auf Fliegen. *Berichte D Bot Ges* 69:387–398

Lall AB, Lord ET & Trouth CO (1982) Vision in the firefly *Photuris lucicrescens* (Coleoptera: Lampyridae): spectral sensitivity and selective adaptation in the compound eye. *J Comp Physiol A* 147:195–200

Laverty T (1994) Bumble bee learning and flower morphology. *Anim Behav* 47:531–545

Levin DA & Berube DE (1972) *Phlox* and *Colias*: the efficiency of a pollination system. *Evolution* 26:242–250

Levin DA & Kerster HW (1968) Local gene dispersal in *Phlox*. *Evolution* 22:130–139

Lewis AC (1986) Memory constraints and flower choice in *Pieris rapae*. *Science* 232:863–865

Lewis AC (1989) Flower visit consistency in *Pieris rapae*, the cabbage butterfly. *J Anim Ecol* 58:1–13

Lewis AC (1993) Learning and the evolution of resources: pollinators and flower morphology. In: Papaj DR & Lewis AC (eds) *Insect learning: ecological and evolutionary perspectives*, pp 219–242. Chapman & Hall, London

Lin J-T & Wu C-Y (1992) A comparative study on the color vision of four Coleopteran insects. *Bull Inst Zool Acad Sinica* 31:81–88

Lindauer M (1963) Allgemeine Sinnesphysiologie. Orientierung im Raum. *Fortschr Zool* 16:58–140

Linhart YB & Mendenhall JA (1977) Pollen dispersal by hawkmoths in a *Lindenia rivalis* Benth. population in Belize. *Biotropica* 9:143

Listabarth C (1996) Pollination of *Bactris* by *Phyllotrox* and *Epurea*: implications of the palm breeding beetles on pollination at the community level. *Biotropica* 28:69–81

Lunau K (1988) Innate and learned behavior of flower-visiting hoverflies: flower-dummy experiments with *Eristalis pertinax* (Scolopi) (Diptera, Syrphidae). *Zool Jb Physiol* 92:487–499

Lunau K (1992) Limits of colour learning in a flower-visiting hoverfly, *Eristalis tenax* L. (Syrphidae, Diptera). *Eur J Neurosci* Suppl no 5:103

Lunau K & Maier EJ (1995) Innate colour preferences of flower visitors. *J Comp Physiol A* 177:1–19

Mallet J, Longino JT, Murawski D, Murawski A & De Gamboa AS (1987) Handling effects in *Heliconius*: where do all the butterflies go? *J Anim Ecol* 56:377–386

Manning JC & Goldblatt P (1997) The *Moegistorhynchus longirostris* (Diptera: Nemestrinidae) pollination guild: long-tubed flowers and a specialized long-proboscid fly pollination system in southern Africa. *Plant Syst Evol* 206:51–69

Martínez del Rio C & Búrquez A (1986) Nectar production and temperature dependent pollination in *Mirabilis jalapa* L. *Biotropica* 18:28–31

McGregor SE (1976) *Insect pollination of cultivated crop plants*, Agriculture Handbook no. 496. Agricultural Research Service, USDA, Washington, DC

Menzel R (1967) Untersuchungen zum Erlernen von Spektralfarben durch die Honigbiene (*Apis mellifica*). *Z vergl Physiol* 56:22–62

Michener CD & Grimaldi DA (1988) The oldest fossil bee: apoid history, evolutionary stasis, and antiquity of social behavior. *Proc Natl Acad Sci USA* 85:6424–6426

Momose K, Yumoto T, Nagamitsu T, Kato M, Nagamasu H, Sakai S, Harrison RD, Itioka T, Hamid AA & Inoue T (1998) Pollination biology in a lowland dipterocarp forest in Sarawak, Malaysia. I. Characteristics of the plant–pollinator community in a lowland dipterocarp forest. *Am J Bot* 85:1477–1501

Murawski DA & Gilbert LE (1986) Pollen flow in *Psiguria warscewiczii*: a comparison of *Heliconius* butterflies and hummingbirds. *Oecologia* 68:161–167

Murphy DD (1984) Butterflies and their nectar plants: the role of the checkerspot butterfly *Euphydryas editha* as a pollen vector. *Oikos* 43:113–117

Nilsson LA, Jonsson L, Ralison L & Randrianjohany E (1997) Angraecoid orchids and hawkmoths in central Madagascar: specialized pollination systems and generalist foragers. *Biotropica* 19:310–318

Olesen JM & Warncke E (1989) Temporal changes in pollen flow and neighbourhood structure in a population of *Saxifraga hirculus* L. *Oecologia* 79:205–211

Papaj DR (1986) Conditioning of leaf-shape discrimination by chemical cues in the butterfly, *Battus philenor*. *Anim Behav* 34:1281–1288

Papaj DR & Prokopy RJ (1989) Ecological and evolutionary aspects of learning in phytophagous insects. *Ann Rev Entomol* 34:315–350

Parmesan C, Singer MC & Harris I (1995) Absence of adaptive learning from the oviposition foraging behaviour of a checkerspot butterfly. *Anim Behav* 50:161–175

Pellmyr O (1985) Flower constancy in individuals of an anthophilous beetle, *Byturus ochraceus* (Scriba) (Coleoptera: Byturidae). *Coleopterists' Bull* 39:341–345

Pellmyr O & Patt J (1986) Function of olfactory and visual stimuli in pollination of *Lysichiton americanum* (Araceae) by a staphylinid beetle. *Madroño* 33:47–54

Pettersson MW (1991) Pollination by a guild of fluctuating moth populations: option for unspecialization in *Silene vulgaris*. *J Ecol* 79:591–604

Pickens LG (1990) Colorimetric versus behavioral studies of face fly (Diptera: Muscidae) vision. *Environ Entomol* 19:1242–1252

Plotkin HC (1979) Learning in a carabid beetle (*Pterostichus melanarius*). *Anim Behav* 27:567–575

Poinar GJ (1994) Bees in fossilized resin. *Bee World* 75:71–77

Proctor M, Yeo P & Lack A (1996) *The natural history of pollination*. Timber Press, Portland, OR

Prokopy RJ, Averill AL, Cooley SS & Roitberg CA (1982) Associative learning in egglaying site selection by apple maggot flies. *Science* 218:76–77

Pyke GH, Pulliam HR & Charnov EL (1977) Optimal foraging: a selective review of theory and tests. *Q Rev Biol* 52:137–154

Ren D (1998) Flower-associated *Brachycera* flies as fossil evidence for Jurassic angiosperm origins. *Science* 280:85–88

Scherer C & Kolb G (1987a) Behavioral experiments on the visual processing of color stimuli in *Pieris brassicae* L. (Lepidoptera). *J Comp Physiol A* 160:645–656

Scherer C & Kolb G (1987*b*) The influence of color stimuli on visually controlled
    behavior in *Aglais urticae* L. and *Pararge aegeria* L. (Lepidoptera). *J Comp Physiol A*
    161:891–898

Schmitt J (1980) Pollinator foraging behavior and gene dispersal in *Senecio* (Compositae).
    *Evolution* 34:934–943

Schmitt J (1983) Density-dependent pollinator foraging, flowering phenology, and
    temporal pollen dispersal patterns in *Linanthus bicolor*. *Evolution* 37:1247–1257

Shimohigashi M & Tominaga Y (1991) Identification of UV, green and red receptors, and
    their projection to lamina in the cabbage butterfly, *Pieris rapae*. *Cell Tissue Res*
    263:49–59

Silberglied RE (1984) Visual communication and sexual selection among butterflies. In:
    Vane-Wright RI & Ackery PR (eds) *Biology of butterflies*, vol. 11, pp 207–223.
    Academic Press, London

Spatz H-C, Emanns A & Reichert H (1974) Associative learning of *Drosophila melanogaster*.
    *Nature* 248:359

Stanton ML (1984) Short-term learning and the searching accuracy of egg-laying
    butterflies. *Anim Behav* 32:33–40

Steiner KE (1998) The evolution of beetle pollination in a South African orchid. *Am J Bot*
    85:1180–1193

Stephens DW (1993) Learning and behavioral ecology: incomplete information and
    environmental predictability. In: Papaj DR & Lewis AC (eds) *Insect learning:
    ecological and evolutionary perspectives*, pp 195–218. Chapman & Hall, London

Sun G, Dilcher DL, Zheng S & Zhou Z (1998) In search of the first flower: a Jurassic
    angiosperm, *Archaefructus*, from Northeast China. *Science* 282:1692–1695

Sutherland JP, Sullivan MS & Poppy GM (1999) The influence of floral character on the
    foraging behaviour of the hoverfly, *Episyrphus balteatus*. *Entomol Exp Appl*
    93:157–164

Swihart CA (1971) Colour discrimination by the butterfly, *Heliconius charitonius* Linn.
    *Anim Behav* 19:156–164

Swihart CA & Swihart SL (1970) Colour selection and learned feeding preferences in the
    butterfly, *Heliconius charitonius* Linn. *Anim Behav* 18:60–64

Thomson, JD & Thomson BA (1992) Pollen presentation and viability schedules in
    animal-pollinated plants: consequences for reproductive success. In: Wyatt R
    (ed) *Ecology and evolution of plant reproduction: new approaches*, pp 1–24. Chapman &
    Hall, New York

Tinbergen N (1968) *Curious naturalists*. Doubleday Anchor, Garden City, NY

Traynier RMM (1986) Visual learning in assays of sinigrin solution as an oviposition
    releaser for the cabbage butterfly, *Pieris rapae*. *Entomol Exp Appl* 40:25–33

Troje N (1993) Spectral categories on the learning behaviour of blowflies. *Z Naturforsch*
    48c:96–104

Turner JRG (1981) Adaptation and evolution in *Heliconius*: a defense of neoDarwinism.
    *Ann Rev Ecol Syst* 12:99–121

Venables BAB & Barrows EM (1985) Skippers: pollinators or nectar thieves? *J Lepid Soc*
    39:299–312

Waller DA & Gilbert LE (1982) Roost recruitment and resource utilization: observations
    on a *Heliconius charitonius* L. roost in Mexico. *J Lepid Soc* 36:178–184

Waser NM (1982) A comparison of distances flown by different visitors to flowers of the
    same species. *Oecologia* 55:251–257

Waser NM, Chittka L, Price MV, Williams NM & Ollerton J (1996) Generalization in pollination systems, and why it matters. *Ecology* 77:1043–1060

Weiss MR (1991) Floral colour changes as cues for pollinators. *Nature* 354:227–229

Weiss MR (1995) Associative colour learning in a nymphalid butterfly. *Ecol Entomol* 20:298–301

Weiss MR (1997) Innate colour preferences and flexible colour learning in the pipevine swallowtail. *Anim Behav* 53:1043–1052

Weiss MR & Lamont BB (1997) Floral color change and insect pollination: a dynamic relationship. *Israel J Plant Sci* 45:185–199

Widen B & Widen M (1990) Pollen limitation and distance-dependent fecundity in females of the clonal gynodioecious herb *Glechoma hederacea* (Lamiaceae). *Oecologia* 83:191–196

Wiklund C (1977) Oviposition, feeding, and spatial separation of breeding and foraging habitats in a population of *Leptidea sinapsis* (Lepidoptera). *Oikos* 28:56–68

Wiklund C, Eriksson T & Lundberg H (1979) The wood white butterfly *Leptidea sinapsis* and its nectar plants: a case of mutualism or parasitism? *Oikos* 33:358–362

Willmott AP & Búrquez A (1996) The pollination of *Merremia palmeri* (Convolvulaceae): can hawk moths be trusted? *Am J Bot* 83:1050–1056

Wilson P & Thomson JD (1991) Heterogeneity among floral visitors leads to discordance between removal and deposition of pollen. *Ecology* 72:1503–1507

Winston ML (1987) *The biology of the honey bee.* Harvard University Press, Cambridge, MA

Yeboah Gyan K & Woodell SRJ (1987) Analysis of insect pollen loads and pollination efficiency of some common insect visitors of four species of woody Rosaceae. *Funct Ecol* 1:269–274

Young HJ (1986) Beetle pollination of *Dieffenbachia longispatha* (Araceae). *Am J Bot* 73:931–944

**10**

# Pollinator individuality: when does it matter?

I have always regretted that I did not mark the bees by attaching bits of cotton wool or eiderdown to them with rubber, because this would have made it much easier to follow their paths.

Charles Darwin, cited by Freeman (1968)

The symposium that stimulated this book arose from the editors' conviction that botanists interested in biotic pollination would benefit from a consideration of recent research on the behavior and the sensory capabilities of flower-visiting animals. We hoped to offer perspectives that would correct misapprehensions, enrich future work, and open new questions. In this chapter, we continue in this evangelistic vein by indulging in long-standing personal interests in the individuality of pollinating animals. Ignoring the uniqueness of individuals will invite regrets like those expressed by Darwin in reviewing his work on the flight patterns of male bumble bees. Although he investigated this question for several years, Darwin never published his observations. Might he have considered his failure to mark the bees a fatal flaw?

Our goals are to outline some of the insights that are made possible by treating pollinators as individuals, and to show possible pitfalls of *not* doing so. Some well-known conclusions regarding pollinator physiology and behavior can be given alternative interpretations by invoking individuality. We hope that this chapter will stimulate more systematic approaches to pollinator individuality.

There are many relevant axes along which individual pollinators may vary, including gross behavioral aspects such as foraging-site preferences, food-plant preferences, and numerous aspects of foraging style (including sampling effort, level of flower constancy, giving-up thresholds, etc.).

[191]

These in turn may be underlain by variation in basic neurophysiological processes such as learning ability (speed, capacity, and duration), sensitivity to interference, efficiency at detecting flowers, etc. There are also multiple causes for observed variations in foraging behavior. These can be genetic, learning-related, age-dependent, or induced by parasites. In what follows, we are mostly concerned with cases where neglecting pollinator individuality may lead to erroneous conclusions.

### Basic observations

### Small foraging areas

Several studies showing that social insects use spatial memory in foraging date back to the penultimate century (e.g. Fabre 1879; Müller 1882). After many decades of detailed research on spatial memory of bees (e.g. Chittka *et al.* 1995; Menzel, this volume), most pollination biologists accept that such memory exists, but most associate it with finding the nest rather than finding food sources. Optimal foraging theory is partially responsible for this (Healy & Hurly, this volume): some adherents of this theory proposed that pollinators forage using essentially the same rules as protozoans. The numbers of places visited during a foraging bout seemed to many biologists too high (often, several thousand flowers must be visited to fill the stomach of a bee) for bees to memorize much detail of the complex flight path.

Yet, if one catches bee workers at a patch of flowers, marks them, and releases them, one will frequently see some of them return to the site (Ribbands 1949; Heinrich 1976; Free 1993). This indicates that at least some individuals have established small foraging areas to which they return for all or most of their feeding. In one study, 37 plants of *Penstemon strictus* were planted in a meadow in a hexagonal pattern with 1.5 m between plants (Thomson *et al.* 1997). We marked bees and followed some of them intensively from 23–28 July 1990. Several bees did all of their foraging in this area; one worker in particular, *Bombus flavifrons* "Blue," worked the array for our entire period of close observations (23 July through 5 August 1990). She would visit the 37 plants (and some of other meadow species that grew interspersed with the *Penstemon*) essentially all day, disappearing for only a few minutes at *c.* half-hour intervals to drop off collected rewards at the nest. Bumble bees of other species have performed comparably on other plants (Thomson *et al.* 1987), but we do not know whether this site fidelity is typical.

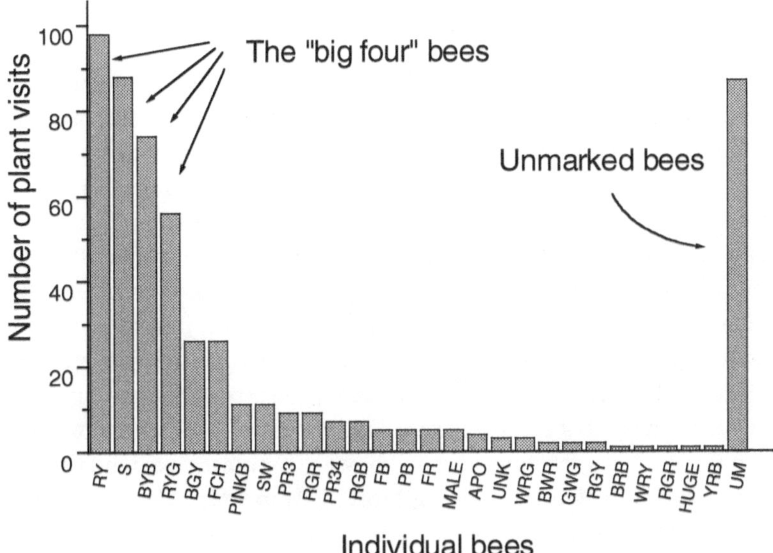


Fig. 10.1. Frequencies of visits in one day to a focal plant of *Penstemon strictus* by a number of marked bumble bees (Williams & Thomson 1998). Bee names mostly indicate painted color marks (e.g., RY = red–yellow), although a few distinctive bees were identified by natural attributes (e.g., HUGE).

In another study of *Penstemon strictus*, Williams & Thomson (1998) videotaped all visits to a single potted plant in a circular array of 27 plants. We had marked visitors on previous days. Four individuals made over half of the plant visits (Fig. 10.1); these bees returned to the focal plant at statistically regular intervals, with mean interarrival times of 5.36, 5.90, 7.07, and 7.91 min. Unmarked bees might have been vagabonds with no site fidelity, site-faithful bees that evaded marking, or site-faithful bees that were new arrivals.

Individual bumble bees may maintain more than one foraging area. Brian (1952) noted that *Bombus agrorum* (now *pascuorum*) workers tended to leave the nest in characteristic compass direction, but that some individuals had more than one departing direction. These bees also came back with different pollens when they left in different directions. Karen Goodell (personal communication) found that certain workers of *B. ephippiatus* collected one of two different *sets* of several pollen species on different trips in a montane Neotropical habitat. The most likely explanation for the covariation of several species is that the bees were going to two different localities, then foraging inconstantly in each place.

### Traplining behavior

If bees do return frequently to foraging areas, they may also tend to visit a set of plants within those areas in a particular, somewhat repeatable circuit (Manning 1956; Heinrich 1976; Thomson *et al.* 1982, 1987, 1997). In fact, such traplining is a case where pollinator individuality manifests itself *par excellence*. In one study, we let bumble bees *(Bombus impatiens)* forage in an arena with six artificial flowers at fixed positions. The nectar rewards were adjusted to bee crop capacity, so that bees had to visit all six flowers (but not more) to fill their stomach once. Each bee was tested individually and encountered an absolutely identical array during 40 successive foraging bouts. Yet, each bee found a unique solution to the problem of linking the six flowers, and used this solution repeatedly (see Fig. 10.2).

Although we lack comparative studies that would indicate how often bumble bees show trapline behavior, or what circumstances tend to elicit it, it seems likely that traplining is most likely to emerge (1) when nectar or pollen rewards are replenished rapidly after being drained by a visitor, and (2) when there are spaces between plants, with sufficient landmarks to allow bees to orient. Bumble bees, especially *Bombus ternarius*, showed clear traplining behavior on scattered plants of *Aralia hispida* in central New Brunswick (Thomson *et al.* 1982); in dense stands of *Solidago* spp. (goldenrods) nearby, however, bees of the same species showed no discernable tendency to repeat their flight paths, although they were using small foraging areas (J. D. Thomson & W. Maddison, unpublished data).

Two aspects of bumble bee traplining are most relevant to this paper. First, although traplines are quite flexible – bees do not slavishly follow a fixed route, but rather add new plants and drop old ones as conditions change – there is a conservative tendency for bees to keep using accustomed flight paths (Thomson 1996) and to keep returning to plants that have been particularly rewarding in the past (Thomson 1988). For example, Manning (1956) described how bees that had been trained to visit potted plants still returned to those locations after the pots had been completely removed. Second, bees return to plants on their traplines at surprisingly brief intervals, *c.* 10 min in both *Penstemon strictus* and *Aralia hispida*.

### Variation in working speed

When following marked pollinators, one is frequently struck by variations in the speed of individuals. Some of this variation is caused by differences in the nectar offerings of plants on which these individuals forage.

For example, bees and butterflies will fly more rapidly when more nectar is available, an observation with several possible explanations (Núñez 1970; Kunze & Chittka 1996). But there is also variation between individuals who are using the same resources at the same time. Some such variation can be explained by size: larger bees are faster fliers (Spaethe *et al.* 2000). In addition, some sensory attributes correlate with size and influence the speed with which bees detect flowers. Spaethe *et al.* (2000) recently found that larger bees have better visuo-spatial resolution, and are therefore substantially more accurate and faster at detecting small flowers. Furthermore, foraging speed is dependent on colony needs in bumble bees (Cartar 1992*a*).

In studies with numerous marked bees, Thomson has frequently encountered a few individuals that seem to fly much faster and to handle inflorescences very quickly. Because such bees are hard to observe for long bouts, they may be underrepresented in certain types of observational data.

Even among the more stolid bees for which data are available, however, there are individual differences in working speed and in other aspects such as flower constancy (Table 10.1). The mean flower-handling times of 17 bees in the 1994 data varied two-fold. Recall that all of these data come from the same plant on a single day. Bees also varied about two-fold in the duration of their plant visits (measured as the mean number of flowers visited per plant visit), but plant-visit durations varied so much within bees that the variation among bees was insignificant. In addition to showing variation among individuals, the data for "Blue" suggest that this bee's foraging tempo slowed over the two weeks she was observed.

### Variation in foraging mode

Different bees may adopt different ways of using flowers. One of the more conspicuous differences involves the type of floral reward – pollen or nectar – being actively sought. On *Penstemon strictus*, for example, most *Bombus* workers enter the large flowers rightside-up and tongue the paired nectaries at the filament bases. These bees usually accumulate small pollen loads, but they never fill their corbiculae, presumably because their honeystomachs fill first. Other bees, mostly *B. bifarius*, ignore the nectaries, turning upside-down to grasp the anthers and sonicate pollen from them. These bees accumulate very large pollen loads. Some bees combine the two behaviors, but most individuals tend to stick with one type of behavior over at least a few days. Still, changes occur; bees

# Bee 1

A
F
H
B
D
E
C

Percent
100
80
60
40
20
0
A B C D E F
Flower
1
2
3
4
5
6
Sequential position

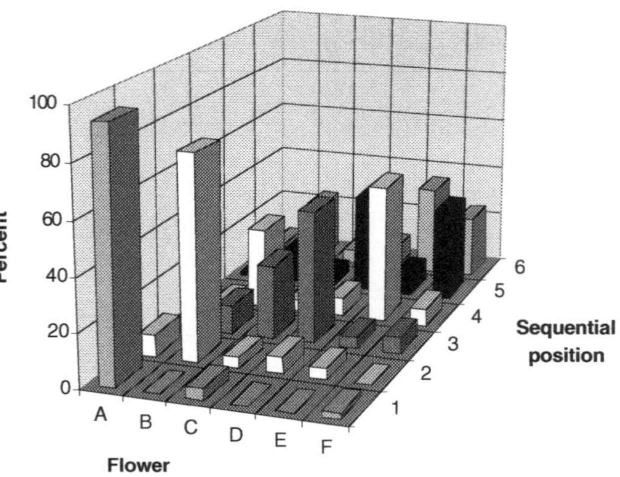

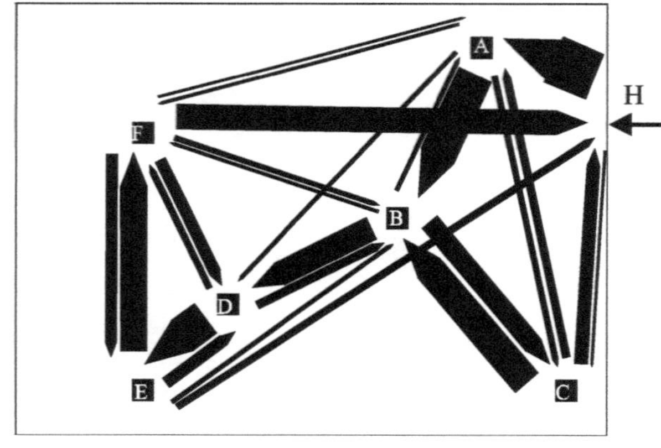

# Bee 2

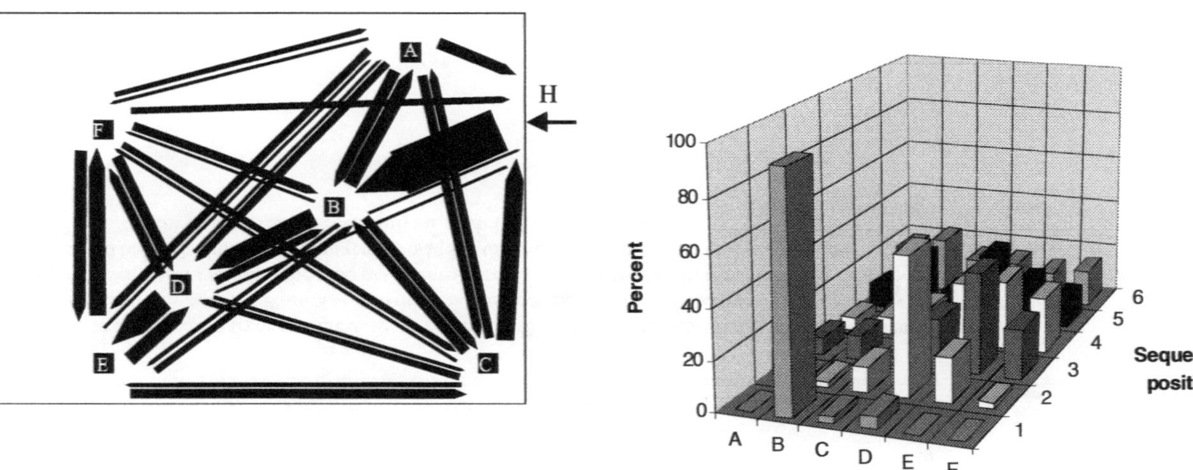


Fig. 10.2. Traplining of two bumble bees in a flight arena (105 cm × 75 cm) with six artificial flowers. The left panels show the arrangement of flowers (black squares), and the frequency of flight paths within it. The arrow on the right marks the location of the nest entrance to the arena. The width of the arrows corresponds to the frequency each trajectory was taken. For example, in the upper left panel, the arrow from the hive to the closest flower corresponds to 38 of a total of 40 foraging bouts. The right panels show how often each flower occupied a given sequential position within a foraging bout. Note that the right panels contains only the first six visits of each foraging bout, whereas the left panels contain the information from all flights. Because bees often revisited flowers they had already emptied, a foraging bout can have more than six visits.

Table 10.1. *Summary of movement patterns of three individual bees foraging in a hexagonal grid of* Penstemon strictus *plants set in a meadow with other plants*

| Bee | Moves to neighbors[a] | | | | Constancy[b] | | | | Summaries[c] | | |
|---|---|---|---|---|---|---|---|---|---|---|---|
| | 1st | 2nd | 3rd | Far | $P$ to $P$ | $P$ to Other | Other to $P$ | Other to Other | Bateman's Index | Other/$P$ | Mean bout |
| "Blue" early[d] | 741 | 56 | 35 | 19 | 712 | 148 | 137 | 19 | −0.10 | 0.21 | 22.0 |
| "Blue" late[d] | 304 | 28 | 11 | 6 | 282 | 74 | 72 | 41 | 0.19 | 0.26 | 55.3 |
| "Red–Blue" | 255 | 33 | 16 | 12 | 293 | 27 | 22 | 5 | 0.22 | 0.09 | 15.2 |
| "Pink" | 285 | 26 | 7 | 12 | 225 | 110 | 115 | 15 | −0.31 | 0.49 | 13.4 |

*Notes:*

[a] Counts of visits to first-nearest, second-nearest, third-nearest, farther-than-third-nearest neighboring *Penstemons*.

[b] Frequencies of transitions between *Penstemons* and other flower species, where $P$ = *Penstemon* and Other = any other species.

[c] Bateman's Index (see Waser 1986) and the ratio of Other to $P$ are given as summaries of constancy. The mean bout length (last column) is the number of plant visits observed before losing the bee; longer bouts indicate slow-moving bees that are easy to follow.

[d] Data for bee "Blue" are divided into early (23–28 July) and late (5 August) periods.

that collected pollen while young may turn to nectar collecting with age, or vice versa. "Blue," for example, accumulated small corbicular loads during all bouts from 23–28 July, but by 5 August was no longer carrying visible loads.

Even though the relative efforts made into pollen and nectar foraging are genetically controlled (Robinson & Page 1989), there is also strong plasticity in the way in which individuals react to colony needs (Cartar 1992a; Plowright *et al.* 1993; Fewell & Bertram 1999). There have been recent attempts to understand such task allocation in bee colonies by self-organization models in which each bee is an automaton that differs from other colony members only in the response threshold to particular stimuli in and outside the nest (Bonabeau *et al.* 1997; Pankiw & Page 2000). Even if these models explain some of the observed behavior, there are potential difficulties, because they neglect the individuality of pollinators beyond their inborn thresholds. All animals encounter a basic difficulty when they set out to perform a novel skill: they generally need to *learn* that skill, even if it has innate components. The investments in learning different types of foraging activities (and the costs of interference when switching) can be substantial (Dukas & Visscher 1994); therefore, we cannot understand task allocation and task switching without quantifying these costs (and how bees perceive them). Surprisingly, however, one review of new breakthroughs in task allocation (Gordon 1996) avoids such terms as "learning" and "memory" altogether.

### Learning-related individuality

Pollinators learn about diverse aspects of their environment (see other chapters of this volume). Because each pollinator's experience is unique, its behavior may also be unique. Much of this experience, however, is beyond the control of the observer. Moreover, each bee's experience (e.g., which flower species it experiences as rewarding) may in part be an epiphenomenon of its decision where to forage (see above) or may simply reflect stochastic processes.

The efficiency and accuracy with which pollinators handle flowers depends substantially on their experience with the respective flower type (Laverty 1994; Chittka & Thomson 1997). Some complex handling skills, such as nectar robbing in *Corydalis cava* (Fumariaceae) by bumble bee queens can take several days to develop (Olesen 1996). But handling efficiency on a given flower type can also be influenced by pollinators' experience on other flower types. Depending on the similarity of motor patterns involved (and depending on the timing of visits to the two flower types), transfer or interference may occur (Chittka & Thomson 1997; Gegear & Laverty, this volume).

In the following paragraphs, we are concerned with the possibility of more formative types of learning, i.e., the possibility that early experience may substantially influence how a pollinator later reacts to flowers. Memory-through-metamorphosis has been invoked as a possible mechanism to determine foraging preferences in specialist bees (Dobson 1994), i.e., the possibility that bees become imprinted on particular scents (such as pollen odor) as larvae, and as adults show a preference for flowers with the same scent. The nervous system is entirely reorganized during metamorphosis; therefore a memory that persists through fundamental rewiring of neuronal circuitry is not trivial. However, Lindauer (1985) earlier found evidence for memory-through-metamorphosis in honeybees. The phenomenon, however, was also shown in grain beetles (Alloway 1972) and fruitflies (Tully *et al.* 1994), so it is not restricted to pollinators.

Does early experience shape the brain, as some studies on humans suggest (Elbert *et al.* 1995)? The mushroom bodies, a prominent structure in the insect brain, are essential in memory formation (Menzel, this volume). Interestingly, the size of the mushroom bodies in honeybees is correlated not only with age, but also with type of activity. Durst *et al.* (1994) showed that foragers have larger mushroom-body volumes than nurse bees of the same age, concluding that mushroom-body size is experience-dependent. The rationale was that more information storage requires more neural substrate (e.g more neurons or dendritic proliferations). However, it was not clear whether the mushroom bodies increase in size as a result of experience, or whether the increased mushroom-body volume is a prerequisite in honeybees to switch from nursing to foraging activities.

To resolve this problem, Fahrbach *et al.* (1998) reared honeybees in an extremely deprived environment (social isolation and complete darkness). Mushroom-body volume increased even when bees collected no foraging experience, suggesting that the observed changes in brain structure served to *prepare* the animal for handling complex information in the context of foraging. But the correlation between brain-region size and storage capacity (or behavioral/cognitive ability) remains to be shown empirically.

Early learning may influence later learning without fundamental changes in brain structure, however. We found that when bees were trained extensively on only a single artificial flower type, they had more difficulty in learning to switch between flower types than did bees that learned to switch without the prior phase of visiting only one flower type (Chittka & Thomson 1997). The effect extended to only a few hundred visits (or a few hours), so it may be marginal during a bee's several-weeks-long foraging career. On the other hand, the training phase also involved only a few hundred flower visits, and therefore was much shorter than what bees may really experience in nature. Some bees may spend the first several days of their life foraging in low-diversity situations – such as flowering trees – visiting tens of thousands of flowers of exactly the same type in rapid succession. Might such bees later have more difficulties in learning new flower types, or in learning to minimize interference when switching between flower types? Or do bees maintain complete flexibility, even if their foraging history includes phases where no flexibility is required?

The skill with which bees solve a particular foraging task depends substantially on their earlier experience with related tasks (Zhang & Srinivasan 1994). If bees are exposed to several flower types, some of which are rewarding and others not, bees are able to extract categories and concepts to predict the profitability of novel flowers (Dukas & Waser 1994, Giurfa *et al.* 1996). Whether or not bees acquire such complex skills depends substantially on the sequence with which different flower types are encountered (Zhang & Srinivasan 1994, Chittka & Thomson 1997). An entirely unresolved question is whether, in nature, this sequence is predominantly determined by the spatial arrangement of different flower types, or whether young bees actually choose to forage in diverse floral patches in order to gain the experience necessary for complex cognitive abilities.

### Effects of genotype

Menzel (1985) claimed that the information available to a foraging bee comes from two sources: its own individual experience, and the "species experience" which is derived from evolutionary history and which is written into the species' genome. The implication here is that the "species memory" is identical in all members of the species. This is strictly true only when there is no genetic variability for the trait in question, either because of constraint, or because selection or drift have eliminated variance in the past. However, recent studies have shown that there is heritable variation in learning speed (Brandes 1988), and several other foraging related traits (see references in Waddington, this volume) which means that the *limitations* of the plasticity discussed elsewhere in this chapter are variable and subject to selection. Thus, just as much as foraging is shaped by individual experience, it is also determined by individual genetic histories.

We hope for more studies of heritable variation of sensory and behavioral traits related to foraging. To confirm a hypothesis that a trait is adaptive, we should ideally show that animals *with* that trait have greater fitness than animals *without* that trait, or with a different quantitative expression of the trait. Is traplining adaptive, for example? Is flower constancy a strategy (Menzel, this volume) or a suboptimal solution (Gegear & Laverty, this volume)? Do bumble bees with red preference perform better on some islands than on the mainland, whereas bumble bees without such preference outcompete those with red preference in European mainland habitats (Chittka *et al.*, this volume)? We need to exploit heritable variation to understand whether the cognitive, behavioral, and sensory attributes of pollinators are truly sitting on narrow adaptive peaks, as many workers assume.

### Parasite-induced changes in forager behavior

Certain parasites may force changes in foraging behavior. Late-instar larvae of conopid flies, which occupy much of the host's abdomen, prevent filling of the honey crop (Schmid-Hempel & Schmid-Hempel 1991); these bees concentrate on pollen foraging. On the other hand, parasitism by the protozoan *Crithidia bombi* is associated with reduced pollen foraging (Shykoff & Schmid-Hempel 1991). These parasites can be common. Shykoff & Schmid-Hempel (1991) found 20.2% and 35.7% infection rates by conopids and *C. bombi*, respectively, in bumble bees in the Swiss Alps. Schmid-Hempel & Stauffer (1998) also found that parasites affected floral preferences and switching behavior, but since both parasite load and experience may correlate with age, these changes might also have been driven directly by experience.

### Effects of age

Cartar (1992*a*) and Dukas & Visscher (1994) found that efficiency increases over roughly the first week of a bee's foraging career (an effect which can likely be attributed to learning what, where, and how to forage, and more complex foraging rules). Dukas & Visscher also observed that foraging efficiency declines later in life, but this effect need not necessarily be related to an age-related decline in cognitive ability. In fact, controlled studies show no effects of age on learning ability in honeybees (Bhagavan *et al.* 1994) or bumble bees (Chittka & Reinhold 1999). Several factors affecting foraging efficiency are potentially correlated with age of foragers, for example parasite load (Schmid-Hempel & Stauffer 1998) or wing wear (Cartar 1992b). Having seen several marked bumble bees die during foraging bouts, one of us (JT) can state with certainty that they slow down greatly as their time runs out. In honeybees, a decline in foraging efficiency with age might also be explained by assuming that seasoned foragers invest more time into scouting for new food sources than into harvesting.

### Problems in neglecting interindividual variance: foraging strategies

One can fall into various misinterpretations by aggregating heterogeneous sets of individuals and therefore obtaining spurious correlations. Here we are dealing with specific, pollination-related manifestations of a general statistical problem. If bees do vary substantially in performance, but are treated statistically as equivalent replicates, the interbee variation can pop out in a variety of spurious relationships. For example, Pyke (1978) hypothesized that optimally foraging bees ought to show area-restricted search, i.e., they should fly shorter distances between plants after they have just received larger than average rewards. Because it is hard to know how much nectar a bee has received, Pyke and many others have substituted the time spent at a flower as a surrogate variable for the amount of reward received. This is reasonable, as it takes more time to extract more nectar (Harder 1986; Kato 1988). Making this substitution, one then tests for area-restricted search by testing for positive correlation between the time spent at one flower and the distance flown to the next. Pyke found this pattern. Although this procedure would be trustworthy for observations of a single bee, suppose that some bees in a population – say, those with tattered wings – work all flowers more slowly, and always tend to fly shorter distances. If one then combines data from fast and slow bees, one could obtain the expected positive correlation, even if no individual bee shows area-restricted behavior (Thomson *et al.* 1982).

Analogous difficulties attend field studies of flower constancy. Here, an attractive hypothesis is that a flower visitor should be more willing to switch to another species of flower after having received little reward. This flexibility would allow individuals to track the relative values of different resources and concentrate on the best ones. If flower-handling time is used as a surrogate for reward, if interbee variation in constancy is correlated with variation in working speed, and if data are pooled across bees, however, spurious correlation can cause the hypothesis to be accepted when it should be rejected, or vice versa.

Problems of this sort arose in a study of constancy in many unmarked bumble bees that were followed for as long as possible as they foraged freely in a meadow with several suitable flower species (Chittka *et al.* 1997; L. Chittka, unpublished data). The authors initially classified flower-handling times into two categories, either above or below the grand median for all bees. In this data set, bees were significantly more likely to switch plant species if their last (several) visits had been shorter than the median, and more likely to stay constant if their last visits had been longer than the median. This seems consistent with the hypothesis that bees switch when they are dissatisfied, but when each bee's visit times were re-scaled by the median for *that* bee's bout (rather than the grand median), the effect disappeared. Further exploration of the data suggested that the heterogeneity causing the spurious correlation arose not so much from interbee variation as from temporal variation. In the morning, all bees handled flowers slowly, presumably because nectar levels were high, and all bees tended to be constant. In the afternoon, visits were shorter and constancy dropped overall, so the relationship between visit length and subsequent constancy could not be clearly attributed to short-term behavioral flexibility. In fact, Chittka *et al.* (1997) resurrected the flexibility hypothesis; they reanalyzed the data within bouts, considering not just the upper and lower halves of the visit times but the upper and lower quartiles. Then, bees *were* more likely to switch following very short visits, and more likely to be constant following very long visits.

This example illustrates not just the danger of spurious correlation but also a reasonable way of handling existing data to avoid problems. Although marking animals is not always feasible, more trustworthy results will be obtained by restricting analyses to comparisons within bees, as well as considering other cryptic sources of heterogeneity (such as time of day). One investigation of flower constancy that apparently did not include such precautions is a study of skippers by Goulson *et al.* (1997). They used exactly the procedure initially tried by Chittka *et al.* (1997), except that they used means rather than medians for dividing the data, and they reached the same initial conclusion. It might be worthwhile to analyze their data further, along the lines of Chittka *et al.* (1997), assuming that the bout lengths are long enough.

### Modes of foraging

As Galen & Plowright (1985) showed, bumble bees that forage for nectar on *Epilobium angustifolium* visit the vertical inflorescences differently from those that seek pollen from the same plant. These authors interpreted their results in terms of reward maximization criteria, as if members of a group of equivalent bees first made a decision to specialize on pollen or nectar, then adjusted their movements accordingly. One would also like to know, however, whether parasitic infections also played a role in the food-type decisions; if so, then the population might be more profitably viewed as comprising heterogeneous groups of infected and uninfected individuals with different behaviors.

From the plant's point of view, it is clear that the adoption of pollen- or nectar-collecting behavior by a visitor can greatly change the fitness value of that visitor to the host-plant (Galen & Plowright 1985; Shykoff & Schmid-Hempel 1991; Wilson & Thomson 1991).

### Familiarity with individual plant characteristics

Ignoring pollinator individuality can lead not only to spurious correlations, it can hamper insight regarding the adaptive problems that animals or plants are "trying" to solve. Knowledge of individuals can lead one to pose questions that would otherwise go unasked. For example, researchers concerned with pollinators' responses to variation in plant phenotypes tend to assume that the plant's visitors are influenced only by the characteristics, such as inflorescence size, that the plant presents at the moment. However, the behavior of bees that return frequently to particular plants might also be sensitive to qualities that the plant displayed previously but no longer does. For example, *Aralia hispida* plants change sex from male to female phases several times during a flowering season. When floral rewards were manipulated in male-phase inflorescences (Thomson 1988), bumble bee visitation increased to the richer inflorescences. When all of the variable male-phase inflorescences were replaced with uniform female ones, simulating the natural sex change, the bees preferentially visited female inflorescences that were located where the richer males had been. This result highlighted an ambiguity in interpreting selection on floral displays in terms of sex allocation theory: nectar secreted by a flower in male phase can increase the visitation rate to that flower in female phase. Should the cost of producing that nectar be considered a male or a female cost?

Even without special subtleties due to sex roles, early flowers can influence visitation rates to later flowers if pollinators show "trapline holdover," as bumble bees sometimes do (Thomson 1988, 1996). This effect could provide adaptive explanations for some aspects of floral biology, such as the tendency of many plants to burst into bloom with many flowers, then to taper off flower production. Here, the early flowers may benefit the plant not only through their own gametes but also by recruiting a faithful set of individual pollinators that will continue to serve the plant through its blooming period (cf. Thomson 1988). Without knowing the site fidelity of individual pollinators, one cannot fully interpret how pollinator-based selection might act on inflorescence architecture.

### Scent-marking at flowers

We have long had indications that bumble bees scent-mark flowers and respond to those marks (e.g., Cameron 1981; Kato 1988; Schmitt & Bertsch 1990), but this evidence has not yet been well incorporated into the thinking of many who study foraging primarily from an energetic point of view. The energetic viewpoint has interpreted bees' decisions at flowers as being driven mostly by direct assessment of rewards gained at a blossom, rather than indirect olfactory assessment of recent visitation. This is partly because the evidence for scent marks has been mostly indirect and partly because the interpretation has been somewhat confusing. Schmitt & Bertsch (1990) review the evidence up to that date for bumble bees and honeybees; they indicate that some chemicals deposited on flowers may serve as attractants that denote rewarding flowers, while others, probably more volatile and short-lived, may serve as repellents that signal bees not to revisit flowers that have recently been drained. Schmitt & Bertsch interpret their results as strong evidence for an attractant role. Conversely, Giurfa & Núñez (1992) found evidence that marks recently left by honeybee foragers act as repellents. More recently, Goulson *et al.* (1998) have reported field evidence from bumble bees for a repellent role, a finding reinforced by experimental application of extracts from bee tarsal glands to flowers (Stout *et al.* 1998). To date, however, it is not clear if bees use more than a single scent to mark flowers, nor whether scent-marking is an active process (Chittka *et al.* 1999). It is equally possible that tarsal secretions are used for adherence of bee feet to flowers, and are used as scent marks only as an epiphenomenon: bees might use the scent marks as repellents if the flowers are known to refill slowly, and as attractant if they remember the flowers as having high refill rates.

Our goal in considering scent marks in this chapter is not to resolve controversies but to show how an individualistic perspective can help clarify how these marks should be interpreted. If one adopts an adaptationist viewpoint of bees as optimal foragers that search widely for food, scent-marking is hard to understand. Of course, it is easy to see that a short-lived repellent mark might be useful in helping an individual avoid revisiting flowers that it has just probed, but it is harder to see how it could be adaptive to leave long-lived attractive marks on rewarding flowers. It would seem to require some special conditions. First, there must be an expectation that the bee who does the marking will return in time to benefit from the mark. This condition is easily met if bees use small foraging areas. Second, and more onerous, the mark must be expected to be of *more* benefit to the bee who left it than to other bees that may also detect it. It will do an individual little good to flag a rich resource if the primary result is to help competing bees exploit that resource. This paradox could be explained by kin selection if most of the visitors to a plant were sisters. In honeybees, which might combine scent-marking of flowers with site-specific dance information in the hive (von Frisch 1967), this may sometimes be the case. In bumble bees, however, these conditions probably do not apply: they lack a site-specific recruitment system (Dornhaus & Chittka 1999); workers range too far, workers per colony are too few, and colony densities are too high (Cumber 1953; Harder 1986), for sibling encounters to be frequent.

On the other hand, if a traplining individual is making a substantial fraction of the visits to a plant (Fig. 10.1), that bee may reap enough benefits from attractive scent marks to offset the possible advantage given to competitors. Some analogous mechanism might help explain a puzzling observation by Williams & Thomson (1998) in the 1994 data mentioned above. Modeling nectar production and removal with some simple assumptions, they estimated that the bees that visited the focal plant most often – i.e., the regular trapliners – gained more reward per plant visit than did the casual visitors that arrived less often. Interestingly, the trapliners achieved their edge not by arriving at times when the plant had more reward overall, but rather by being better at selecting the flowers that had not been visited recently by others. Positing scent cues does not in itself dispel the puzzle, for the casual visitors presumably have as much access to scent cues as the trapliners do. Conceivably, the trapliners simply pay more attention to these cues for some reason; an interesting alternative is that bees can leave some private cues that are not accessible to others. Individual-specific trail marks are known in some species of ants (Maschwitz *et al.* 1986). In laboratory tests, scent marks left by bees on artificial flowers have also been shown to be more efficient in repelling the individual which left them than other bees (Giurfa 1993), but whether this effect holds up in the field remains to be shown. If it does, trapliners that return at regular intervals might be able to make the best use of marks left by themselves and those left by other bees.

### Problems in neglecting interindividual variance: pollinator sensory physiology

Many physiologists have treated all variance between individuals as noise, and eliminated it by averaging the responses from several animals. A typical example is a study by Peitsch *et al.* (1992), who measured the color-receptor wavelength positions of several species of Hymenoptera. They found that differences between species, although slight, exceeded differences between individuals of the same species, and concluded that variance between data from animals of one species was entirely caused by measurement error. This may be correct; however it would also be worthwhile to take the possibility seriously that there might be *real* (i.e., heritable) variation between individuals. While such variation may be a nuisance for the physiologist trying to extract smooth functions from noisy data, it is a resource for the evolutionary biologist interested in predicting how animals will respond to directional selection.

A more serious (and common) error in physiological work is caused by equating intraindividual variance with interindividual variance. Many authors regarded it is legitimate to take repeated measurements from the same individual animal, and treat these as if they had taken independent measurements from different animals. In fact, numbers of individuals tested are not even available in some behavioral or physiological studies of honeybees; instead, only total numbers of choices or measurements are given. The result is that the numbers of observations, in such studies, is often drastically inflated. It is trivial to most biologists that one cannot obtain a sample size of 150 leaf diameters by measuring 3 leaves 50 times over. Yet, this is precisely what some physiologists do in their data analyses. This is especially dangerous when comparisons between groups of animals are performed. For example, Vorobyev *et al.* (1999) tested honeybees' ability to detect artificial "flowers" of different colors on a green background. They used the total $n$ of choices (270) as a basis for the conclusion that white was more easily detectable than gray, but the number of individuals tested (which should have been used for statistics) for white flowers was only three! Clearly, within-individual behavior is noisy, and therefore one needs several data points from each animal. However, behavioral variance across individuals can be large, particularly in honeybees whose experience before and between experiments is outside the control of the experimenter. Thus, once each animal's behavior is quantified (if necessary, with several tests), only a single data point per animal may be used for comparison between groups of individuals (see, e.g., Chittka & Thomson 1997; Chittka *et al.* 1997).

Because interindividual differences were regarded as noise, many authors pooled data from individuals without testing for heterogeneity. This can be hazardous. For example, Scherer & Kolb (1987) tested innate floral color preferences of *Pieris* butterflies. They found that colors both in the blue and the red part of the spectrum were preferred, and conjectured that a neuronal mechanism summing up the responses from blue and red receptors might drive this behavior. This mechanism is simple and therefore attractive, but there are alternative explanations. In essence, Scherer & Kolb (1987) used average *group* behavior to deduce the neuronal mechanisms implemented in *individuals*. When only pooled data are presented, it is equally possible that group behavior is caused by some individuals preferring red, others blue. If this were the case, no single butterfly would use summed inputs from two receptors to search for flowers. Furthermore, pooling the data masks possible sequence effects. Say, for example, that butterflies tend to visit blue first (bypassing red flowers), then red (bypassing the blue flowers it has already found unrewarding); such a pattern would also suggest a different model of neuronal control.

### Recommendations

Pollination biologists ought to consider the ways in which pollinator individuality may affect their interpretations of both pollinator behavior and pollinator-driven selection on plants. In addition to documenting variation among pollinators in characteristics such as constancy and the mode and tempo of working flowers, future studies should concentrate first on how these traits covary across individuals, and second on how the variation and covariation change as individuals gain experience.

For any interpretation of how pollinators respond to variation among plants, and thereby exert selection pressure, it is important to know whether individual pollinators show site fidelity or traplining behavior. This is particularly important with respect to scent-marking behavior. In particular, we need to go beyond existing studies, which show only that *some* individuals trapline in some situations. We need to know whether this behavior is typical, and we need to know what circumstances promote it.

In our opinion, an interesting unanswered question in pollinator foraging ecology is, "How do individual animals choose foraging areas?" For example, do they preferentially forage in areas with rich spatial detail, so as to facilitate memorization of particularly rewarding patches (Cohen & Keasar 2000)? Does spatial memory develop "passively" as animals move between flowers based on simple foraging rules, or do animals first establish a "cognitive map" of their home range within which they then place the coordinates of profitable foraging sites (Menzel, this volume)? Heinrich (1979, p. 114) states that "Young bees wander about a great deal before settling down," but we know little about how their experiences while wandering affect their ultimate decisions to settle. We also do not know if the wandering phase simply serves searching for the most rewarding flowers, or whether pollinators "deliberately" visit many different flower types to extract more complex foraging rules, including, for example, categorization of food types, or optimal decision rules for when to leave floral patches. Perhaps bees whose early experiences have favored flower-constant behavior will preferentially choose foraging areas with monospecific stands of flowers that make it easy to be constant. Relationships of this sort would necessarily color our mechanistic interpretations of behavioral patterns, yet we tend to ignore them. Focusing on flower visitors as individuals – with individual histories of learning about the world – can be a useful corrective.

## References

Alloway TM (1972) Retention of learning through metamorphosis in the grain beetle (*Tenebrio molitor*). Am Zool 12:471–477

Bhagavan S, Benatar S, Cobey S & Smith BS (1994) Effect of genotype but not of age or caste on olfactory learning performance in the honeybee, *Apis mellifera*. Anim Behav 48:1357–1369

Bonabeau E, Theraulaz G, Deneubourg JL, Serge A & Camazine S (1997) Self-organisation in social insects. *Trends Ecol Evol* 12:188–193

Brandes C (1988) Estimation of heritability of learning behaviour in honeybees (*Apis mellifera capensis*). *Behav Genet* 18:119–132

Brian AD (1952) Division of labour and foraging in *Bombus agrorum* Fabricius. *J Anim Ecol* 21:223–240

Cameron SA (1981) Chemical signals in bumblebee foraging. *Behav Ecol Sociobiol* 9:257–260

Cartar RV (1992*a*) Adjustment of foraging effort and task switching in energy-manipulated wild bumblebee colonies. *Anim Behav* 44:75–87

Cartar RV (1992*b*) Morphological senescence and longevity: an experiment relating wing wear and life span in foraging wild bumble bees. *J Anim Ecol* 61:225–231

Chittka L & Reinhold H (1999) Towards an individual-based approach to insect learning. In: Elsner N & Eysel U (eds) *Göttingen Neurobiology Report*, p 257. Thieme Verlag, Berlin

Chittka L & Thomson JD (1997) Sensori-motor learning and its relevance for task specialization in bumble bees. *Behav Ecol Sociobiol* 41:385–398

Chittka L, Kunze J & Geiger K (1995) The influences of landmarks on distance estimation of honeybees. *Anim Behav* 50:23–31

Chittka L, Gumbert A & Kunze J (1997) Foraging dynamics of bumble bees: correlates of movement within and between plant species. *Behav Ecol* 8:239–249

Chittka L, Williams NM, Rasmussen H & Thomson JD (1999) Navigation without vision: bumblebee orientation in complete darkness. *Proc R Soc Lond B* 266:45–50

Cohen D & Keaser T (2000) Prediction of future intake rates: anticipation and learning by foraging bees. *International Behavioral Ecology Congress, Zürich, Abstract Volume*, p 35

Cumber RA (1953) Some aspects of the biology and ecology of humble-bees bearing upon the yields of red-clover seed in New Zealand. *NZ J Sci Technol* 34:227–240

Dobson HE (1994) Floral volatiles in insect biology. In: Bernays EA (ed) *Insect–plant interactions*, vol. 5, pp 48–81. CRC Press, Boca Raton FL

Dornhaus A & Chittka L (1999) Evolutionary origins of bee dances. *Nature* 401:38

Dukas R & Visscher PK (1994) Lifetime learning by foraging honeybees. *Anim Behav* 48:1007–1012

Dukas R & Waser NM (1994) Categorization of food types enhances foraging performance of bumblebees. *Anim Behav* 48:1001–1006

Durst C, Eichmüller S & Menzel R (1994) Development and experience lead to increased volume of subcompartments of the honeybee mushroom body. *Behav Neural Biol* 62:259–263

Elbert T, Pantev C, Wienbruch C, Rockstroh B & Taub E (1995) Increased cortical representation of the fingers of the left hand in string players. *Science* 270:305–307

Fabre J-H (1879) *Souvenirs entomologiques*. Delagrave, Paris

Fahrbach SE, Moore D, Capaldi EA, Farris SM & Robinson GE (1998) Experience-expectant plasticity in the mushroom bodies of the honeybee. *Learn & Mem* 5:115–123

Fewell JH & Bertram SM (1999) Division of labor in a dynamic environment: response by honeybees (*Apis mellifera*) to graded changes in colony pollen stores. *Behav Ecol Sociobiol* 46:171–179

Free JB (1993) *Insect pollination of crops*. Academic Press, San Diego, CA

Freeman RB (1968) Charles Darwin on the routes of male humble bees. *Bull Br Mus Nat Hist Ser* 3:177–189

Frisch K von (1967) *The dance language and orientation of bees.* Belknap Press, Cambridge, MA

Galen C & Plowright RC (1985) Contrasting movement patterns of nectar-collecting and pollen-collecting bumble bees (*Bombus terricola*) on fireweed (*Chamaenerion angustifolium*) inflorescences. *Ecol Entomol* 10:9–17

Giurfa M (1993) The repellent scent-mark of the honeybee *Apis mellifera ligustica* and its role as communication cue during foraging. *Insectes Soc* 40:59–67

Giurfa M & Núñez J (1992) Honeybees mark with scent and reject recently visited flowers. *Oecologia* 89:113–117

Giurfa M, Eichmann B & Menzel R (1996) Symmetry perception in an insect. *Nature* 382:458–461

Gordon DM (1996) The organization of work in social insect colonies. *Nature* 380:121–124

Goulson D, Ollerton J & Sluman C (1997) Foraging strategies in the small skipper butterfly, *Thymelicus flavus*: when to switch? *Anim Behav* 53:1009–1016

Goulson D, Hawson SA & Stout JC (1998) Foraging bumblebees avoid flowers already visited by conspecifics or by other bumblebee species. *Anim Behav* 55:199–206

Harder LD (1986) Influences on the density and dispersion of bumble bee nests (Hymenoptera: Apidae). *Holarctic Ecol* 9:99–103

Heinrich B (1976) The foraging specializations of individual bumblebees. *Ecol Monogr* 2:105–128

Heinrich B (1979) *Bumblebee economics.* Harvard University Press, Cambridge, MA

Kato M (1988) Bumblebee visits to *Impatiens* spp.: pattern and efficiency. *Oecologia* 76:364–370

Kunze J & Chittka L (1996) Bees and butterflies fly faster when plants feed them more nectar. In: Elsner N & Schnitzler H (eds) *Göttingen Neurobiology Report 1996*, p 109. Thieme Verlag, Stuttgart

Laverty TM (1994) Bumble bee learning and flower morphology. *Anim Behav* 47:531–545

Lindauer M (1985) The dance language of honeybees: the history of a discovery. In: Hölldobler B & Lindauer M (eds) *Experimental behavioral ecology*, pp 129–140. Fischer Verlag, Stuttgart

Manning A (1956) Some aspects of the foraging behaviour of bumble-bees. *Behaviour* 9:164–201

Maschwitz U, Lenz S & Buschinger A (1986) Individual specific trails in the ant *Leptothorax affinis* (Formicidae: Myrmicinae). *Experientia* 42:1173–1174

Menzel R (1985) Learning in honeybees in an ecological and behavioral context. In: Hölldobler B & Lindauer M (eds) *Experimental behavioral ecology*, pp 55–74. Fischer Verlag, Stuttgart

Müller H (1882) Versuche über die Farbenliebhaberei der Honigbiene. *Kosmos Jahrg.* 6

Núñez JA (1970) The relationship between sugar flow and foraging and recruiting behaviour of honeybees (*Apis mellifera* L.). *Anim Behav* 18:527–538

Olesen JM (1996) From naiveté to experience: bumblebee queens (*Bombus terrestris*) foraging on *Corydalis cava* (Fumariaceae). *J Kans Entomol Soc* Suppl 69:274–286

Pankiw T & Page RE (2000) Response thresholds to sucrose predict foraging division of labor in honeybees. *Behav Ecol Sociobiol* 47:265–267

Peitsch D, Fietz A, Hertel H, de Souza J, Ventura DF & Menzel R (1992) The spectral input systems of hymenopteran insects and their receptor-based colour vision. *J Comp Physiol A* 170:23–40

Plowright RC, Thomson JD, Lefkovitch LP & Plowright CMS (1993) An experimental study of the effect of colony resource level manipulation on foraging for pollen by worker bumble bees (Hymenoptera: Apidae). *Can J Zool* 71:1393–1396

Pyke GH (1978) Optimal foraging: movement patterns of bumblebees between inflorescences. *Theor Popul Biol* 13:72–98

Ribbands CR (1949) The foraging method of individual honeybees. *J Anim Ecol* 18:47–66

Robinson GE & Page RE (1989) Genetic determination of nectar foraging, pollen foraging, and nest-site scouting in honeybee colonies. *Behav Ecol Sociobiol* 24:317–323

Scherer C & Kolb G (1987) Behavioral experiments on the visual processing of color stimuli in *Pieris brassicae* L. (Lepidoptera). *J Comp Physiol* 160:645–656

Schmid-Hempel R & Schmid-Hempel P (1991) Endoparasitic flies, pollen-collection by bumblebees and a potential host–parasite conflict. *Oecologia* 87:227–232

Schmid-Hempel P & Stauffer H-P (1998) Parasites and flower choice of bumblebees. *Anim Behav* 55:819–825

Schmitt U & Bertsch A (1990) Do foraging bumblebees scent-mark food sources and does it matter? *Oecologia* 82:137–144

Shykoff JA & Schmid-Hempel P (1991) Incidence and effects of four parasites in natural populations of bumble bees in Switzerland. *Apidologie* 22:117–125

Spaethe J, Tautz J & Chittka, L (2000) Foraging economics in bumble bees: do larger bees do a better job? In *International Behavioral Ecology Congress, Zürich, Abstract Volume*, p 185

Stout JC, Goulson D & Allen JA (1998) Repellent scent-marking of flowers by a guild of foraging bumblebees (*Bombus* spp.). *Behav Ecol Sociobiol* 43:317–326

Thomson JD (1988) Effects of variation in inflorescence size and floral rewards on the visitation rates of traplining pollinators of *Aralia hispida*. *Evol Ecol* 2:65–76

Thomson JD (1996) Trapline foraging by bumblebees. I. Persistence of flight-path geometry. *Behav Ecol* 7:158–164

Thomson JD, Maddison WP & Plowright RC (1982) Behavior of bumble bee pollinators of *Aralia hispida* Vent. (Araliaceae). *Oecologia* 54:326–336

Thomson JD, Peterson S & Harder LD (1987) Response of traplining bumble bees to competition experiments: shifts in feeding location and efficiency. *Oecologia* 71:295–300

Thomson JD, Slatkin M & Thomson BA (1997) Trapline foraging by bumble bees. II. Definition and detection from sequence data. *Behav Ecol* 8:199–210

Tully T, Cambiazo V & Kruse L (1994) Memory through metamorphosis in normal and mutant *Drosophila*. *J Neurosci* 14:68–74

Vorobyev M, Hempel de Ibarra N, Brandt R & Giurfa M (1999) Do "white" and "green" look the same to a bee? *Naturwiss* 86:592–594

Waser, NM (1986) Flower constancy: definition, cause and measurement. *Am Nat* 127:593–603

Williams NM & Thomson JD (1998) Trapline foraging by bumble bees. III. Temporal patterns of visitation and foraging success at single plants. *Behav Ecol* 9:612–621

Wilson P & Thomson JD (1991) Heterogeneity among floral visitors leads to discordance between removal and deposition of pollen. *Ecology* 72:1503–1507

Zhang SW & Srinivasan MV (1994) Prior experience enhances pattern discrimination in insect vision. *Nature* 368:330–332

11

# Effects of predation risk on pollinators and plants

Almost all pollination studies neglect the possible effects of predation on flower visitors. Various authors have even claimed that predation is too infrequent to influence pollinator behavior. It is tempting to dismiss the role of predation because it is rarely observed. In the past two decades, however, ecologists have learned to appreciate the central role that predation risk plays in animal behavior and ecology, mostly through a variety of measures animals take to minimize predation. Studies on a wide variety of animals from zooplankton to mammals have suggested that predation risk affects: diurnal patterns of activity; choice of diet, habitat, food patches, and food type; ways of handling food items; social organization; choice of nest sites; and various physiological factors such as diurnal and seasonal levels of fat reserves and respiration patterns (Price *et al.* 1980; Lawton 1986; Bernays & Graham 1988; Lima & Dill 1990; Clark 1993; Martin 1995; Lima 1998*a, b*; Ydenberg 1998).

Are flower-visiting animals really immune to predation, or does the prevailing view about the unimportance of predation in pollination systems merely reflect researchers' inattention? In this chapter, I shall review some of the literature and argue that pollination ecologists have mostly overlooked a central factor influencing pollinator traits and pollination systems. Specifically, I ask: (1) Are there significant levels of predation on pollinators? (2) How might predation affect pollinator traits? And, (3) how might predation influence pollinator–plant interactions?

## Are there significant levels of predation on pollinators?

I watched these wasps at work all through that afternoon, and soon became absorbed in finding out exactly what was happening in this

> busy insect town. ... Now and then a wasp would fly out and, after half
> an hour or longer, return with a load, which was then dragged in.
> Every time I examined the prey, it was a Honeybee. No doubt they
> captured all these bees on the heath ... A rough calculation showed that
> ... on a sunny day like this several thousands bees fell victims to this
> large colony of killers. Tinbergen (1958, p. 21)

Niko Tinbergen's account of bee predation by the infamous bee-wolf, *Philanthus triangulum* Fabr., is just one of numerous anecdotal reports of apparently significant levels of predation on flower-visiting animals. I shall begin this section with a brief listing of the major predators of pollinators, proceed to review published data about predation on pollinators, and then evaluate the importance of predation, emphasizing its indirect effects on pollinators.

### Major predators of flower-visiting animals

I shall focus here on predators that attack pollinators outside the nest, as these are most relevant to pollination ecology. I shall also include within my broadly defined "predator" category parasitoids that attack pollinators outside the nest. Parasitoids eventually kill their hosts, though there is some time delay between successful egg-laying and host death. Predation and parasitism inside nests, especially those of social insects, have been recently addressed elsewhere (Morse & Nowogrodzki 1990; Godfray 1994; Schmid-Hempel 1998). Outside the nest, predation on pollinators may occur at flower patches, where various ambush predators wait for flower-visiting insects, on the way between flower patches and the nest, or at the nest entrance. While some predators are confined mostly to one of the above locations, others are rather opportunistic.

The most frequent ambush predators that sit on or near flowers are crab spiders (Thomisidae) (Morse 1981, 1986), predaceous bugs (Hemiptera) (Balduf 1939; Greco & Kevan 1995), and praying mantids (Mantidae) (Canon 1990). Bee-wolves (*Philanthus* spp.) (Evans & O'Neill 1988) and other wasps (Evans & Eberhard 1970; De Jong 1990) commonly hunt for insects that are perching on flowers or flying near the colony. Orb-weaving spiders catch insects moving among plants (Robinson & Robinson 1970; Caron & Ross 1990). Pollinators are also attacked by robber flies (Lavigne 1992; Rabinovich & Corley 1997) and dragonflies (Fry 1983).

Many bird species are opportunistic predators of various pollinators. Among the more significant predators of bees are bee-eaters (Meropidae),

Old and New World flycatchers (Muscicapidae and Tyrannidae), swifts (Apodidae), swallows (Hirundinidae), shrikes (Laniidae), jacamars (Galbulidae), and drongos (Dicruridae) (Davies 1976; Fry 1983; Ambrose 1990). Nocturnal pollinators such as moths and hawkmoths are attacked by bats (Svensson *et al.*, 1999). Various raptors are probably the most significant hummingbird predators, although information about hummingbird predation is scant (Miller & Gass 1985). Among parasitoids, conopid flies (Conopidae) are a major source of bumble bee mortality (Schmid-Hempel *et al.* 1990), and phorid flies (Phoridae) are known to attack honeybees in the New World tropics (Knutson & Murphy 1990).

### Attack and predation rates on pollinators

Throughout this chapter, "attack" will mean an approach by a predator attempting to capture an individual. A successful attack results in predation, while an unsuccessful attack means that the potential prey escapes either harmlessly, or with some injury that may diminish fitness. Although there are numerous studies of predation on pollinators, only a few allow quantitative estimates of predation rates.

Morse (1979, 1981, 1986) presented detailed predation rates of crab spiders (*Misumena vatia*) on insect visitors to the common milkweed (*Asclepias syriaca*). In that study, a bumble bee (*Bombus vagans*) had a 14% chance of being attacked each day by a spider while visiting flowers. The daily attack probability for a honeybee (*Apis mellifera*) was 8%. The probability of predation per day was much smaller, 1.4% and 3% for a bumble bee and honeybee, respectively (Fig. 11.1).

Morse's estimates are much smaller than mortality rates of forager bumble bees and honeybees estimated from observations of foragers at observation hives. In such studies, the daily rates of non-returning workers range between 3% to 7% for bumble bees (Rodd *et al.* 1980; Goldblatt & Fell 1987) and 8% to 10% for honeybees (Wolf & Schmid-Hempel 1989; Dukas & Visscher 1994) (Fig. 11.1). On the one hand, Morse's data underestimate overall predation because they include only one predator. On the other hand, observations at the nest overestimate predation because worker disappearance may be caused by other factors, such as disorientation (Rodd *et al.* 1980).

Simonthomas & Simonthomas (1980) reported on the collapse of a local honeybee business at the Dakhla oasis in Egypt that was probably caused by high levels of predation by the bee-wolf, *Philanthus triangulum*. They cite similar cases in Western Europe. *Philanthus triangulum* is considered a

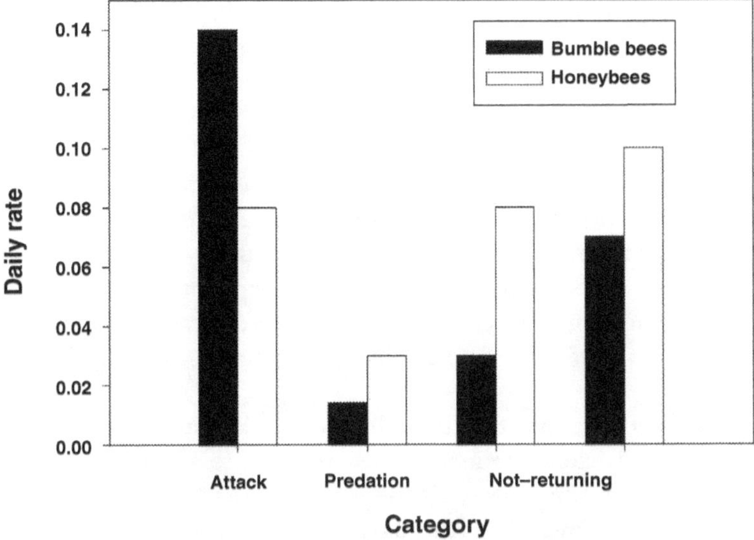



Fig. 11.1. The daily probabilities of attack and predation by crab spiders on a worker bumble bee and a worker honeybee (data from Morse 1986), and two values each for bumble bees and honeybees of overall daily disappearance rates of workers observed at the hive. (Data from Rodd *et al.* 1980; Goldblatt & Fell 1987; Wolf & Schmid-Hempel 1989; Dukas & Visscher 1994.)

honeybee specialist, although it occasionally takes other bees of similar size (Evans & O'Neill 1988).

The North American "bumblebee-wolf," *P. bicinctus,* mostly specializes on bumble bees. An aggregation of approximately 200 bumblebee-wolves observed by Gwynne (1981) was estimated to prey on a total of over 7500 bumble bees (Evans & O'Neill 1988). Armitage (1965) recorded 154 attacks by bumblebee-wolves on worker bumble bees foraging on clover and goldenrod flowers, 23% of which resulted in predation.

Hymenoptera predominate in the diet of most bee-eaters (Meropidae); honeybees are the most frequent species taken, although the European bee-eater (*Merops apiaster*) seems to prefer bumble bees (Fry 1983). Fry (1983) estimated that about 33% of the predation attempts by bee-eaters were successful.

Conopid flies (Conopidae) parasitized 12.7% of bumble bee workers in eastern Canada (MacFarlane & Pengelly 1974) and an average of 13.2% in Switzerland (Schmid-Hempel *et al.* 1990). In the latter study, the frequency of parasitation increased from 0% before June to a peak of 35% in

July. Other studies found up to 20% parasitism by conopids in Switzerland (Shykoff & Schmid-Hempel 1991) and in the northeastern United States (Heinrich *et al.* 1977).

### Indirect effects of predation

Documented low rates of predation may inform us little about the importance of predation. First, even relatively low rates of predator attack are significant because of the severe consequences: injury or death. Second, observed predation rates already represent an equilibrium under which animals typically have taken various measures to decrease probabilities of attack and capture by predators (Lima 1990; Lima & Dill 1990; Ydenberg 1998).

Although experiments with pollinators are lacking, ants have been experimentally shown to modify their foraging to avoid patches with higher densities of predatory ants (Fig. 11.2) (Nonacs & Dill 1990, 1991), or to forage less at times when phorid parasitoids are more active (Folgarait & Gilbert 1999; see also Feener 1988). Hymenopteran flower visitors presumably have similar capabilities.

### How might predation affect pollinator traits?

### Assessment of predation risk and avoidance learning

Gould (1987) trained honeybees to avoid landing at certain locations. He used a radially symmetrical mechanical flower with six petals, but allowed the bees to land on only one "correct" petal, either the lower right or lower left one. Solenoids mounted behind each of the petals caused the petals to flick forward, scaring off any bee that landed on a "wrong" petal. Although naïve bees strongly preferred landing on the bottom petal, trained bees showed significant preference for landing on the correct petal. Gould's study may illustrate a bee's ability to avoid landing at dangerous locations, but the alternative that bees simply learned to land where they had been rewarded cannot be rejected. My own preliminary experiments with a similar mechanical flower strongly suggested that honeybees showed avoidance learning: during calibration of the solenoids' power, I found that a violent flicking of the petals caused the bees to avoid the flower altogether rather than a specific petal as I had wished. Some of these bees returned to the flower vicinity and hovered near the petals but failed to land. Other well-controlled experiments clearly documented avoidance learning by honeybees (Abramson 1986; Smith *et al.*

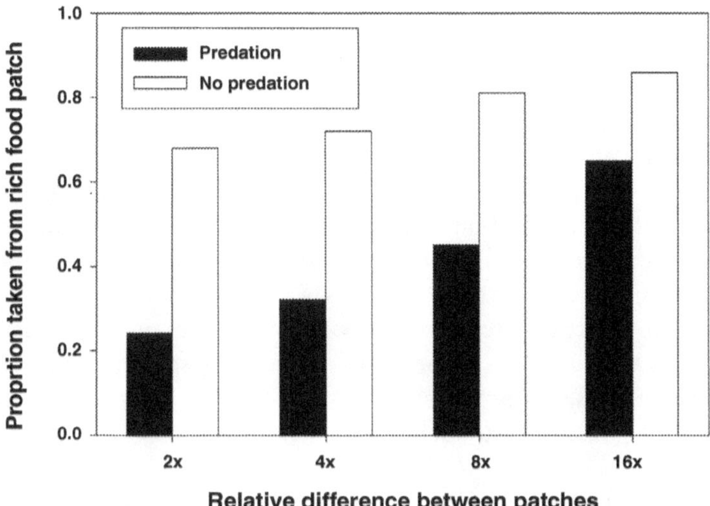


Fig. 11.2. Patch use by 12 *Lasius pallitarsis* ant colonies in response to relative food quality and predation risk. The rich-food patch had food concentration 2–16 times higher than the poor-food patch (the *x* axis). Experimental sessions either included a predator at the rich-food patch or had no predators at either patch. The ants varied the proportion of food taken from the rich-food patch in response to its relative quality and the predation risk. The effects of relative food quality and predation treatments were statistically significant ($p < 0.01$). (Data from Nonacs & Dill 1991.)

1991) as well as other insects (Quinn *et al.* 1974; Dukas 1999; Liu *et al.* 1999). Nevertheless, I highlight Gould's data, because further elaboration of his protocol may illustrate that bees can learn to avoid landing on one plant where they encountered a predation attempt while they keep visiting neighboring safe plants.

Morse (1986) concluded that bees in his study failed to either bypass flowers containing crab spiders or leave flowers with non-attacking spiders faster than spider-free flowers. The low resolution of the insect eye (Land 1981) and limited attention (Dukas 1998*a*; Dukas & Kamil 2000) can probably explain pollinators' failure to notice motionless, cryptic crab spiders or other ambushing predators. Still, the studies on crab spiders, bee-eaters, and bee-wolves reviewed above indicate that the success rates of these bee predators range between 10% to 30%. This means that bees can at least potentially respond to failed predation attempts by altering behavior in an attempt to reduce subsequent attacks.

Recently, Grostal & Dick (1999) reported that herbivorous spider mites

(*Tetranychus urticae*) avoided leaves where predation on conspecifics had occurred, and Dixon & Agarwala (1999) found that pea aphids (*Acyrthosiphon pisum*) responded to odor tracks left by predatory ladybird larvae (*Adalia bipunctata*). Bees and other flower visitors may also perceive predator activity indirectly through sensing predator odor, an attack on another individual, or injured conspecifics. In honeybees, components of the sting and mandibular gland pheromones deter conspecifics (Free 1987; Balderrama *et al.* 1996). Honeybees may employ such pheromone-based information to avoid locations of high predator activity.

The ability to perceive predators, however, does not imply that pollinators must always alter their behavior in response to predator presence. The response to predation risk should reflect long-term costs and benefits, which I discuss below.

### How should pollinators respond to perceived predation?

Anti-predatory adaptations have the potential to increase fitness, either through increased lifetime reproduction in solitary species or increased worker's lifetime contribution to colony growth in social insects. Bees seem to have some obvious anti-predatory traits: most species possess stings and many have aposematic coloration (Schmidt 1990). Sympatric bumble bee species even segregate into groups with similar color patterns, which are better explained by geographical than taxonomic affiliation, suggesting Müllerian mimicry (Plowright & Owen 1980). Similarly, many noxious butterflies have aposematic coloration; these butterflies and their non-poisonous mimics appear in complexes of Müllerian and Batesian mimicry (Gilbert 1983). In addition to morphological adaptations, bees and other pollinators probably possess less apparent behavioral traits that help them decrease predation risk.

Many times, alternative feeding options with different rates of food intake also have characteristic mortality rates. Such foraging–mortality tradeoffs can be evaluated with models that consider the lifetime outcomes of the available alternatives. The foraging–mortality tradeoff faced by a social bee can be approximated by the ratio $g/\mu$, where $g$ is the rate of food collection and $\mu$ is the mortality rate during foraging (Clark & Dukas 1994; see also Gilliam & Fraser 1987; Dukas & Edelstein-Keshet 1998). Maximization of this ratio would result in maximizing the lifetime contribution of the bee to her colony growth. Using this approximation, we can conclude that a social bee collecting food in flower patch 1 at the rate of 1 unit per trip and facing an expected predation rate of 0.01 per trip should prefer the less rewarding flower patch 2 with half the expected predation rate ($\mu = 0.005$) if the reward rate in patch 2 is larger than half the reward rate in patch 1. For example, with the values given, the expected lifetime contribution in patch 1 is 100 food units, so if patch 2 (with $\mu = 0.005$) offers 0.6 food units per trip, it should be preferred because the expected lifetime contribution of a forager in that patch is 120 food units. Although the bee gains food in patch 2 at only 60% the rate of patch 1, her lifetime contribution is 20% higher due to increased expected lifespan. (For more elaborate models, see Clark & Dukas 1994; Dukas & Edelstein-Keshet 1998.)

To illustrate the power of a simple anti-predatory response, consider the following example based on Morse's (1986) data. Suppose a bumble bee had just successfully escaped a crab spider attack. Should she return to the same patch on subsequent foraging trips, or switch to another area? The bumble bee's probability of being attacked on the same inflorescence containing the crab spider may be as high as 0.08 compared to an attack probability by crab spiders of only 0.0004 on a randomly chosen inflorescence (Morse 1986, Table 5). Given the 200-fold difference in attack probability by crab spiders, the bee should probably attempt to avoid the inflorescence where she has been attacked even if this implies switching to a patch with somewhat lower reward rates.

Although sensitivity to predation risk should be ubiquitous among pollinators, one can readily imagine cases where short- or long-term fitness considerations might lead animals to ignore predators (Ydenberg & Dill 1986). For example, Cartar (1991) compared the response to a model predator of several bumble bees that foraged for a colony with either ample or depleted honey supply. The workers from honey-rich colonies were three times more likely to flee from the model predator than the bees from the honey-depleted colonies.

So far, I have assumed that the predators cannot alter their behavior as well. In reality, however, the interactions between pollinators and predators should be analyzed with a game-theory approach, where the predators are allowed to respond to the prey and vice versa (reviewed by Sih 1998). For example, Craig *et al.* (1996) suggested that the spider *Nephila clavipes* produces golden webs that attract more bees than any other web color (see also Craig & Bernard 1990; Craig & Freeman 1991; Craig 1994*a*; Blackledge & Wenzel 1999). The arms race between predators and pollinators would not lead to pollinators ignoring predators, but it would likely render anti-predator measures less effective due to the counter-adaptations by the predators.

### Direct response to predation risk

Anti-predator behavioral adaptations of pollinators may be expressed all or most of the time, or expressed only in response to immediate predation risk (Dukas 1998*b*). Only the latter requires a capacity to assess levels of predation risk in space and time. That is, even if a pollinator cannot assess predation risk, due to sensory or cognitive limitations (Dukas 1998*a*), it can still possess evolved behaviors that reduce predation. The ability to respond selectively to perceived predation may be more efficient than a relatively fixed behavioral adaptation, because the former allows the expression of potentially costly anti-predator tactics only when necessary. However, selective response requires a pollinator to perceive the presence of a threat in appropriate circumstances.

Three types of direct response to a perceived heightened predation risk are increasing vigilance, avoiding subsequent visits to the same plant species or location, or adjusting foraging parameters in a way that can reduce the probability of subsequent attack or capture. Increasing predator vigilance can be achieved even through a fundamental sensitization mechanism, which most organisms possess (Jennings 1906; Papaj & Prokopy 1989). This means that after an attack, a pollinator would more readily initiate escape in response to stimuli such as sudden movement. I know of no systematic study documenting increased vigilance due to sensitization or learning in pollinators after exposure to threat. However, evidence presented by Armitage (1965) is rather suggestive.

Armitage compared wing wear of bumble bees captured by the bumblebee-wolf *Philanthus bicinctus* to wing wear of bumble bees he collected at the same flowers that were used as hunting grounds by the wasps. In two bumble bee species, the wasps' prey was strongly biased in favor of less worn (presumably younger) bees (Fig. 11.3). The wasps probably preferred or had higher success rates with younger, inexperienced bees. Gradual accumulation of experience may enhance foraging success of honeybees (Dukas & Visscher 1994), so such long-term experience might increase specific antipredatory behavior as well.

Suggestive field evidence for the ability of insects to avoid an area where they have encountered a predation attempt comes from two butterfly studies (Singer & Wedlake 1981; Mallet *et al.* 1987) that looked at the

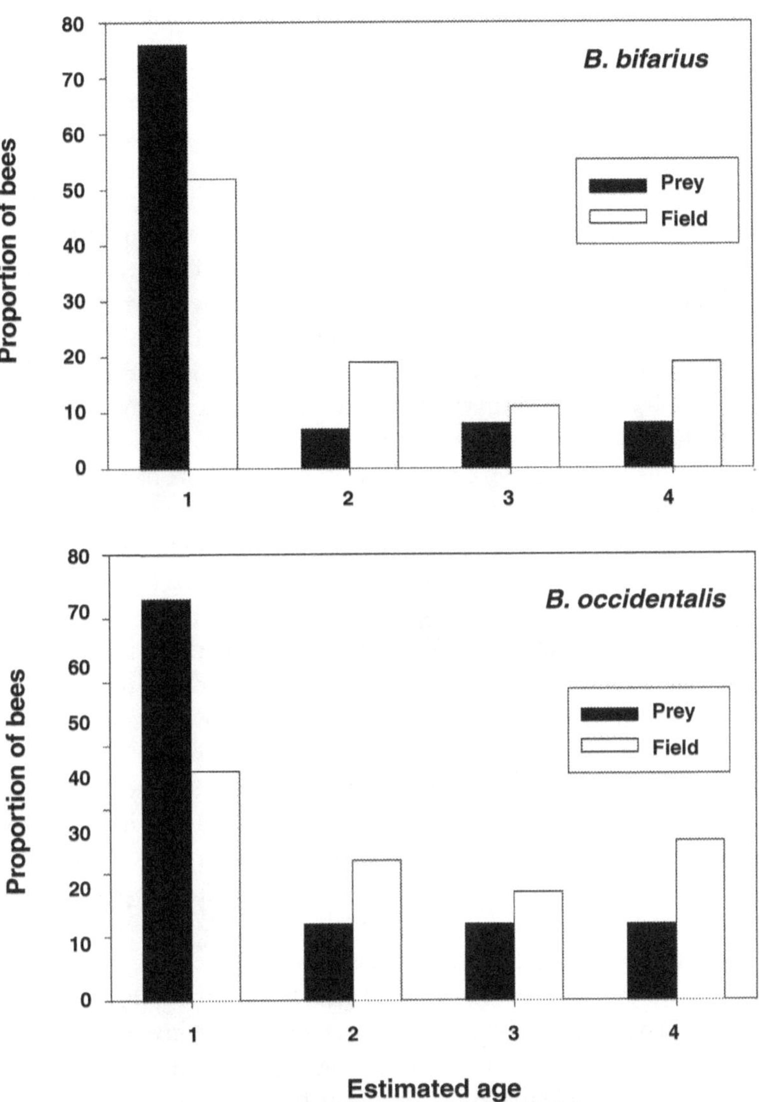


Fig. 11.3. The percentage of worker bumble bees of four age categories (1 = youngest, 4 = oldest) that were either captured by bee wolves ("prey") or collected at the same flowers ("field"). The field sample represents the available age distribution, while the prey reflects the age bias in the wasps' catch. The bias towards young workers of both bee species was statistically significant ($p < 0.001$). The total number of bees was 71 prey and 64 field *Bombus bifarius*, and 72 prey and 92 field *B. occidentalis*. (Data from Armitage 1965.)

effect of handling butterflies, which the butterflies probably perceived as a failed predation attempt. *Heliconius* are long-lived butterflies that repeatedly return to the same individual food and host-plants over a long period of time. Some species also maintain gregarious roosting sites, which they frequent every night (Gilbert 1975; Mallet 1986). Mallet *et al.* (1987) compared the percentage of individually recognized *Heliconius* butterflies that were re-sighted or re-captured at a given vine two days after the butterflies were either captured and marked, or just observed on the vine. Only 15% of the handled butterflies returned to the same vine two days later, compared with a 61% return rate two days after the observation alone. To eliminate the possibility of increased death rate of marked individuals, Mallet *et al.* also observed the butterflies at their roosting sites. They found that 30% of the handled butterflies and 13% of the observed butterflies did not return to the roosting sites on the same night of capture or observation, but that the proportions of non-returning butterflies converged after a few nights, with only 2% and 4% of the handled and observed butterflies, respectively, never returning to roost. These suggestive results invite carefully designed experiments using either butterflies or bees. An advantage of using honeybees for such study is the ability to mark bees upon emergence and later conduct long-term experiments, which include observations on the individually recognized foragers as they exit and enter the hive and at their foraging site (see Dukas & Visscher 1994).

Thomson (this volume) reviews evidence that, under some conditions, individual bumble bees visit individual plants repeatedly, in sequences termed traplines. If the bees can learn about specific routes and the rich patches along such routes, they can probably also learn to bypass hazardous locations (Craig 1994*b*).

In addition to modifying individual behavior, social bees may alter colony-level activity or the distribution of foragers. Korelov (1948, cited by Fry 1983) reported that flight activity near a honeybee hive was reduced from 90 to 4 bees per minute when bee-eaters preyed on bees at the hive entrance. It is feasible that honeybees, which recruit hive-mates to food-rich areas through dance (von Frisch 1967; Seeley 1996), quit recruitment when encountering predation near the hive, or reduce recruitment to a specific patch where they have experienced predation attempts.

Another way of responding to perceived predation risk is to modify foraging activity, such as reducing activity during riskier times. Another behavioral change may be to reduce meal size or the food load carried to the nest. In birds, increased body mass reduces maneuverability; birds are sensitive to this, as they maintain lower fat reserves under heightened predation risk (Metcalfe & Ure 1995; Veasey *et al.* 1998). Similarly, at least one insect study indicates strong effects of body mass on flight performance: Berrigan (1991) measured lift production in the flesh fly (*Neobellieria bullata*), finding that, compared to immature females, sexually mature females had 40% less lift than lighter immatures. Such decrease in flight performance can strongly decrease the fly's ability to escape from approaching predators.

The life span of worker honeybees decreased when weights were attached (Wolf & Schmid-Hempel 1989; Schmid-Hempel 1991). Although these studies focused on work load and physiological wear, increased predation due to decreased flight maneuverability may have contributed additional mortality.

If mass-dependent flight maneuverability is critical for successfully escaping predators, should flower-visiting animals stay lighter under heightened predation? If predation occurs at flowers, perhaps decreasing the food load is the optimal strategy. However, if predation occurs predominantly during flight between the nest and flowers, the alternative of reducing the number of trips and increasing load per trip may result in higher fitness. This issue requires further evaluation.

Many bee species concentrate either in aggregations of solitary individuals or in social nests. Although foraging close to the nest may increase the net rate of food intake, the relative predation rate at flowers and on the way between the nest and flowers may alter the optimal patch distance (Dukas & Edelstein-Keshet 1998). For example, if most of the predation occurs during flight, bees should show a stronger preference for patches closer to their nest than predicted from energetic considerations alone. That is, in addition to floral traits, pollination levels and patterns of pollen flow are likely to be affected by the spatial distribution of colonial pollinators (most bees) and the patterns of predation (Dukas & Edelstein-Keshet 1998).

### General behavioral adaptations to predation

#### The timing of activity

Predation risk typically varies among seasons and times of day, so pollinators may shift their activity season away from that of a major predator. For example, Schmid-Hempel *et al.* (1990) suggested that parasitoids were responsible, at least in part, for the very early activity period of some of the bumblebee species in their sites. These early-season bumble bee species escape the main activity season of the parasitoids, although the same seasonal pattern may be explained solely by resource competition with other sympatric bumble bee species. Similarly, Svensson *et al.* (1999) suggested that the dusk activity of winter moths in Scandinavia occurred at the time of lower predation risk (see also Andersson *et al.* 1998).

Flower specialization and flower constancy

Predation risk may also vary among food plants and spatial locations. The reasons for such inter-host or location variance are that animals may escape predation while occupying locations that (1) are not searched by the predators, (2) are inaccessible to predators, or (3) diminish efficient predator search (Price *et al.* 1980). For example, Geitzenauer & Bernays (1996) examined predation by paper wasps (*Polistes arizonensis*) on tobacco budworms (*Heliothis virescens*) occurring on sunflower (*Helianthus annuus*) and groundcherry (*Physalis pubescens*). They found higher predation on sunflower, perhaps because the caterpillars were more conspicuous on sunflower (Fig. 11.4). Such results have two relevant implications for pollinator behavior. First, if generalist predators are more attracted to one plant species than another due to a greater ease of capturing herbivores, the pollinators of that plant may also incur higher predation rates than those of the other plant. Second, various floral characteristics may generate distinct attack and predation rates on pollinators on different flowers.

Indeed, Morse (1981) suggested differences in attack and predation rates by crab spiders on three sympatric flowers. Spiders on pasture rose (*Rosa carolina*) attacked visiting bumble bees two times more frequently than spiders on common milkweed and goldenrod (*Solidago juncea*), but the success rate of the spiders was more than four times lower on the rose than either milkweed or goldenrod. Honeybees in that study visited only milkweed and goldenrod. The spiders attacked them four times more often on milkweed than goldenrod, with success rates of 7% on milkweed and 0% on goldenrod (Fig. 11.5). Although highly suggestive, Morse's comparative data do not allow calculation of the predation rate per pollinator visiting each of the three flower species.

In two studies, pollinator specialization has been associated with increases in long- (Strickler 1979) and short-term (Laverty & Plowright

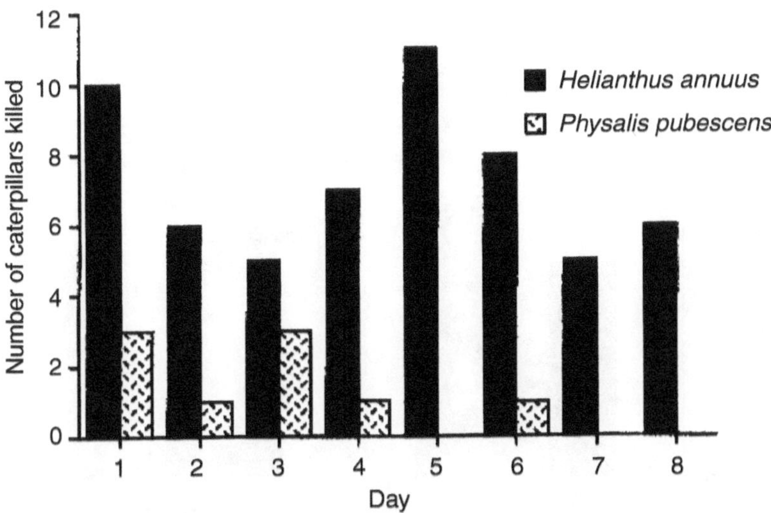


Fig. 11.4. The number of tobacco budworms (*Heliothis virescens*) captured by paper wasps (*Polistes arizonensis*) on the two host-plants, sunflower (*Helianthus annuus*) and groundcherry (*Physalis pubescens*) on eight different days. The two host-plant species were set up in a paired-choice test with one caterpillar on each; data from 20 wasps were recorded on each day. Capture rate was significantly different between the two host species ($p < 0.001$). (Figure from Geitzenauer & Bernays 1996.)

1988) foraging efficiency. Furthermore, Minckley *et al.* (1994) suggested that synchronization between a specialist bee and its host flower was a key factor justifying specialization. Yet predation may also have played a role in selecting for specialization. Bernays (1989; Bernays & Graham 1988) suggested that herbivorous insects with greater host-plant breadth were more vulnerable to generalist predators; similar effects might influence diet breadths of pollinators. At the least, specialist bees may possess innate escape movements that are most efficient for their specific host flower and allow higher escape rates from predators than on other flowers. In addition, the specialization on one plant species may have been in part due to less predation on that species.

Experience with a single plant species may allow individual generalist pollinators to perfect inter-flower and inter-plant movements, to become more familiar with the flowers and hence more likely to notice ambushing predators, and to acquire refined escape movements. These may reduce predation risk for flower-constant individuals.

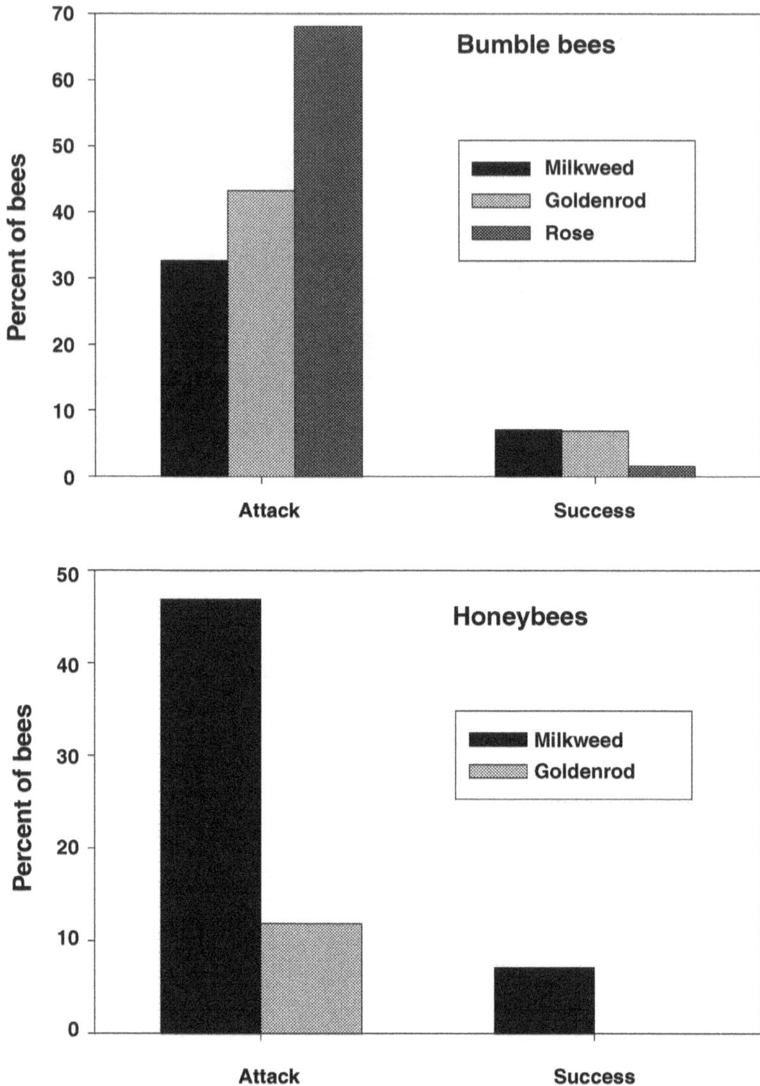



Fig. 11.5. A comparison among three sympatric flowers of the percentage of bees within attack range of crab spiders that were attacked, and the percentage of attacked bees that were captured. Note that no honeybees visited the rose flowers in this study, and that the success rate of spiders attacking honeybees on goldenrod was zero. (Data from Morse 1981.)

## How might predation influence plant characteristics through pollinator–plant interactions?

I argued above that variation in predation risk among seasons and times of day may affect pollinators' activity. This in turn could have selected for changes in flowering times. For example, selection for early activity by bumble bees (driven by lesser activity of conopid parasitoids) may have selected for earlier blooming of bumble bee flowers. Similarly, the dusk activity of moths may have selected for flowers that open and secrete nectar at dusk.

If predation rates do vary among plant species, this factor may influence floral choices by pollinators, in addition to other factors such as the quality and quantity of reward. Furthermore, differential predation rates may have contributed to specialization and flower constancy by pollinators. These behavioral patterns allow higher rates of intraspecific pollen transfer than would be possible with generalist, inconstant flower visitors.

### Floral traits

Various floral traits may reduce predation rates on pollinators. Although it is likely that the effect of many of these traits on predation is incidental, it is conceivable that some evolved due to their effect on lower predation. For example, young *Acacia* flowers appear to emit a volatile chemical signal that deters ant-guards, allowing pollinating bees safe passage (Willmer & Stone 1997). Other possibilities are outlined in the sections below.

#### Flower shape and size

Morse (1981) noted various floral characteristics that affected predation activity by crab spiders (see 'Flower specialization and flower constancy', above). On the large nectarless rose flowers, spiders ambushed at the centers of the flowers near the stamens, which predictably received insect visits. In contrast, on milkweed and goldenrod, the spiders could not occupy a spot guaranteed to receive insect visits. As a result, spiders on rose attacked 68% of the visiting bumble bees, but spiders on milkweed and goldenrods attacked 33% and 43% of the visiting bumble bees, respectively. (Note, however, that the *predation* rates (i.e., successful attacks) on bumble bees were actually higher on milkweed and goldenrod than on rose.)

The rose flowers, however, had a major disadvantage as hunting grounds for crab spiders. Each nectarless flower attracted insect visitors for only a few hours during the morning because the pollen was mostly removed by noon. Consequently, the spiders had to move every day to a new flower, a procedure that decreased their daily hunting duration compared to the hunting duration on goldenrod and milkweed (Morse 1979, 1981). These are examples for how floral characteristics affect the activity of an ambush predator on different flowers. Have some floral traits, such as a flower's lifespan, size, or color, evolved due to a negative effect on predation?

### Nectar availability

The tarnished plant bug, *Lygus lineolaris,* is rarely attacked by the braconid parasitoid (*Leiophron pallipes*) on *Oenothera, Daucus, Amaranthus,* and *Solidago,* but is parasitized at rates of up to 40% on *Erigeron* species (Price *et al.* 1980). Apparently, the parasitoids are more attracted to *Erigeron* flowers, which provide higher quality nectar. Analogous three-level interactions are possible for flower–pollinator–predator interactions. For example, might bumble bees receive more conopid attacks at flowers with nectar accessible to the flies? Has the evolution of concealed nectar been driven in part due to its negative effect on pollinators' parasitoids?

### Spur length

The celebrated match between long-spurred orchids and long-tongued hawkmoths (Nilsson 1988, 1998) has traditionally been attributed to a coevolutionary race between the flowers and pollinators, although an alternative version assuming that the long-tongued hawkmoths existed before the long-spurred orchids is also feasible. According to either scenario, the plants with longer spurs receive higher rates of pollen transfer to stigmas because the hawkmoths' proboscis bases are more likely to contact the floral sexual organs when the hawkmoths insert their tongues more deeply into the spurs. The open question is what factor(s) selected for the hawkmoths' long tongues. Two answers are feasible: the traditional answer is that longer tongues are associated with greater net rate of energy gain and fitness while feeding on flowers with long spurs. A non-mutually-exclusive alternative suggested recently by Wasserthal (1993, 1997, 1998) is that tongues longer than spurs also allow hawkmoths to oscillate sideways while hovering at flowers, and that this "swing hovering" decreases predation. Field measurements (Inouye 1980) and controlled laboratory experiments (Harder 1983) indicated that bumble bees with longer mouthparts should prefer long-tube flowers, which typically contain more nectar. These studies, however, do not preclude a role for predation risk in the evolution of long nectar-extracting mouthparts in hawkmoths or other taxa. Hence Wasserthal's innovative suggestion deserves critical experimental evaluation.

## Conclusions

The available evidence suggests significant levels of predation on pollinators. Hence the effect of predation risk should be integrated into future studies on pollinator behavior and pollinator–plant ecology and evolution. Currently, there are only few detailed accounts on attack and predation rates on pollinators, so further quantitative studies will be useful. Perhaps more important would be studies evaluating how predation affects pollinator behavior and plants. Throughout this chapter, I presented numerous ideas for the necessary experiments.

## Acknowledgements

I thank R. Ydenberg, L. Dill, B. Roitberg, L. Chittka, and J. Thomson for discussions and comments on the manuscript, and W. Pasayten for helping me crystallize some important ideas. My research was supported by grants from the National Institutes of Health and US Department of Agriculture. During writing, I was partially supported by an NSERC operating grant to B. Roitberg.

## References

Abramson CI (1986) Aversive conditioning in honeybees (*Apis mellifera*). *J Comp Psychol* 100:108–116

Ambrose JT (1990) Birds. In: Morse RA & Nowogrodzki R (eds) *Honey bee pests, predators, and diseases*, pp 243–160. Cornell University Press, Ithaca, NY

Andersson S, Rydell J & Svensson MGE (1998) Light, predation and the lekking behaviour of the ghost swift *Hepialus humuli* (L.) (Lepidoptera: Hepialidae). *Proc R Soc Lond B* 265:1345–1351

Armitage KB (1965) Notes on the biology of *Philanthus bicinctus* (Hymenoptera: Sphecidae). *J Kans Entomol Soc* 38:89–100

Balderrama N, Núñez J, Giurfa M, Torrealba J, De Albornoz EG & Almeida LO (1996) A deterrent response in honeybee (*Apis mellifera*) foragers: dependence on disturbance and season. *J Insect Physiol* 42:463–470

Balduf WV (1939) Food habits of *Phymata pennsylvanica americana* Melin (Hemip.). *Can Entomol* 71:66–75

Bernays EA (1989) Host range in phytophagous insects: the potential role of generalist predators. *Evol Ecol* 3:299–311

Bernays EA & Graham M (1988) On the evolution of host specificity in phytophagous arthropods. *Ecology* 69:886–892

Berrigan D (1991) Lift production in the flesh fly, *Neobellieria* (= *Sarcophaga*) *bullata* Parker. *Funct Ecol* 5:448–456

Blackledge TA & Wenzel JW (1999) Do stabilimenta in orb webs attract prey or defend spiders? *Behav Ecol* 10:372–376

Caron DM (1990) Other insects. In: Morse RA & Nowogrodzki R (eds) *Honey bee pests, predators, and diseases*, pp 156–176. Cornell University Press, Ithaca, NY

Caron DM & Ross KG (1990) Spiders and pseudoscorpions. In: Morse RA & Nowogrodzki R (eds) *Honey bee pests, predators, and diseases*, pp 177–187. Cornell University Press, Ithaca, NY

Cartar RV (1991) Colony energy requirements affect response to predation risk in foraging bumble bees. *Ethology* 87:90–96

Clark CW (1993) Dynamic models of behavior: an extension of life history theory. *Trends Ecol Evol* 8:205–209

Clark CW & Dukas R (1994) Balancing foraging and antipredator demands: an advantage of sociality. *Am Nat* 144:542–548

Craig CL (1994*a*) Limits to learning: effects of predator pattern and colour on perception and avoidance-learning by prey. *Anim Behav* 47:1087–1099

Craig CL (1994*b*) Predator foraging behavior in response to the perception and learning by its prey: interactions between orb-spinning spiders and stingless bees. *Behav Ecol Sociobiol* 35:45–52

Craig CL & Bernard GD (1990) Insect attraction to ultraviolet-reflecting spider webs and web decorations. *Ecology* 71:616–623

Craig CL & Freeman CR (1991) Effects of predator visibility on prey encounter: a case study on aerial web weaving spiders. *Behav Ecol Sociobiol* 29:249–254

Craig CL, Weber RS & Bernard GD (1996) Evolution of predator–prey systems: spider foraging plasticity in response to the visual ecology of prey. *Am Nat* 147:205–229

Davies NB (1976) Prey selection and the search strategy of the spotted flycatcher (*Muscicapa striata*): a field study on optimal foraging. *Anim Behav* 25:1016–1033

De Jong D (1990) Insects: hymenoptera (ants, wasps, and bees). In: Morse RA & Nowogrodzki R (eds) *Honey bee pests, predators, and diseases*, pp 135–155. Cornell University Press, Ithaca, NY

Dixon AFG & Agarwala BK (1999) Ladybird-induced life-history changes in aphids. *Proc R Soc Lond B* 266:1549–1553

Dukas R (1998*a*) Constraints on information processing and their effects on behavior. In: Dukas R (ed) *Cognitive ecology*, pp 89–127. University of Chicago Press, Chicago

Dukas R (1998*b*) Evolutionary ecology of learning. In: Dukas R (ed) *Cognitive ecology*, pp 129–174. University of Chicago Press, Chicago

Dukas R (1999) Ecological relevance of associative learning in fruit fly larvae. *Behav Ecol Sociobiol* 45:195–200

Dukas R & Edelstein-Keshet L (1998) The spatial distribution of colonial food provisioners. *J Theor Biol* 190:121–134

Dukas R & Kamil AC (2000) The cost of limited attention in blue jays. *Behav Ecol* 11:502–506

Dukas R & Visscher PK (1994) Lifetime learning by foraging honeybees. *Anim Behav* 48:1007–1012

Evans HE & Eberhard JW (1970) *The wasps*. University of Michigan Press, Ann Arbor, MI

Evans HE & O'Neill KM (1988) *The natural history and behavior of North American beewolves*. Cornell University Press, Ithaca, NY

Feener DH (1988) Effects of parasites on foraging and defense behavior of a termitophagous ant, *Pheidole titanis* Wheeler (Hymenoptera: Formicidae). *Behav Ecol Sociobiol* 22:421–427

Folgarait PJ & Gilbert LE (1999) Phorid parasitoids affect foraging activity of *Solenopsis richteri* under different availability of food in Argentina. *Ecol Entomol* 24:163–173

Free JB (1987) *Pheromones in social bees*. Chapman & Hall, London

Frisch K von (1967) The dance language and orientation of bees. Belknap Press, Cambridge, MA

Fry CH (1983) Honeybee predation by bee-eaters, with economic considerations. *Bee World* 64:65–78

Geitzenauer HL & Bernays EA (1996) Plant effects on prey choice by a vespid wasp, *Polistes arizonensis*. *Ecol Entomol* 21:227–234

Gilbert LE (1975) Ecological consequences of a coevolved mutualism between butterflies and plants. In: Gilbert LE & Raven PH (eds) *Coevolution of animals and plants*, pp 210–240. University of Texas Press, Austin, TX

Gilbert LE (1983) Coevolution and mimicry. In: Futuyma DJ & Slatkin M (eds) *Coevolution*, pp 263–281. Sinauer, Sunderland, MA

Gilliam JF & Fraser DF (1987) Habitat selection under predation hazard: test of a model with foraging minnows. *Ecology* 68:1856–1862

Godfray HCJ (1994) *Parasitoids: behavioral and evolutionary ecology*. Princeton University Press, Princeton, NJ

Goldblatt JW & Fell RD (1987) Adult longevity of workers of the bumble bees *Bombus fervidus* (F.) and *Bombus pennsylvanicus* (DeGeer) (Hymenoptera: Apidae). *Can J Zool* 65:2349–2353

Gould JL (1987) Honey bees store learned flower-landing behaviour according to time of day. *Anim Behav* 35:1579–1581

Greco CF & Kevan PG (1995) Patch choice in the anthophilous ambush predator *Phymata americana*: improvement by switching hunting sites as part of the initial choice. *Can J Zool* 73:1912–1917

Grostal P & Dick M (1999) Direct and indirect cues of predation risk influence behavior and reproduction of prey: a case for acarine interactions. *Behav Ecol* 10:422–427

Gwynne DT (1981) Nesting biology of the bumblebee wolf *Philanthus bicinctus* Mickel (Hymenoptera: Sphecidae). *Am Midl Nat* 105:130–138

Harder LD (1983) Flower handling efficiency of bumble bees: morphological aspects of probing time. *Oecologia* 57:274–280

Heinrich B, Mudge PR & Deringis PG (1977) Laboratory analysis of flower constancy in foraging bumblebees: *Bombus ternarius* and *B. terricola*. *Behav Ecol Sociobiol* 2:247–265

Inouye DW (1980) The effect of proboscis and corolla tube lengths on patterns and rates of flower visitation by bumblebees. *Oecologia* 45:197–201

Jennings HS (1906) *Behavior of the lower organisms*. Indiana University Press, Bloomington, IN

Knutson LV & Murphy WL (1990) Insects: Diptera (flies). In: Morse RA & Nowogrodzki R (eds) *Honey bee pests, predators, and diseases*, pp 120–134. Cornell University Press, Ithaca, NY

Land MF (1981) Optics and vision in invertebrates. In: Autrum H (ed) *Handbook of sensory physiology*, vol. VII/6B, *Invertebrate visual centers*, pp 472–592. Springer-Verlag, Berlin

Laverty TM & Plowright RC (1988) Flower handling by bumblebees: a comparison of specialists and generalists. *Anim Behav* 36:733–740

Lavigne RJ (1992) Ecology of *Neoaratus abludo* Daniels (Diptera: Asilidae) in south Australia, with notes on *Neoaratus pelago* (Walker) and *Neoaratus rufiventris* (Macquart). *Proc Entomol Soc Wash* 94:253–262

Lawton JH (1986) The effect of parasitoids on phytophagous insect communities. In: Waage J & Greathead D (eds) *Insect parasitoids*, pp 265–287. Academic Press, London

Lima SL (1990) Evolutionary stable antipredator behavior among isolated foragers: on the consequences of successful escape. *J Theor Biol* 14:377–389

Lima SL (1998*a*) Nonlethal effects in the ecology of predator–prey interactions. *BioSci* 48:25–34

Lima SL (1998*b*) Stress and decision making under the risk of predation: recent developments from behavioral, reproductive, and ecological perspectives. *Adv Study Behav* 27:215–291

Lima SL & Dill LM (1990) Behavioral decisions made under the risk of predation: a review and prospectus. *Can J Zool* 68:619–640

Liu L, Wolf R, Roman E & Heisenberg M (1999) Context generalization in *Drosophila* visual learning requires the mushroom bodies. *Nature* 400:753–756

MacFarlane RP & Pengelly DH (1974) Conopidae and Sacrophagidae (Diptera) as parasites of adult Bombinae (Hymenoptera) in Ontario. *Proc Entomol Soc Ontario* 105:55–59

Mallet J (1986) Gregarious roosting and home range in *Heliconius* butterflies. *Nat Geog Res* 2:198–215

Mallet J, Longino JT, Murawski D, Murawski A & Simpson de Gamboa A (1987) Handling effects on *Heliconius*: where do all the butterflies go? *J Anim Ecol* 56:377–386

Martin TE (1995) Avian life history evolution in relation to nest sites, nest predation, and food. *Ecol Monogr* 65:101–127

Metcalfe NB & Ure SE (1995) Diurnal variation in flight performance and hence potential predation risk in small birds. *Proc R Soc Lond B* 261:395–400

Miller RS & Gass CL (1985) Survivorship in hummingbirds: is predation important? *Auk* 102:175–178

Minckley R., Wcislo WT, Yanega D & Buchmann SL (1994) Behavior and phenology of a specialist bee (*Dieunomia*) and sunflower (*Helianthus*) pollen availability. *Ecology* 75:1406–1419

Morse DH (1979) Prey capture by the crab spider *Misumena calycina* (Arcaneae: Thomisidae). *Oecologia* 39:309–319

Morse DH (1981) Prey capture by the crab spider *Misumena vatia* (Clerck) (Thomisidae) on three common native flowers. *Am Midl Nat* 105:358–367

Morse DH (1986) Predation risk to insect foraging at flowers. *Oikos* 46:223–228

Morse RA & Nowogrodzki R (1990) *Honey bee pests, predators, and diseases*. Cornell University Press, Ithaca, NY

Nilsson LA (1988) The evolution of flowers with deep corolla tubes. *Nature* 334:147–149

Nilsson LA (1998) Deep flowers for long tongues. *Trends Ecol Evol* 13:259–260

Nonacs P & Dill LM (1990) Mortality risk vs. food quality trade-offs in a common currency: ant patch preferences. *Ecology* 71:1886–1892

Nonacs P & Dill LM (1991) Mortality risk versus food quality trade-offs in ants: patch use over time. *Ecol Entomol* 16:73–80

Papaj DR & Prokopy RJ (1989) Ecological and evolutionary aspects of learning in phytophagous insects. *Ann Rev Entomol* 34:315–350

Plowright RC & Owen RE (1980) The evolutionary significance of bumble bee color patterns: a mimetic interpretation. *Evolution* 34:622–637

Price PW, Bouton CE, Gross P, McPheron BA, Thompson JN & Weis AE (1980) Interactions among three trophic levels: influence of plants on interactions between insect herbivores and natural enemies. *Ann Rev Ecol Syst* 11:41–65

Pyke GH (1979) Optimal foraging in bumblebees: rule of movement between flowers within inflorescences. *Anim Behav* 27:1167–1181

Quinn WG, Harris WA & Benzer S (1974) Conditioned behavior in *Drosophila melanogaster. Proc Natl Acad Sci USA* 71:708–712

Rabinovich M & Corley JC (1997) An important new predator of honeybees: the robber fly *Mallophora ruficauda* Wiedemann (Diptera, Asilidae) in Argentina. *Am Bee J* 137:303–306

Robinson MH & Robinson B (1970) Prey caught by a sample population of the spider *Argiope argentata* (Araneidae) in Panama: a year's census data. *Zool J Linn Soc* 49:345–357

Rodd FH, Plowright RC & Owen RE (1980) Mortality rates of adult bumble bee workers (Hymenoptera: Apidae). *Can J Zool* 58:1718–1721

Schmid-Hempel P (1991) The ergonomics of worker behavior in social hymenoptera. *Adv Study Behav* 20:87–134

Schmid-Hempel P (1998) *Parasites in social insects.* Princeton University Press, Princeton, NJ

Schmid-Hempel P, Muller C, Schmid-Hempel R & Shykoff JA (1990) Frequency and ecological correlates of parasitism by conopid flies (Conopidae, Diptera) in populations of bumblebees. *Insectes Soc* 37:14–30

Schmidt JO (1990) Hymenopteran venoms: striving toward the ultimate defense against vertebrates. In: Evans DL & Schmidt JO (eds) *Insect defenses,* pp 387–419. State University of New York Press, Albany, NY

Seeley TD (1996) *The wisdom of the hive.* Harvard University Press, Cambridge, MA

Shykoff JA & Schmid-Hempel P (1991) Incidence and effects of four parasites in natural populations of bumble bees in Switzerland. *Apidologie* 22:117–125

Sih A (1998) Game theory and predator prey response races. In: Dugatkin LA & Reeve HK (eds) *Game theory and animal behavior,* pp 221–238. Oxford University Press, New York

Simonthomas RT & Simonthomas AMJ (1980) *Philanthus triangulum* and its recent eruption as a predator of honeybees in an Egyptian oasis. *Bee World* 61:97–107

Singer MC & Wedlake P (1981) Capture does affect probability of recapture in a butterfly species. *Ecol Entomol* 6:99–121

Smith BH, Abramson CI & Tobin TR (1991) Conditional withholding of proboscis extension in honeybees (*Apis mellifera*) during discriminative punishment. *J Comp Psychol* 105:345–356

Strickler K (1979) Specialization and foraging efficiency of solitary bees. *Ecology* 60:988–1009

Svensson MGE, Rydell J & Brown R (1999) Bat predation and flight timing of winter moths, *Epirrita* and *Operophtera* species (Lepidoptera, Geometridae). *Oikos* 84:193–198

Tinbergen N (1958) *Curious naturalists*. Basic Books, New York

Veasey JS, Metcalfe NB & Houston DC (1998) A reassessment of the effect of body mass upon flight speed and predation risk in birds. *Anim Behav* 56:883–889

Wasserthal LT (1993) Swing-hovering combined with long tongue in hawkmoths, an antipredator adaptation during flower visits. In: Barthlott W, Naumann CM, Schmidt-Loske K & Schuchmann KL (eds) *Animal–plant interactions in tropical environments*, pp 77–86. Zoologisches Forschungsinstitut, Bonn

Wasserthal LT (1997) The pollinators of the Malagasy star orchids *Angraecum sesquipedale, A. sororium* and *A. compactum* and the evolution of extremely long spurs by pollinator shift. *Bot Acta* 110:343–359

Wasserthal LT (1998) Deep flowers for long tongues. *Trends Ecol Evol* 13:458–459

Willmer PG & Stone GN (1997) How aggressive ant-guards assist seed-set in *Acacia* flowers. *Nature* 388:165–167

Wolf TJ & Schmid-Hempel P (1989) Extra loads and foraging life span in honeybee workers. *J Anim Ecol* 58:943–954

Ydenberg RC (1998) Behavioral decisions about foraging and predator avoidance. In: Dukas R (ed) *Cognitive ecology*, pp 343–378. University of Chicago Press, Chicago

Ydenberg RC & Dill LM (1986) The economics of fleeing from predators. *Adv Study Behav* 16:229–249

12

# Pollinator preference, frequency dependence, and floral evolution

## Pollinator responses to frequency – definitions and importance

Frequency-dependent selection (FDS) occurs when the relative fitness of a genotype or phenotype is a function of its frequency in the population (Wright 1948; Clarke 1962). In behavioral ecology, FDS usually indicates that the identity of the fittest genotype (or phenotype) reverses at some intermediate frequency (Heino *et al.* 1998). When rare genotypes have an advantage, in this narrow sense of FDS, such selection will result in stable polymorphic equilibria (Clarke & O'Donald 1964). This "negative" FDS has captured the interest of many evolutionary biologists (Ayala & Campbell 1974).

An example of a floral polymorphism believed to be maintained by FDS is heterostyly, a suite of floral traits including reciprocal style- and stigma-length polymorphisms. These polymorphisms can increase the amount of pollen carried to alternative phenotypes, causing rare morphs to have increased outcross mating opportunities (Heuch 1979; Eckert *et al.* 1996). Such selection arises purely from the architecture of the sexual organs, even if pollinators forage randomly among phenotypes. Levin (1972), however, suggested that behavioral preferences of the pollinators themselves might induce FDS among floral traits.

During the 1960s and early 1970s, a number of studies suggested that behavioral preferences were frequency-dependent (Ayala & Campbell 1974). Allen & Clarke (1968) showed that predators, especially birds, selected proportionately more of the most common prey types in a color-varying prey population, even if energetic rewards were equivalent for the different types. Preferential predation on common forms should lead to a rare-morph advantage, resulting in stable polymorphism for the trait concerned (Clarke & O'Donald 1964). This behaviorally induced stability excited many behavioral ecologists, producing an explosion of studies on frequency-dependent choices of prey, mates, and hosts in natural populations [reviewed by Allen 1988; O'Donald & Majerus 1988; Barrett 1988; see Clarke (1962) for shell-pattern polymorphisms in *Cepea* snails and Turner (1977) for wing-pattern mimicry in tropical butterfly species].

Levin (1972) suggested that pollinators choose disproportionately the most common floral types in plant populations, even if nectar rewards are the same. Whereas predators cause mortality selection in prey, pollinators cause fecundity selection in plants. Because he expected pollinator visitation rate to be positively related to plant fecundity and fitness, Levin predicted that pollinator behavior would lead to common-morph advantage among floral types, i.e., "positive" FDS.

Levin (1972) tested the hypothesis of frequency-dependent pollinator foraging using arrays of two shape morphs of *Phlox*. He defined relative morph fitness as the mean proportion of outcrossed seeds, quantified as the proportion of heterozygotes among progeny derived from the recessive morph. At low morph frequency, he found the deficit of heterozygous progeny predicted by FDS. This inference assumed that a decrease in the proportion of heterozygote progeny accurately reflected a decrease in the total number of overall outcross matings, however, whereas it could alternatively reflect an increase in the proportion of matings between similar morphs, i.e., assortative visitation (Kay 1978). Therefore, Levin's results did not conclusively show that pollinators prefer common morphs.

Why is it important to test whether pollinator-induced FDS is an important selective force acting in plant populations? A common-morph advantage will tend to fix alleles (Thomson 1984); therefore, FDS could generate stabilizing selection, severely constraining floral evolution in animal-pollinated plants. Alternatively, if pollinator preferences are reversed, a rare-morph advantage should promote phenotypic diversity. We need to identify the circumstances that can lead to this reversal. Pollinator-driven FDS is also implicated in other selective processes, such as the evolution of mimetic pollination systems and the convergence of floral signals of different plant species (Roy & Widmer 1999).

The aim of this paper is to review the evidence both for frequency-dependent pollinator foraging behavior and for FDS in plant populations. Do pollinators really select in a frequency-dependent way, and if so, why? Does pollinator preference induce FDS of sufficient strength to account for patterns of monomorphism or polymorphism among floral traits?

## Are pollinator foraging patterns frequency-dependent? The results of laboratory experiments

I consider here experiments that offered arrays of artificial flowers under controlled conditions to foraging pollinators. The provisioning behavior of bumble bee workers allowed researchers to study long foraging sequences without satiation of the experimental subjects.

### Analyzing data to test for FDS

Typically, laboratory experiments test bees' preferences on arrays of two or more colors of artificial flowers. The frequencies of the colors are varied to see if frequency affects the proportion of visits a color morph receives. Frequency-dependent (common or rare morph) and frequency-independent (bias to one morph independent of frequency) preferences can occur simultaneously. Figure 12.1 shows the range of potential relationships between frequency and visitation, along with the resultant expected fitness relationships.

Gendron (1987) discussed statistical tests for frequency dependence Comparing observed and expected numbers of visits with goodness of fit tests is suitable only for very small data sets, due to the risk of Type I errors. Two authors have developed methods specifically for analyzing frequency-dependent behavior.

Manly's model (1973) can be used to measure selection when morph frequencies change during an experiment, and this model is frequently used for predator–prey experiments. However, it does not fit some types of data well, as the probability of a morph being selected is constrained to be a linear function of frequency (Gendron 1987), and may be invalid if selectivity varies with learning (Greenwood & Elton 1979). I have adopted Greenwood & Elton's (1979) model,

$$F = \frac{(VA)^b}{(VA)^b + (1-A)^b},$$

which relates the availability of a morph in a two-morph system ($A$) to those eaten ($F$) using two parameters, frequency-independent ($V$) and frequency-dependent ($b$) components. This model is constrained by being

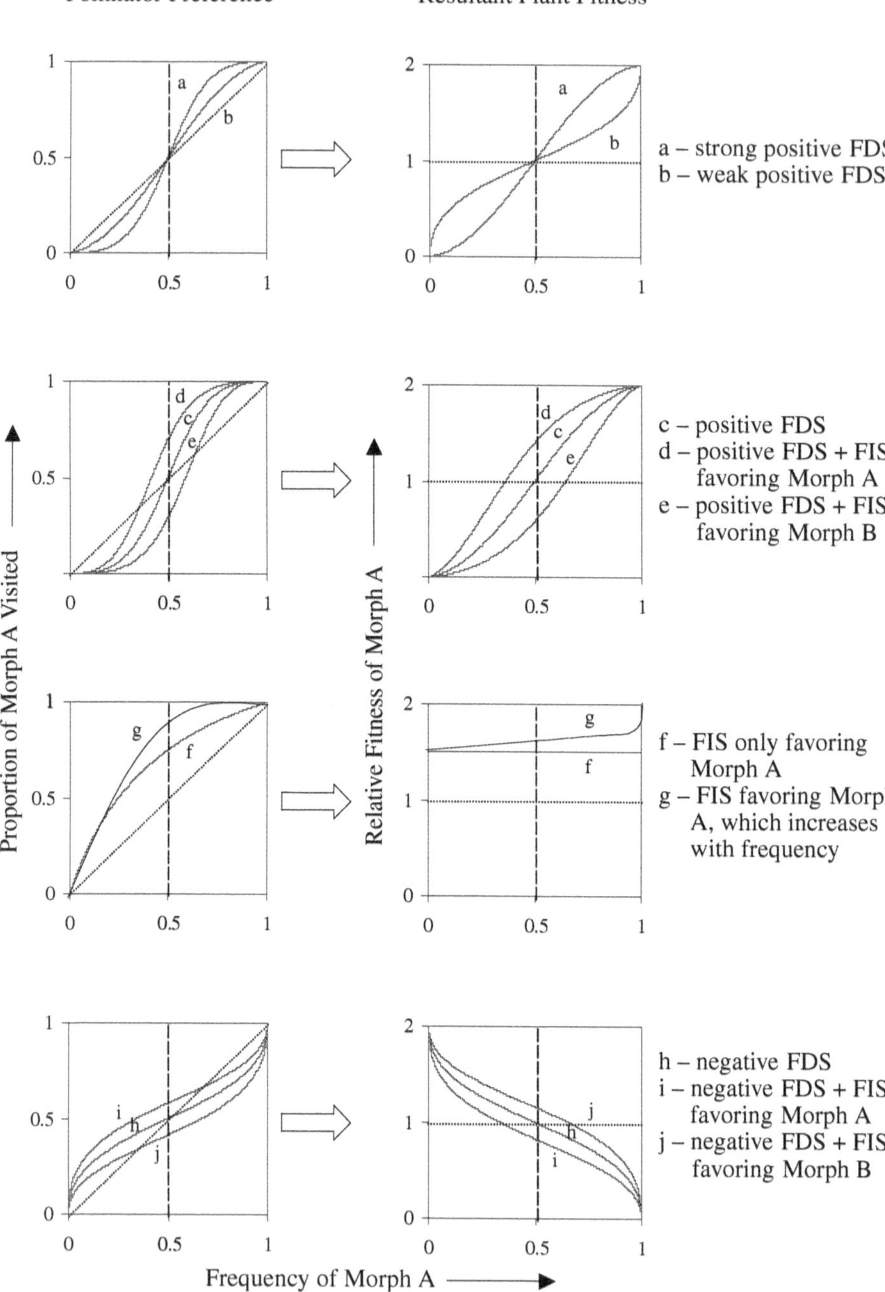

Pollinator Preference
Resultant Plant Fitness
Proportion of Morph A Visited
Relative Fitness of Morph A
Frequency of Morph A
a – strong positive FDS
b – weak positive FDS
c – positive FDS
d – positive FDS + FIS
favoring Morph A
e – positive FDS + FIS
favoring Morph B
f – FIS only favoring
Morph A
g – FIS favoring Morph
A, which increases
with frequency
h – negative FDS
i – negative FDS + FIS
favoring Morph A
j – negative FDS + FIS
favoring Morph B

forced through the origin at extreme morph frequencies (Gendron 1987), but it fits most data well and is appropriate to pollinator–plant experiments where apparent morph frequencies do not change.

I have reanalyzed published data sets, using the Greenwood & Elton model where possible. I have found it essential to log-transform data before analysis, due to non-normality of variances. Analyses can be greatly influenced by outliers, so bootstrap estimates of parameters and standard errors are appropriate. With small data sets, power is limited, but I have examined correlations between morph frequency and deviations of observed from expected values.

### Bumble bee behavior on single artificial flowers

Both Real (1990) and I (Smithson & Macnair 1996, 1997*a*, *b*) explored frequency-dependent behavior by bumble bees using a "bee-board" – a large, rigid, plastic sheet drilled with small wells to hold "nectar." Colored discs representing flowers are placed under selected holes. Flower color, density, positioning, and reward content are easily manipulated, and the experiments are easily replicable. One shortcoming is the lack of inflorescence structure, which may affect behavior. Table 12.1 summarizes results from this method.

Significant frequency dependence emerged in 11 of 13 experiments with flowers of two colors offered to bumble bees at different frequencies (Table 12.1). In 9 of 14 experiments, bees expressed significant preferences for one color. In 6 of the 8 experiments where both colors were equally rewarding, bees preferred the common color. Altering the relative amount of reward provided by each color did not significantly affect frequency-independent preference, although variances were greatly increased (Smithson 1995). However, a significant change in frequency-independent preference was recorded when the reward provided by one color was more variable than the other (Table 12.1, $t = 8.242$, $p < 0.001$). Other authors have found

Fig. 12.1 Opposite. Diagram showing various types of frequency-dependent pollinator preferences amongst two morphs (A and B) and the resultant selection regimes that could be induced in plant populations assuming a simple and positive relationship between preference and plant fitness. Lines a and b contrast different strengths of common morph preference and resultant positive FDS. Lines c, d, and e show the effects of FDS and FIS acting simultaneously. Line f shows FIS only, with line g showing FIS increasing with morph frequency – this could be considered as FDS in a broad sense (see text). Lines h, i, and j show rare-morph preference and resultant negative FDS. The dotted line represents the line of equal preference or fitness for the two morphs, and the dashed line shows 50% morph A.

Table 12.1. *Results of experiments testing for frequency-dependent and frequency-independent choice in artificial flower experiments using bumble bee workers; all experiments used flowers placed on a bee-board, and all experiments contained two flower colors, yellow and blue, which were tested for frequency-dependence (except for Smithson & Macnair 1997a, where three flower colors were present but only two, blue and purple, were tested)*

| Reward[a] | Reward schedule[b] | Frequency dependence[c] | Frequency independence[c] | Source |
|---|---|---|---|---|
| Y | =, constant, 2 | 1.39*** | 0.92 | Real 1990 |
| Y | <>, variable, 2 | 1.22* | 0.28*** | Real 1990 |
| Y | =, variable, 2 | 1.18 | 0.86 | Real 1990 |
| Y | =, constant, 2 | 1.83*** | 0.32*** | Smithson & Macnair 1996 |
| Y | =, constant, 5 | 1.49* | 0.63 | Smithson & Macnair 1996 |
| Y | =, constant, 2 | 2.00*** | 0.79 | Smithson & Macnair 1997a |
| Y | =, constant, 2 | 2.09*** | 1.04 | Smithson & Macnair 1997a |
| Y | =, constant, 2 | 1.11*** | 0.83*** | Smithson & Macnair 1997b |
| Y | =, constant, 5 | 1.03 | 0.85* | Smithson & Macnair 1997b |
| N | =, rewardless | 0.75** | 0.63** | Smithson & Macnair 1997b |
| Y | <>, constant, 2/5 | 1.88*** | 0.52*** | Smithson 1995 |
| Y | <>, constant, 2/5 | 1.62*** | 0.70*** | Smithson 1995 |
| Y | <>, constant, 1/5 | 2.15*** | 0.64*** | Smithson 1995 |
| Y | <>, constant, 5/2 | 1.76*** | 0.65*** | Smithson 1995 |

*Notes:*
[a] Reward indicates whether sucrose solution was present (Y) in flowers tested or not (N).
[b] Rewards equal in both color types (=); rewards not equal in both color types (<>); rewards the same within each reward type (constant); one or both color types had variable rewards (variable). Values in the reward schedule column indicate amount of sucrose added; two values are given if color types have different reward levels.
[c] The Greenwood & Elton (1979) model was fitted to data in all cases to test for selection (see text). The degree of frequency dependence ($b$) and its significance (* $p < 0.05$, ** $p < 0.01$, *** $p < 0.001$) are given; a $b$ value $<1$ indicates preference for the rare morph; $b > 1$ indicates common-morph preference. The value and significance of frequency-independence ($V$) is indicated the same way. Frequency independence increases with departure from unity.

stronger learned responses to patterns of empty and full flowers (Perez & Waddington 1996; Waddington, this volume). Neither reward schedule (Smithson 1995) nor variability (Table 12.1, $F_{2,64} = 0.38$, NS) significantly affected the strength of frequency dependence, although that strength tended to decrease with increasing variability. Other experiments testing the effects of density and total reward also found no significant effect on the strength of frequency dependence (Smithson & Macnair 1996, 1997a). However, where neither color morph contained reward, behavioral patterns were reversed, and bees showed significant rare-morph preference (Smithson & Macnair 1997*b*).

Visitation order was correlated with frequency dependence: when bees displayed common-morph preference, they visited the same colors sequentially, but when they displayed rare-morph preference, they tended to visit morphs disassortatively (Smithson & Macnair 1996, 1997*a*, *b*). Frequency-dependent behavior increased with experience, being weak in the first 50 flowers visited, developing to its maximum extent over 50–100 flowers visited, and changing little thereafter; however, the level of assortative or disassortative visitation remained constant over all flower visits (Smithson & Macnair 1996).

Bumble bees forage with unequivocal frequency-dependence on arrays of simple artificial flowers that vary in color. Variation in flower size produced no FDS (A. Smithson unpublished data), but other floral traits such as shape or fragrance have not been tested for a frequency-dependent response, nor have other pollinator types been tested.

## Why do bumble bees visit artificial flowers in a frequency-dependent way?

Tinbergen (1960) suggested that common-morph preference arises because foragers can maximize feeding efficiency by concentrating on a single food type. Bumble bee workers, however, show common corolla color morph preference even when rewards and handling times are identical. Real (1990) suggested behavioral constraints as the cause: the subjectively perceived probability and the actual probability of encountering a morph may differ, such that low probability events are overestimated; alternatively, a bee may repeatedly switch its attention from one morph to the other (the shifting attention hypothesis *sensu* Dawkins 1971*a*). However, neither of these suggestions fully explains all the behavioral patterns observed. Chittka & Thomson (1997) trained bees on artificial flowers of different colors and morphologies. With respect to handling flowers, avoiding and correcting errors, and initial traveling times between flowers, bees trained on only one color-morphology combination outperformed bees trained on two combinations. I have noted a reduction in traveling time as the common-morph preference develops. Both observations suggest efficiency advantages for pollinator specialization even when floral morphologies are identical; further experiments are required.

Mechanistically, the link between visitation order and common-morph preference implies that short-term memory affects frequency dependence (Menzel 1999, this volume). The increase of frequency dependence with experience suggests that continued reinforcement on common morphs consolidates long-term memory, resulting in continued preference. Predators that develop "search images" while learning to distinguish cryptic prey from the background will display frequency dependence if those images interfere with each other (Dawkins 1971*b*). Although interference has been shown for short-term memory (Menzel 1979), the importance of interference for common-morph preference by pollinators is unclear.

Why do rare unrewarding morphs receive more pollinator visits? Ultimately, this is expected as a consequence of efficient avoidance of empty flowers by pollinators. Bumble bees may learn to avoid unrewarding flowers by making a particular number of test visits (Heinrich 1975). If this mechanism applies, the task of learning to avoid all unrewarding morphs will increase as the number of morphs increases. A bee will have to make more sampling visits. Furthermore, the numbers of each unrewarding morph sampled should be independent of that morph's frequency in the population. Dukas & Real (1993) did not find the predicted increase in sampling visits with morph number. I found that the number of flowers sampled of unrewarding morphs did depend on the frequencies of those morphs (Spearman's $r_s = 0.60$ and $0.52$, $p < 0.001$, for two unrewarding morphs that varied in frequency). Further, the equal-sampling hypothesis does not predict disassortative visitation. Dukas & Real (1993) suggested that unrewarding morphs are not memorized during sampling, but if this is correct, rare-morph preference would not occur (Ferdy *et al.* 1998). An alternative hypothesis (Smithson & Macnair 1997*b*) suggests that sampling an unrewarding morph causes negative reinforcement stored in short-term memory, increasing the likelihood subsequently of sampling a different morph, thereby causing disassortative visitation. Data from humans suggest that the time to search for a particular target type does not increase as the number of alternative stimuli increases, as long as the alternatives vary only in one way, e.g., color or shape. If the alternative stimuli vary in more than one way, e.g., both color and shape, it takes significantly longer to find the target (Treisman & Gelade 1980; see also Gegear & Laverty this volume).

The mechanisms behind pollinator sampling of unrewarding morphs and consequent negative FDS require further experimentation, particularly with respect to interactions among several traits. In nature, many unrewarding species are rare orchids, so if increasing phenotypic variability within a population can increase the total number of pollinator visits to the species, this may have important conservation implications (Ferdy *et al.* 1998).

## Bumble bee behavior on artificial inflorescences

Although bumble bees may show frequency dependence on simple artificial flowers, they might not exert frequency-dependent selection in real populations. First, patterns of behavior might be affected by the organization of flowers into inflorescences: encountering many flowers clustered together could affect a bee's perception of frequency. Second, even if visitation rates are well correlated with female function through seed set (Waser & Price 1981), they may be weakly coupled to other components of fitness such as outcrossing through male function (Stanton *et al.* 1986, 1989). In particular, if pollinators visit more flowers per inflorescence on one morph, more pollen will be transferred between flowers within inflorescences of that species (Klinkhamer & de Jong 1993; Harder *et al.*, this volume). This will cause geitonogamous selfing in self-compatible species, and may block receipt of compatible pollen in self-incompatible species; it will also reduce the amount of pollen available for outcross seed paternity (de Jong *et al.* 1993; Harder & Barrett 1995). Thus, differences in the numbers of flowers visited per inflorescence between morphs can affect final reproductive success.

Artificial inflorescence experiments attempt to bridge the gap between bee-board experiments and pollinator behavior in the field; some have been used to test optimality models (Cartar & Abrahams 1996). We used artificial inflorescences to test both for frequency dependence and differences in the numbers of flowers visited per inflorescence between morphs (A. Smithson & L. Gigord, unpublished data). We made artificial inflorescences from green plastic rods 40 cm long and 1 cm in diameter, the top of which held 10 "flowers" arranged spirally around the rod 1.5 cm apart. Flowers were colored card stock "corollas," with central holes that gave access to wells inside the rod, into which we pipetted sucrose solution. We conducted two experiments in a cage, with inflorescences of purple and yellow arranged randomly on a grid at frequencies of 50 yellow:25 purple and 50 purple:25 yellow. Worker bumble bees (*Bombus terrestris*) from a captive colony foraged singly.

For the array with 50 yellow:25 purple, 7 out of 11 bees showed a preference for yellow. Overall results indicated a common-morph preference ($G = 42.92$, $p < 0.001$). For 50 purple:25 yellow, 3 out of 11 bees showed a preference for purple, while 4 showed a preference for yellow; heterogeneity between individuals caused deviations from expected ratios. This suggests common-morph preference combined with frequency-independent preference for yellow.

After removing data from revisited inflorescences, we found that, for most of the bees, morphs differed significantly in the number of flowers visited per inflorescence for both experiments, although overall the difference was significant only for the 50 yellow:25 purple array ($F = 19.52$, $p < 0.001$). The relationship between visitation rate and the mean number of flowers visited per inflorescence was strong, non-linear, and positive: a regression analysis explained 79% of the variance. The relationship did not differ across experiments or color, and was not influenced by revisiting empty inflorescences, as there was no difference between morphs in the proportion of revisits (A. Smithson & L. Gigord, unpublished data).

These results are consistent with a relationship between inflorescence learning and flower learning. If bumble bees interpret each flower as an individual learning event, then preference for a morph should also result in an increased number of flowers visited per inflorescence. In other unpublished experiments, we found a positive correlation between the number of flowers visited on unrewarding artificial inflorescences and with visitation rate. These results show that although pollinators may visit plants in a frequency-dependent way, fitness relationships are unlikely to be predicted by visitation rate alone. Other components of fitness may also depend on morph frequency.

### Do pollinators forage in a frequency-dependent way under field conditions?

For plant species that produce rewards (nectar or pollen), four studies have quantified pollinator visitation rates for alternative floral phenotypes under field conditions and over a range of morph frequencies (Table 12.2). One study used the natural floral polymorphism observed in plant populations, and three studies included, either wholly or partly, manipulative experiments containing arrays of alternative floral phenotypes. Plants varied either in corolla color or in the presence of a corolla spot.

Table 12.2. *The effects of frequency dependence and frequency independence on pollinator choice between different morphs in five plant species*

| Plant | Variable trait | Pollinators | Reward[a] | Replicates[b] | Frequency dependence[c] | Frequency Independence[c] | Source |
|---|---|---|---|---|---|---|---|
| *Clarkia gracilis* | Corolla spot | Solitary bees Social bees | Y | 5 (3) | $r_s = 0.63$ | $\chi^2 = 8.66$ | Jones 1996a |
| *Ipomoea purpurea* | Corolla color | Bumble bees | Y | 24 (5) | $1.27^{**}$ | $0.91^{**}$ | Epperson & Clegg 1987 |
| *Delphinium nelsonii* | Corolla color | Hummingbirds Bumble bees | Y | 6 (2) | $r_s = -0.88^*$ | $\chi^2 = 32.64^{***}$ | Waser & Price 1981 |
| *Cirsium palustre* | Corolla color | Bumble bees | Y | 21 (6) | 1.02 | $0.89^*$ | Mogford 1978 |
| *Tolumnia variegata* | Scent production | Bee sp. | N | 6 (2) | $F = 0.28$ | $F = 0.27$ | Ackerman *et al.* 1997 |

*Notes:*

[a] Reward is either present (Y) or absent (N).

[b] The numbers of replicates are given as the total number, with the number of different frequencies considered in parentheses.

[c] Where possible, frequency dependence and frequency independence are tested as for Table 12.1 using the Greenwood & Elton (1979) model. Where numbers of replicates were too low, frequency independence gives the result of a $\chi^2$ test for heterogeneity and frequency dependence the result of testing deviations of expected values for rank correlation with morph frequency (see text for discussion). In one case, the authors' F-statistics are given.

For the two studies with larger sample sizes, researchers found significant preferences for the common corolla colors in *Ipomoea purpurea*, and overall preference for one corolla color over another in both *I. purpurea* and *Cirsium palustre* (Table 12.2). For the two studies with smaller sample sizes, there is no indication of non-random pollinator visitation for *Clarkia gracilis*, but non-randomness is suggested for *Delphinium nelsonii*. There was no common-morph advantage in either data set; indeed, the marginally significant pattern for *D. nelsonii* suggests negative rather than positive frequency dependence. However, in this species one of the four data points shows a particularly large deviation from expected values accounts for most of the correlation; conclusions from such small sample sizes must be tentative. Using meta-analysis (Kirby 1993), I combined all results for the rewarding species to test the hypothesis of positive frequency dependence. The result did not support the hypothesis (average effect size $= 0.17$, $t = 0.42$, $p = 0.71$); the experiments were significantly heterogeneous (heterogeneity $\chi^2 = 11.32$, $p = 0.01$).

One unrewarding species has been studied with respect to frequency-dependent pollinator visitation – an epiphytic orchid, *Tolumnia variegata*. This species produces no nectar but relies on deception (without mimicry) of naïve pollinators for pollen transfer. Such species are termed "non-model deceptive" (Dafni 1986). *Tolumnia variegata* is polymorphic for the presence or absence of scent production. By manipulating morph frequencies in adjacent trees, Ackerman *et al.* (1997) found that pollinators showed neither overall preference for a morph nor frequency-dependent selection in favor of the rare morph.

Overall, a survey of the literature does reveal significant common-morph preference in at least one rewarding plant species, but this is weak; predictions are not upheld in three other rewarding species. For an unrewarding species, pollinators did not show significant preference for either morph type. Of course, these studies involved different species of animals.

### From individual behavior to selection – frequency-dependence in plant populations

### Evidence for FDS in rewarding plant populations

Four published studies of rewarding plant populations provide data relating the relative reproductive success of alternative morphs to morph frequency (Table 12.3). I have excluded floral size traits from consideration, along with Levin's (1972) data set for reasons discussed above. The studies include those from Table 12.2, with the exception of *Cirsium palustre*, and include an additional study on *Phlox pilosa* (Levin & Kerster 1970). For each study, I calculated relative fitness for one of the morphs considered, then used rank correlations to test for a relationship between relative reproductive success and morph frequency. I predicted that, for rewarding species, there should be a positive correlation between morph frequency and relative fitness (positive FDS). Note that this method tests for FDS in a broad sense; an indication of reversal of fitness advantage at some frequency must also be observed to implicate FDS in its stricter sense.

None of the published studies indicates a significant relationship between morph frequency and relative reproductive success (Table 12.3). In all three cases where seed set is considered, there are positive trends but the correlations are not significant. Outcrossed seed set in *Ipomoea purpurea* suggests a tendency for a negative correlation between relative reproductive success and morph frequency, but this result differs from that of the original authors (Epperson & Clegg 1987). They analyzed the deviation from expected for each data point individually; results must therefore be regarded as tentative. The only experiment to consider outcross seed paternity found a positive but non-significant relationship between relative reproductive success and morph frequency. Meta-analysis of these data was not appropriate due to the diversity of fitness indices used and the small sample sizes for some studies. Overall, results from rewarding species fail to support the hypothesis that differences in pollinator visitation rates between morphs induce positive FDS in plant populations.

### Evidence for FDS in unrewarding plant populations

Only one previously published study has measured FDS in an unrewarding plant species: again, this is *Tolumia variegata* (Table 12.3) (Ackerman *et al.* 1997). For an unrewarding species, we expect a negative correlation between morph fitness and relative reproductive success (negative FDS). However, the pollinators did not discriminate strongly between different scent morphs; there were no significant effects of frequency on male or female reproductive success (Ackerman *et al.* 1997).

In 1996, I attempted a test of the negative FDS hypothesis using

Table 12.3. *The impact of morph frequency on relative reproductive success in six plant species*

| Plant | Reward[a] | Replicates[b] | Fitness index[c] | Frequency dependence[d] | Source |
|---|---|---|---|---|---|
| *Clarkia gracilis* | Y | 7 | MO, FS | MO: $r_s = +0.43$<br>FS: $r_s = +0.58$ | Jones 1996*b* |
| *Ipomoea purpurea* | Y | 4 | FO | $r_s = -0.63$ | Epperson & Clegg 1987 |
| *Delphinium nelsonii* | Y | 12 | FS | $r_s = +0.36$ | Waser & Price 1981 |
| *Phlox pilosa* | Y | 11 | FS | $r_s = +0.32$ | Levin & Kerster 1970 |
| *Tolumnia variegata* | N | 6 | MR, FS | 1993: F = 0.00<br>1994: F = 0.61 | Ackerman *et al.* 1997 |
| *Dactylorhiza sambucina* | N | 17 | MR | $r_s = -0.44^*$ | A. Smithson, unpublished data |

*Notes:*
[a] Indicates whether rewards are present (Y) or not (N).
[b] Populations or arrays studied.
[c] MO = outcrossed seed paternity, FO = outcrossed seed set, MR = pollen removal, FS = total seed set.
[d] Frequency dependence was tested where possible as the rank correlation between relative reproductive success and morph frequency. Where not possible, the author's *F*-statistics are given. Significances are indicated as for Table 12.1.

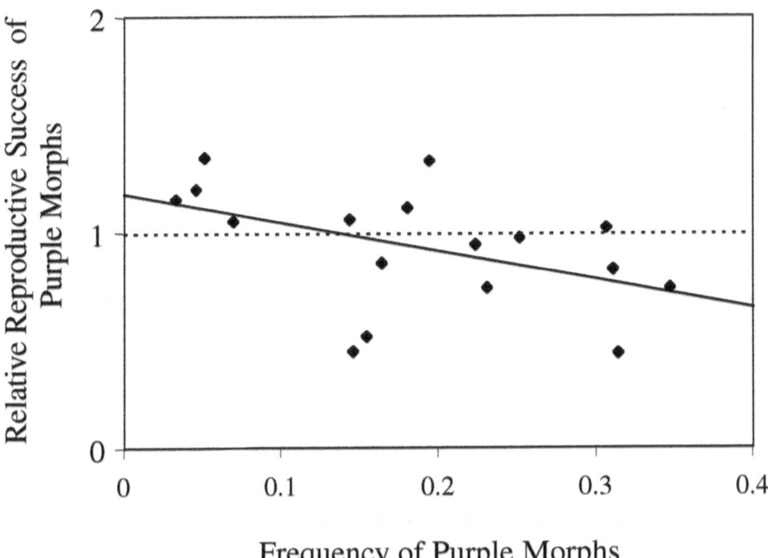

Frequency of Purple Morphs

Fig. 12.2. The relationship between morph frequency and relative reproductive success for two morphs of the unrewarding orchid Dactylorhiza sambucina. The dashed line indicates the line of equal morph fitness and the solid line the least-squares regression line ($y = 1.177 - 1.320\,x$).

*Dactylorhiza sambucina*, a non-model, deceptively pollinated orchid (Nilsson 1980). Widespread in mainland Europe, it has a dramatic corolla-color dimorphism, with both yellow and purple morphs present throughout its range (Tutin *et al.* 1980). Reproductive success through both fruit set and pollinia removal is pollinator-limited; pollinators are newly emerged queen bumble bees (Nilsson 1980). I measured morph frequencies and relative reproductive success in 17 populations of *D. sambucina* in southern France in May 1996. I sampled 20 individuals of each morph from each population (or all individuals in a population for rare morphs) and measured relative reproductive success as the number of pollinia removed per plant, (corrected for inflorescence size). Environmental parameters (altitude, location, substrate) affected neither mean morph frequency nor relative reproductive success, nor were there significant effects of population size (A. Smithson, unpublished data). Relative reproductive success and morph frequency were negatively correlated (Table 12.3, Fig. 12.2). The regression line relating relative reproductive success and morph frequency shows a reversal in fitness advantage at a frequency of 13.4% purple morphs (Fig. 12.2). This predicted equilibrium morph frequency approximates the actual mean frequency of the populations studied (18.96% ± 4.52%), but the strength of the FDS selection recorded – as indicated by the slope of the regression line – is comparatively weak (1.18). It could be argued that specific mimicry could produce the observed relationship. However, we have no evidence of specific mimics being present in the populations, and co-flowering species varied greatly among populations. To date, we have not found density to significantly interact with morph frequency or absolute reproductive success or relative reproductive success (L. Gigord & A. Smithson, unpublished data). Overall, the *D. sambucina* data currently support the hypothesis that there is a reproductive advantage for rare corolla color variants of unrewarding plant species; further manipulative experiments are under way to test competing hypotheses.

### Frequency-dependent selection by pollinators – fact or fiction?

Bumble bee workers under controlled conditions unequivocally forage in frequency-dependent ways in response to variation in corolla color, showing a common-morph preference when flowers contain rewards but a rare-morph preference when they do not. Field experiments, however, showed that pollinators preferred common corolla morphs in only one of four rewarding species. No significant correlations between morph frequency and relative reproductive success have been found in rewarding plant species, although there are suggestions of a negative relationship for a study on an unrewarding species. How may the contradictions in these results be explained?

### Does pollinator behavior in the field differ from that observed in the laboratory?

One potential cause of a difference between laboratory and field behavior is that the frequency dependence observed in bumble bees is not found in other pollinator types. However, FDS has been demonstrated in many animals, including birds (Allen 1988) and a range of predatory, parasitic, and herbivorous arthropods (Sherratt & Harvey 1993). Although further experiments are required, given that frequency-dependent behavior is widespread and common, many polylectic pollinators may respond to variation in floral traits such as color in a frequency-dependent way.

Laboratory experiments typically use highly distinct color morphs. It could be argued that natural floral polymorphisms are often less distinctive to the eyes of their pollinators, and thus induce weaker frequency-dependence. Alternatively, perhaps overall preference for a morph is so strong that frequency dependence is difficult to detect. Pollinators might not respond to floral traits other than color in a frequency dependent way. In experiments using three color morphs of rewarding artificial flowers, and testing for FDS between two that are relatively similar to the eyes of bumble bees, Smithson & Macnair (1997*b*) still detected significant positive FDS, but this was weaker than shown on yellow and blue flowers that contrasted more strongly (Table 12.1). Overall bias is important in determining the frequency at which preference switches in favoring one morph over another (Fig. 12.1, Table 12.1), so if this bias is strong, FDS may be detectable only at more extreme morph frequencies than those used in field experiments. Further experiments are required to test pollinators' responses to other types of floral traits.

It could also be argued that the sample sizes in the above studies may be too small. If pollinators in the wild forage as they do under laboratory conditions, we expect strong preferences, which should have been detectable at the sample sizes used. However, it is certainly noteworthy that experiments with significant effects are usually the ones with the higher sample sizes!

It is possible that pollinators might react differently to natural polymorphisms because real morphs differ not only in color but also in other traits, due to pleiotropy. Pleiotropic effects of variation in corolla color have been demonstrated in several cases (e.g., Waser & Price 1981), but not in others (e.g., Jones 1996*b*). Clearly, for floral traits like corolla color, pleiotropic effects will depend on the biochemical pathways involved. Raguso (this volume) suggests that because floral scent compounds may be produced by the same biochemical pathways as some floral pigments, pleiotropy may be common in characters important to pollinators. Such pleiotropic effects on pollinator behavior need evaluation.

### Do the predicted fitness relationships differ from those observed in the field?

A classic example of FDS is Batesian mimicry in butterflies, in which the fitness of the unpalatable model is inversely related to its abundance relative to that of its palatable mimic (Turner 1977). Both predator and prey are mobile, and the predator either consumes the prey totally or it does not. Contrast this with the type of selection that pollinators exert on immobile plants. The constant spatial remixing of available types does not occur as it does in butterflies. Pollinators consume only part of a plant; a visit usually depletes, but rarely fully consumes, the nectar or pollen. Nectar reward may be replenished by the plant, but slowly. Reproductive success is not an absolute form of selection, and pollinators may affect many aspects of overall fitness. I argue that these fundamental differences between predator–prey and plant–pollinator systems may cause a substantial difference in the frequency dependence expected in plant–pollinator systems.

First, if pollinator visitation levels are high, pollinators will deplete the rewards. Laboratory experiments show that varying the relative amounts of nectar present in two morphs of flowers has a weaker effect on frequency-dependent behavior by bumble bees than changing availability through presence or absence of nectar (Table 12.1). This suggests that differences in reward production between two morphs will affect FDS weakly as long as flowers are not emptied by pollinators, i.e., visitation rates are low. As soon as some flowers are empty, preference patterns will change. Depletion of one morph would lead to strong frequency-independent selection, and depletion of both morphs may to lead to negative FDS. Thus, selection patterns will be expected to vary both temporally and spatially according to factors that determine pollinator abundance and overall visitation rates to a species.

Second, as noted above, if morph frequencies are constant, a preference for common corolla colors develops initially over 100 flower visits and changes little subsequently. However, if morph frequencies of rewarding species fluctuate from patch to patch, patch-hopping pollinators may develop common-morph preference only weakly because of a lag in pollinator response to the morph frequency in the current patch. Alternatively, pollinator preference might be strongly influenced by the morph frequency in the first patch visited. Color morphs are patchily distributed within plant populations (Epperson & Clegg 1986), potentially making FDS much weaker than expected.

Third, the number of flowers visited per inflorescence increases with visitation rate, both for rewarding and unrewarding inflorescences. This relationship leads to the expectation that, while selection by visitation rate alone may be positively frequency-dependent for rewarding species, selection through selfing rates and outcross male and female reproductive success may be negatively frequency-dependent. Final reproductive success may not be predicted by visitation rates alone.

For rewarding plants, positive FDS is thus not necessarily expected, either for pollinator behavior or for selection in plant populations. Furthermore, results for different fitness indices may conflict. Pollinator response to the frequency of different plant species, as opposed to different morphs, also suggests that FDS may go in different directions, depending on the fitness index considered (Stout *et al.* 1998). Pollinator abundance is likely to be crucial, however, in determining the likelihood of positive FDS – it is more likely if pollinators are rare, because visitation rate will be a more important variable than the number of flowers visited per inflorescence in determining fitness.

Things are different for plants that present neither nectar or pollen to visitors. These receive few visits. Pollinators soon locate alternative rewards (Nilsson 1980), thereby limiting the reproductive success of the unrewarding plants (Gill 1989). Visits also do not change the reward status of unrewarding plants. I expect pollinator preferences for unrewarding plants to be independent of variation in morph frequency from patch to patch. Because the flowers are empty, pollinators should move long distances between visits, so they will not stay long enough in a patch to detect shifts in morph frequency (Pyke 1978). Also, when visitors sample only a few flowers per plant, there is little scope for the number of flowers visited per inflorescence to differ between morphs. In populations of unrewarding plant species without mimicry, therefore, negative FDS should act straightforwardly on floral variants for such traits as color.

### What can we predict about the evolutionary dynamics of floral polymorphisms?

Does FDS by pollinators retard floral evolution in rewarding plant populations by putting rare, novel morphs at a disadvantage? I suggest that only where pollinator visits are rare and reproductive success is pollinator-limited will such a disadvantage handicap the spread of a novel mutant. In other cases, pollinator bias for particular colors, and effects such as differences in nectar production, will be more important in determining whether a novel morph spreads. Where pollinator visitation rates are high, negative FDS may be exerted through selection on selfing rates and male function. To test these hypotheses, we need experiments that vary overall visitation rates and assess relative reproductive success.

Does FDS maintain floral polymorphisms in unrewarding plant species? Both behavioral and field data suggest that it does. We need further experiments to distinguish negative FDS from non-specific mimicry as the agent for the maintenance of the high levels of floral diversity found in unrewarding plant species (Heinrich 1975; Cropper & Calder 1990).

## Acknowledgements

I thank Luc Gigord and Jean-Baptiste Ferdy for much discussion and for working with me on artificial inflorescence experiments, Fran Harper for assistance with field research, and NERC for a research fellowship.

## References


Ackerman JD, Meléndez-Ackerman EJ & Salguero-Faria J (1997) Variation in pollinator abundance and selection on fragrance phenotypes in an epiphytic orchid. *Am J Bot* 84:1383–1390

Allen JA (1988) Frequency-dependent selection by predators. *Phil Trans R Soc Lond B* 319:485–503

Allen JA & Clarke B (1968) Evidence for apostatic selection by wild passerines. *Nature* 220:501–502

Ayala FJ & Campbell CA (1974) Frequency-dependent selection. *Ann Rev Ecol Syst* 5:115–138

Barrett JA (1988) Frequency-dependent selection in plant-fungal interactions. *Phil Trans R Soc Lond B* 319:473–483

Cartar RV & Abrahams LM (1996) Risk sensitive foraging in a patch departure context: a test with worker bumblebees. *Am Zool* 36:447–458

Chittka L & Thomson JD (1997) Sensori-motor learning and its relevance for task specialization in bumble bees. *Behav Ecol Sociobiol* 41:385–398

Clarke B (1962) Natural selection in mixed populations of two polymorphic snails. *Heredity* 17:319–345

Clarke B & O'Donald P (1964) Frequency-dependent selection. *Heredity* 19:201–206

Cropper SC & Calder DM (1990) The floral biology of *Thelymitra epipactoides* (Orchidaceae), and the implications of pollination by deceit on the survival of this rare orchid. *Plant Syst Evol* 170:11–27

Dafni A (1986) Floral mimicry – mutualism and unidirectional exploitation of insects by plants. In: Juniper B & Southwood R (eds) *Insects and the plant surface*, pp 81–90. Edward Arnold, London

Dawkins M (1971a) Shifts of "attention" in chicks during feeding. *Anim Behav* 19:575–582

Dawkins M (1971b) Perceptual changes in chicks: another look at the "search image" concept. *Anim Behav* 19:566–574

de Jong T, Waser NM & Klinkhamer PGL (1993) Geitonogamy: the neglected side of selfing. *Trends Ecol Evol* 8:321–325

Dukas R & Real LA (1993) Learning constraints and floral choice behavior in bumblebees. *Anim Behav* 46:637–644

Eckert CG, Manicacci D & Barrett SCH (1996) Frequency-dependent selection on morph ratios in tristylous *Lythrum salicaria* (Lythraceae). *Heredity* 77:581–588

Epperson BK & Clegg MT (1986) Spatial-autocorrelation analysis of flower color polymorphisms within substructured populations of morning glory (*Ipomoea purpurea*). *Am Nat* 128:840–858

Epperson BK & Clegg MT (1987) Frequency-dependent variation for outcrossing rate among flower-color morphs of *Ipomoea purpurea*. *Evolution* 41:1302–1311

Ferdy JB, Gouyon PH, Moret J & Godelle B (1998) Pollinator behavior and deceptive pollination: learning process and floral evolution. *Am Nat* 152:696–705

Gendron RP (1987) Models and mechanisms of frequency-dependent predation. *Am Nat* 130:603–623

Gill DE (1989) Fruiting failure, pollinator inefficiency, and speciation in orchids. In: Otte D & Endler JA (eds) *Speciation and its consequences*, pp 458–481. Sinauer, Sunderland, MA

Greenwood JJD & Elton RA (1979) Analysing experiments on frequency-dependent selection by predators. *J Anim Ecol* 48:721–737

Harder L & Barrett SCH (1995) Mating cost of large floral displays in hermaphrodite plants. *Nature* 373:512–515

Heino M, Metz JAJ & Kaitala V (1998) The enigma of frequency-dependent selection. *Trends Ecol Evol* 13:367–370

Heinrich B (1975) Bee flowers: a hypothesis on flower variety and blooming times. *Evolution* 29:325–334

Heuch I (1979) Equilibrium populations of heterostylous plants. *Theor Popul Biol* 15:43–57

Jones KN (1996a) Pollinator behavior and postpollination reproductive success in alternative floral phenotypes of *Clarkia gracilis* (Onagraceae). *Int J Plant Sci* 157:733–738

Jones KN (1996b) Fertility selection on a discrete floral polymorphism in *Clarkia* (Onagreaceae). *Evolution* 50:71–79

Kay QON (1978) The role of preferential and assortative pollination in the maintenance of flower colour polymorphisms. In: Richards AJ (ed) *The pollination of flowers by insects*, pp 175–190. Academic Press, London

Kirby KN (1993) *Advanced data analysis with Systat*. Van Nostrand Reinhold, New York

Klinkhamer PGL & de Jong TJ (1993) Attractiveness to pollinators – a plant's dilemma. *Oikos* 66:180–184

Levin DA (1972) Low frequency disadvantage in the exploitation of pollinators by corolla variants in *Phlox*. *Am Nat* 106:453–460

Levin DA & Kerster HW (1970) Phenotypic dimorphism and populational fitness in *Phlox*. *Evolution* 24:128–134

Manly BFJ (1973) A linear model for frequency-dependent selection by predators. *Res Popul Ecol* 14:137–150

Menzel R (1979) Behavioral access to short-term memory in bees. *Nature* 281:368–369

Menzel R (1999) Memory dynamics in the honeybee. *J Comp Physiol A* 185:323–340

Mogford DJ (1978) Pollination and flower colour polymorphism, with special reference to *Cirsium palustre*. In: Richards AJ (ed) *The pollination of flowers by insects*, pp 191–199. Academic Press, London

Nilsson LA (1980) The pollination ecology of *Dactylorhiza sambucina* (Orchidaceae). *Bot Notiser* 133:367–385

O'Donald P & Majerus MEN (1988) Frequency-dependent sexual selection. *Phil Trans R Soc Lond B* 319:571–586

Perez SM & Waddington KD (1996) Carpenter bee (*Xylocopa micans*) risk indifference and a review of nectarivore risk-sensitivity studies. *Am Zool* 36:435–446

Pyke GH (1978) Optimal foraging: movement patterns of bumblebees between inflorescences. *Theor Popul Biol* 13:72–98

Real LA (1990) Predator switching and the interpretation of animal choice behavior: the case for constrained optimisation. In: Hughes RN (ed) *Behavioral mechanisms of food selection*, pp 1–21. Springer-Verlag, Heidelberg

Roy BA & Widmer A (1999) Floral mimicry: a fascinating yet poorly understood phenomenon. *Trends Plant Sci* 4:325–330

Sherratt TN & Harvey IF (1993) Frequency-dependent food selection by arthropods: a review. *Biol J Linn Soc* 48:167–186

Smithson A (1995) Frequency-dependent selection amongst floral variants through the foraging behavior of bumblebees, *Bombus terrestris*. PhD thesis, University of Exeter, UK

Smithson A & Macnair MR (1996) Frequency-dependent selection by pollinators: mechanisms and consequences with regard to behavior of bumblebees *Bombus terrestris* (L.) (Hymenoptera: Apidae). *J Evol Biol* 9:571–588

Smithson A & Macnair MR (1997a) Density-dependent and frequency-dependent selection by bumblebees *Bombus terrestris* (L.) (Hymenoptera: Apidae). *Biol J Linn Soc* 60:401–417

Smithson A & Macnair MR (1997b) Negative frequency-dependent selection by pollinators on artificial flowers without rewards. *Evolution* 51:715–723

Stanton ML, Snow AA & Handel SN (1986) Floral evolution: attractiveness to pollinators increases male fitness. *Science* 232:1625–1627

Stanton ML, Snow AA, Handel SN & Bereczky J (1989) The impact of a flower-color polymorphism on mating patterns in experimental populations of wild radish (*Raphanus raphanistrum* L.). *Evolution* 43:335–346

Stout JC, Allen JA & Goulson D (1998) The influence of relative plant density and floral morphological complexity on the behavior of bumblebees. *Oecologia* 117:543–550

Thomson V (1984) Polymorphism under apostatic and aposematic selection. *Heredity* 53:677–686

Tinbergen L (1960) The natural control of insects in pinewoods. I. Factors influencing the intensity of predation by songbirds. *Arch Neerland Zool* 13:265–343

Treisman A & Gelade G (1980) A feature-integration theory of attention. *Cognit Psychol* 12:97–136

Turner JRG (1977) Butterfly mimicry: the genetical evolution of an adaptation. *Evol Biol* 10:163–206

Tutin TG, Heywood VH, Burges NA, Moore DM, Valentine DH, Walters SM & Webb DA (1980) *Flora Europaea*, vol. 5, *Alismataceae to Orchidaceae*. Cambridge University Press, Cambridge

Waser NM & Price MV (1981) Pollinator choice and stabilizing selection for flower color in *Delphinium nelsonii*. *Evolution* 35:376–390

Wright S (1948) On the roles of directed and random changes in gene-frequency in the genetics of natural populations. *Evolution* 2279–294

## 13

# Pollinator-mediated assortative mating: causes and consequences

A typical animal pollinator forages non-randomly among plants in a community, using floral cues to recognize the available options. The tendency of individual foragers to restrict their visits to a subset of the available flowering species increases the proportion of pollen grains that arrive on appropriate stigmas. Pollinators partition themselves among plants in several ways, with the common result of assortative mating according to floral type. First, I discuss the evolutionary implications of assortative mating, in light of recent models that emphasize its importance for species divergence, then review the ways in which pollinator behavior contributes to assortative mating among floral types. Finally, I consider how the different forms of non-random pollinator behavior might influence floral evolution and plant speciation.

There is a long-standing tradition of thought that visitation by different pollinators drives divergence of floral form and provides reproductive isolation among incipient plant species (reviewed by Waser, this volume). However, pollinators rarely specialize completely on a single floral type (plant species or distinct phenotype within a species), leading some investigators to question the role of pollinators in the radiation of the angiosperms, and to suggest that floral evolution is largely decoupled from plant speciation (Waser 1998; Chittka *et al.* 1999). None the less, the remarkable radiation of angiosperms in parallel with pollinators (Grimaldi 1999), and findings that plant families with animal pollination are more speciose than those with abiotic pollination (Dodd *et al.* 1999), suggest that animal pollination was a key innovation in flowering plant evolution. Exactly how pollinators might contribute to speciation and diversification in plants is currently debated (see Waser, this volume), with discussion centered on the importance of pollinators as agents of reproductive isolation and whether divergence can occur despite ongoing gene flow (Grant 1994; Waser 1998).

Insights from recent models of animal speciation (Dieckmann & Doebeli 1999; Kondrashov & Kondrashov 1999) may help to resolve this debate. Application of the modeling results to plants suggests that even moderate pollinator specialization can be important to the initiation or maintenance of divergence in floral form and the process of plant speciation, because such behavior causes some degree of assortative mating among the plants (Jones 1997). Plants have many different mechanisms to promote assortative mating (Levin 1978), manipulation of pollinator behavior being but one. However, genetically simple changes in floral traits can alter pollinator choices significantly (Schemske & Bradshaw 1999); consequently assortative mating may evolve relatively easily via this mechanism. Without single-handedly conferring complete reproductive isolation, pollinators may nevertheless advance plant diversification in their role as agents of assortative mating, acting in concert with other factors as explained below.

## Evolutionary significance of assortative mating

Positive assortative mating is the non-random pairing of individuals that are more closely alike than the average in one or more phenotypic traits (Lincoln *et al.* 1998). For quantitative traits, assortative mating inflates the variance of the trait in a population by increasing the coupling of alleles with similar effects (Lynch & Walsh 1998, p 154). The traits most likely to be affected in animals are those used for mate choice or species recognition, including color pattern, scent, shape, or size. In plants, the affected traits are those used by pollinators to find and recognize flowers.

When intermediate phenotypes have low relative fitness ("disruptive selection"), assortative mating is selectively favored over random mating. If such selection acts directly on a mating trait, assortment with respect to that trait can evolve quickly. The frequency of intermediate phenotypes drops for two reasons: lower production of intermediate offspring due to assortative mating, and selection against those that do arise. Ultimately, indirect selection favors the evolution of reproductive isolation between extreme phenotypes (Kondrashov & Kondrashov 1999). That speciation theoretically can occur in sympatry by this mechanism is not terribly controversial, but the condition that the mating trait itself be the trait under direct disruptive selection is quite restrictive. There are a few clear examples of disruptive selection by pollinators on floral traits, such as in *Polemonium* (Galen *et al.* 1987) and *Ipomopsis* (Campbell *et al.* 1997), but not as many as might be expected if this were an important mechanism of divergence in plants (Wilson & Thomson 1996; Goulson & Jerrim 1997).

A more likely general scenario is that disruptive selection acts on some other ecologically important trait – for example, one involved in resource use or acquisition. Phenotypic divergence may reduce intraspecific competition, with extreme phenotypes favored over intermediate ones. When mating is random with respect to an ecological trait, disruptive selection on the trait increases the phenotypic variance but does not produce the evolutionary branching that would allow further divergence (Dieckmann & Doebeli 1999). The key to divergence in such a scenario is a genetic correlation between the trait under disruptive selection and a trait that promotes assortative mating. In an animal-based illustration of their sympatric speciation model, with disruptive selection on body size and mate choice based on color, Kondrashov & Kondrashov (1999) suggest thinking of the process as "the recruitment of colour for providing reproductive isolation between individuals of the opposite size."

When there is genetic variance in a mating trait, correlations between the mating trait and other characters occur readily via drift or inbreeding, especially for polygenic traits. The correlations usually are temporary, soon broken up by recombination. However, selection against intermediate ecological phenotypes strengthens the genetic correlation, as recombinants would be more likely to mate with the opposite ecological type due to their mating phenotype, if mating is at all assortative with respect to the mating trait. In addition to selection against recombination, assortative mating itself reduces heterozygosity at the mating loci, further constraining the ability of recombination to break up the developing gene complexes. A positive-feedback loop can form, with disruptive selection, assortative mating, and genetic correlation strengthening each other, culminating in genetically isolated groups that are distinct for both mating and ecological characters, according to two different recent models (Dieckmann & Doebeli 1999; Kondrashov & Kondrashov 1999).

In cases where floral traits are genetically correlated with locally adapted ecological races ("ecotypes"), assortative pollinator foraging improves pollen transfer between mates adapted to similar ecological circumstances. When the post-pollination barriers to hybridization are complete, such "pollen targeting" increases pollination efficiency by reducing gamete wastage. When the barriers are partial, so that hybridization occurs but hybrid offspring have low relative fitness or are ill-suited to either environment of parental ecotypes, pollen targeting increases average offspring quality by reducing the frequency of hybrid offspring. Local mating within plant populations contributes to population structuring, with some divergence of genetic neighborhoods (Turner *et al.* 1982; see Waser, this volume). When local adaptation results, assortative mating among neighborhoods helps to maintain co-adapted gene complexes. Overall, selection for assortative mating is likely to be very common in plants, with the strength of selection depending on the relative fitness of hybrid or intermediate offspring.

In general, we might expect those lineages that can attain assortative mating relatively easily to be more diverse and speciose than others. In angiosperms, animal pollination provides several additional mechanisms (beyond floral phenology and pollen–pistil interactions, for example) for assortative mating via pollen targeting.

### Pollinator behavior that results in assortative mating

Faced with multiple types of flowers in a plant community, an individual pollinator typically associates more strongly with a subset of flower types than would be expected based on the frequency of the plants in the community. To the extent that these associations are based on innate or fixed preferences, they should be consistent across individuals within pollinator taxa (Waser 1986). That they often are less than completely consistent within a pollinator species indicates the potential for learning, modification of foraging preferences based on individual experiences, and perhaps different innate preferences among individuals. Not only may different individuals preferentially visit different types of flowers, but the same individual may exhibit different preferences over time. Such "labile preference" is considered adaptive given the dynamic nature of floral resource availability (Heinrich 1976). There also can be further specialization by individuals due to constraints on information processing, particularly memory retrieval. Such specialization results in visitation to fewer flower types than would be predicted if foraging were optimal (Chittka *et al.* 1999; but see Menzel, this volume). These various types of non-random pollinator behavior all contribute to the widespread phenomenon of flower constancy, broadly defined, in which pollinators tend to restrict their visits to one or a few types of flowers within a foraging bout, i.e., to specialize to some extent, at least in the short term. It is useful to analyze separately these different kinds of behavior that make up flower constancy, as their implications for plant evolution are not the same.

Different pollinator taxa may display different innate preferences for certain flower types, based on the morphological fit between animal and flower, for example, or on differing sensitivity to various stimuli such as floral color or scent (see Grant 1994 for review; Giurfa *et al.* 1995). The traditional view – that divergent selection by different pollinator taxa drives floral differentiation and promotes reproductive isolation via assortative mating (Grant 1949) – assumes that pollinator preferences are reasonably consistent over time, as innate preferences are likely to be. At least a few cases appear to fit this scenario, notably in the species-rich Cape flora of South Africa, where some pollinators have highly specialized morphology that "fits" with similarly specialized flowers. For example, several plant taxa in the Cape region have very long, narrow corolla tubes accessed by flies with exceedingly long proboscides (Goldblatt *et al.* 1995; Johnson 1996; see Johnson & Steiner 2000 for review). However, other more common pollinator taxa such as bees and hummingbirds often are quite generalized and opportunistic (Waser *et al.* 1996), as experience may override innate preferences (Giurfa *et al.* 1995). Investigation of the extent to which pollinator taxa partition floral resources in the same way over replicate communities from year to year (e.g., Cripps & Rust 1989) and of how the presence or absence of one pollinator taxon influences resource partitioning by the rest (e.g., Inouye 1978; Laverty & Plowright 1985), would help to determine the potential for innate preferences characteristic of pollinator species to impose consistent selection on floral traits.

Within the range of floral types suitable for a given pollinator species, different pollinator individuals may make different choices (Heinrich 1976). Labile preference and constancy due to behavioral constraints both come into play, as individuals learn from experience but often apparently do not make optimal use of the information gained. Potential explanations for flower constancy involving behavioral constraints have been reviewed thoroughly in this volume (Gegear & Laverty; Menzel) and by Chittka *et al.* (1999).

In addition to varying preferences among individual pollinators, preferences might be expected to vary over the course of a day as reward distributions change or over a season as different plant species become available. Thus, labile preference can be examined by testing for homogeneity of preference among foraging bouts of an individual over time, or among bouts across individuals (see Jones 1997 for methods). Behavioral constraints seem to be especially important in the quick succession of choices within a foraging bout, for which short-term memory may be more important than long-term memory (Chittka *et al.* 1999; but see Menzel, this volume). For this, analyses of the sequences of plants visited within foraging bouts are appropriate (Bateman 1951; Waser 1986; Jones 1997).

While it is useful to distinguish floral resource partitioning between versus within pollinator taxa for comparative purposes (see next section), the two processes certainly are not mutually exclusive. Partitioning between taxa is easy to test by counting visitors to the various floral types, but differences among individuals within pollinator taxa may be superimposed on overall species preferences; thus pollinator-mediated assortative mating may be stronger than counts of visitors might suggest (e.g., Fulton & Hodges 1999; see Thomson & Chittka, this volume). Even in a reasonably clear case of partitioning of pollinator taxa between "hummingbird-pollinated" *Mimulus cardinalis* and "bee-pollinated" *M. lewisii*, pollinator preferences were not completely consistent within taxa (Schemske & Bradshaw 1999). Bumble bees collecting pollen were observed most often on red or orange (rather than pink) flowers, whereas increasing carotenoid (yellow pigment) concentration in petals showed a strong negative correlation with bee visitation overall, as bees foraged for nectar on most bouts on the experimental arrays. Assortative mating in the *Mimulus* complex appears to result from partitioning of floral resources both among and within pollinator taxa, a likely scenario for sympatric plant species pairs in general.

### Consequences for plants

#### Partitioning between pollinator taxa

Innate preferences by different pollinator taxa may produce consistent selection on floral traits. Selection may shape suites of characters that work together (such as flower shape, size, color, scent, and timing of bloom) to attract and reward certain pollinator types as the interaction becomes more specialized. Such coordination among traits may make it difficult to switch to a different type of pollinator, as a mutant in one floral character probably would not do well in the context of the other characters, especially in the presence of the original pollinators. Thus the traditional scenario of floral divergence being driven by adaptation to different pollinators probably requires geographic isolation between plant populations served by different pollinators (Wilson & Thomson 1996; Chittka *et al.* 1999). However, if secondary contact occurs following allopatric divergence, and both pollinator types are present in the zone of contact, then assortative mating via partitioning of the incipient plant species between pollinator taxa may be important for the maintenance of the two distinct lineages (e.g., Fulton & Hodges 1999; Schemske & Bradshaw 1999). If there is no selection against hybrids, such assortative mating probably will not suffice to prevent hybridization and introgression, as it is unlikely that pollinators will be completely constant. The more divergence there has been between populations, the greater the probability that there will be selection against hybrids, through reduced viability, pollinator attraction, or fecundity. This scenario resembles an advanced case of divergence in sympatry, as outlined earlier, in that there is already a strong genetic correlation between mating characters and traits responsible for low hybrid fitness. The combination of the genetic correlation, assortative mating, and selection against hybrids may suffice to keep the divergent lineages genetically distinct. Modeling of such situations is in order, to determine the strength and consistency of selection and assortative mating needed to maintain distinct lineages, along the lines of the animal-based models of Servedio (2000).

It seems very unlikely that partitioning among pollinator taxa would be important for initiating divergence within a population experiencing disruptive selection. Plants within a population have much less opportunity to develop different combinations of floral traits that appeal to innate preferences of different pollinators, compared to an allopatric situation. It is conceivable that variation in one key trait, such as length of a nectar spur, could initiate divergence by effectively restricting which visitor taxa transfer pollen (Fulton & Hodges 1999) and by causing a correlation in trait value (spur length) among mates, i.e., assortative mating. When spur length is correlated with trait(s) under disruptive selection, additional mechanisms that strengthen assortative mating and enable divergence, such as shifts in flowering phenology, would be favored. However, even in this case with variation in a key mating trait, competitive exclusion of alternative floral morphs by the preferred morph of the most effective or frequent pollinator seems the most likely outcome. Different populations with dissimilar pollinator communities may have different competitive outcomes, and thus diverge in allopatry, but gene flow and competition make such divergence within populations unlikely.

Partitioning of floral resources between pollinator taxa may serve to eliminate some members of a plant community from a pollen-transfer pool, but seems unlikely by itself to isolate single species. Even in relatively specialized plant–pollinator communities in the Cape region of South Africa, distantly related plants share pollinators, but often place pollen on different parts of pollinator bodies, thereby furthering assortative mating via mechanical isolation (Dobzhansky 1937; Grant 1949, 1994). For example, *Lapeirousia silenoides* (Iridaceae) and *Pelargonium sericifolium* (Geraniaceae) are both pollinated by long-tongued flies of the genus *Prosoeca* (Nemestrinidae) and exhibit remarkable convergence of floral form (Goldblatt *et al.* 1995). The iris deposits pollen on the head and dorsal side of the flies, whereas the geranium places it ventrally. Flowers of both species have zygomorphic symmetry, which apparently helps to orient the flies "correctly" on the flowers (Goldblatt *et al.* 1995).

In summary, partitioning of floral resources among pollinator taxa may help to maintain distinct lineages when plant populations that have diverged in allopatry come into secondary contact, providing reinforcement of any post-pollination barriers to hybridization. However, this form of pollinator-mediated assortative mating is unlikely to aid in the initial divergence of lineages in sympatry, as gene flow and competition make it difficult to maintain multiple co-adapted gene complexes in a population. In general, partitioning among pollinator taxa seems much more likely to be only part of a suite of mechanisms of assortative mating, rather than to provide by itself complete reproductive isolation of sympatric plant species.

### Partitioning within pollinator taxa

Relatively short-term specialization by individuals has quite different consequences than fixed preferences according to pollinator type. Variation in a single floral trait may suffice to cause assortative mating via behavioral constancy or labile preference, if the trait is used as a recognition cue by foragers; differences in groups of traits are likely to induce stronger constancy (see Gegear & Laverty, this volume). Thus, evolutionarily labile traits such as petal size or color can induce assortative mating without appealing particularly to one kind of pollinator over another. One forager might prefer darker flowers because the first flowers it visited were dark and had plentiful nectar. Another forager might visit the same flowers and find them unrewarding relative to light ones and consequently develop a preference for light flowers. Over several seasons of experiments with freely foraging bumble bees (primarily *Bombus appositus*) visiting randomized arrays of two colors of snapdragons (yellow and white or yellow and red *Antirrhinum majus*: Scrophulariaceae), I found a great deal of heterogeneity of preference among foraging bouts and assortative mating with respect to flower color (Jones 1997; Jones & Reithel 2001).

When the same kinds of pollinators visit incipient plant species, there is little opportunity for disruptive (in sympatry) or divergent (in allopatry) selection directly on floral traits. Plants may diverge in traits that are more or less neutral for pollinator attraction, such as petal color pattern in *Clarkia* (Jones 1996). This idea may help to resolve Ollerton's "paradox of plant–pollinator systems," which hinges in part on the questionable supposition that "specialization to a taxonomically narrow array of pollinators would appear to be a prerequisite for the evolution of floral novelty" (Ollerton 1996). Divergence in floral recognition cues may be an effective way to improve pollen targeting and assortative mating regardless of changes in pollinator communities, when the cues are not specifically tuned to particular pollinator taxa.

To the extent that foragers in an area are systematic, tending to return to patches at regular intervals, the relative rewards of different flower morphs in a patch fluctuate somewhat predictably (Possingham 1988). Extremely systematic foraging – trapline foraging – is known in several kinds of bees (Heinrich 1976; Kadmon 1992; Thomson 1996; Thomson & Chittka, this volume) and hummingbirds (Gill 1988). Labile preference and behavioral constancy should thus tend to "even out" pollinator service, as previously under-exploited flower types are discovered and preferentially visited for awhile. Short-term specialization therefore would be more likely to maintain multiple floral morphs in a population than would fixed pollinator preferences. Following the chain of logic through to the speciation process, individual specialization should be more likely than partitioning among pollinator taxa to help initiate divergence in a plant population under disruptive selection.

In cases of secondary contact of plant populations following divergence in allopatry, individual pollinator specialization seems at least as likely as partitioning among pollinator taxa to maintain genetically distinct lineages. Both types of non-random pollinator foraging provide the necessary assortative mating, but the individual specialization mechanism does not require the presence of multiple kinds of pollinators in order for each incipient plant species to be competitive. For example, an

*Ipomopsis* hybrid zone appears to fit an "advancing wave" model, in which traits characteristic of hummingbird-favored *I. aggregata* have an overall advantage due to the far greater and more reliable abundance of hummingbirds than hawkmoths; therefore *I. aggregata* is predicted to spread at the expense of *I. tenuituba* (Campbell *et al.* 1997). If some hummingbirds occasionally favored *I. tenuituba*, it would seem to have a much better chance of local persistence, but there is no evidence of constancy by hummingbirds foraging among these *Ipomopsis* species and hybrids (see Waser, this volume).

Short-term pollinator specialization seems especially relevant for the establishment of new hybrid or polyploid species in sympatry with parental species. Such lineages – allopolyploids, for example – typically occur in initially low frequencies, and thus face the disadvantage of being minority cytotypes (Levin 1983). Persistence of a new hybrid or polyploid lineage is very unlikely, unless it has some means of reproductive isolation from parental species (Thompson & Lumaret 1992). Assortative mating via pollinator specialization is a possibility when hybrids have distinct flowers, as is the case for allopolyploids such as *Tragopogon mirus* (Cook & Soltis 1999) and the autopolyploid *Heuchera grossulariifolia* (Segraves & Thompson 1999). For example, most pollinators distinguished between diploid and tetraploid *H. grossulariifolia* plants, with several insect species visiting the two floral types at significantly different frequencies, thus providing incomplete partitioning with respect to pollinator taxa (Segraves & Thompson 1999). Whether individual pollinators within species exhibited further specialization is not clear from the study; such specialization could promote stronger assortative mating than suggested by the species preference differences. Because labile preferences and behavioral constancy can result in occasional specialization on generally undervisited floral types, and because hybrid or polyploid flowers are unlikely to be different enough from both parental types to be visited by different pollinator taxa, partitioning within rather than between pollinator taxa seems more likely to aid in the establishment of new hybrid or polyploid species. Polyploidy is considered a prominent mechanism of speciation in plants (Soltis & Soltis 1993) and hybrid formation is quite common, especially in outcrossing perennials: naturally occurring hybrids make up 5%–22% of the species in five biosystematic floras, according to a recent survey (Ellstrand *et al.* 1996).

Pollinator specialization may be harder to come by when hybrids are present and serve as a bridge between species that otherwise are different enough to induce some level of pollinator-mediated assortative mating (Hodges *et al.* 1996; Rieseberg & Carney 1998). This is the case in hybrid swarms of *Baptisia* in Texas, where pollinator constancy to the parental species is strong enough to depress production of F₁ hybrids, but where pollinators move freely between hybrids and parental species, at least in randomized arrays (Leebens-Mack & Milligan 1998). Other mechanisms of assortative mating must maintain the distinct lineages, such as poor competitiveness of pollen from hybrid plants. Pollen competition has been documented in at least two cases (Carney *et al.* 1996; Klips 1999) to be a partial isolating mechanism between hybridizing lineages (a form of intrinsic reproductive barrier, discussed more fully by Waser, this volume).

In summary, when selection favors assortative mating, the easiest way to accomplish this may be to diverge enough in a floral recognition character, such as petal pigmentation pattern, to induce individual pollinator constancy. Short-term specialization may help to maintain a low-frequency floral type or one that is relatively undervisited overall. Pollinator-mediated assortative mating, together with selection against backcrossing (usually very strong in polyploids), may suffice to isolate a new hybrid or polyploid lineage from parental lineages, thus increasing the chance of speciation (Grant 1971; Rieseberg 1997; Wolfe *et al.* 1998; Milne *et al.* 1999); its importance in this regard remains to be established. This task will be facilitated by studies testing for individual pollinator specialization (i.e., constancy in the narrow sense and heterogeneity of preference) as well as partitioning among pollinator taxa with respect to the different plant lineages; the former behavioral mechanism is more likely and generally less studied by botanists than the latter.

## Conclusions

Floral traits generally may be under selection both to increase pollinator service and to improve targeting of pollen to compatible mates; this is not a new idea (Raven 1972; Heinrich 1975). However, pollen targeting becomes a more important issue if it plays a key role in the process of speciation, as well as serving to reinforce existing barriers to hybridization and reduce gamete wastage. Floral traits that are used as recognition cues by pollinators, such as color and scent, may evolve primarily to improve pollen targeting rather than to increase pollinator visitation.

Floral differences sufficient to induce some degree of pollinator constancy and thus assortative mating can be much more subtle than those that change the taxonomic identity of pollinators, particularly to different "syndromes." For purposes of assortative mating, the divergent groups do not need to adapt to different pollinator taxa, and no selection directly on floral traits is required. Instead, the species or incipient species need to accomplish the relatively easy task of diverging sufficiently for individual pollinators to use their variable trait(s) as a basis for constancy or labile preference, thereby improving pollen targeting. The pool of pollinators is thus partitioned on a much finer scale than by taxa.

Assortative mating due to flower constancy by pollinators may be a critical step toward reproductive isolation between ecologically divergent incipient species. In light of recent simulations indicating expanded conditions under which assortative mating may induce reproductive isolation, it is reasonable to envision scenarios in which plant divergence is propelled by assortative pollen flow arising from modified floral signals, often without a change in pollinator identity. I have presented several scenarios for pollinator-mediated divergence in sympatry and reinforcement of differences evolved in allopatry, not because I think they are ubiquitous, but rather to prompt tests of these often overlooked possibilities as mechanisms of diversification and maintenance of species boundaries in angiosperms.

## Acknowledgements

Many thanks to R. Abbot, P. Hecht, K. Holsinger, P. Soltis, J. Wakeley, N. Waser, and especially D. Baum for helpful discussions of these ideas. This work was made possible by a National Science Foundation POWRE (Professional Opportunities for Women in Research and Education) Visiting Professorship, hosted by D. Baum at Harvard University.

## References

Bateman AJ (1951) The taxonomic discrimination of bees. *Heredity* 5:271–278

Campbell DR, Waser NM & Melendez-Ackerman EJ (1997) Analyzing pollinator-mediated selection in a plant hybrid zone: hummingbird visitation patterns on three spatial scales. *Am Nat* 149:295–315

Carney SE, Hodges SA & Arnold ML (1996) Effects of differential pollen-tube growth on hybridization in the Louisiana irises. *Evolution* 50:1871–1878

Chittka L, Thomson JD & Waser NM (1999) Flower constancy, insect psychology, and plant evolution. *Naturwiss* 86:361–377

Cook LM & Soltis PS (1999) Mating systems of diploid and allotetraploid populations of *Tragopogon* (Asteraceae). I. Natural populations. *Heredity* 82:237–244

Cripps C & Rust RW (1989) Pollen foraging in a community of *Osmia* bees
     (Hymenoptera: Megachilidae). *Environ Entomol* 18:582–589
Dieckmann U & Doebeli M (1999) On the origin of species by sympatric speciation.
     *Nature* 400:354–357
Dobzhansky T (1937) *Genetics and the origin of species*, 1st edn. Columbia University Press,
     New York
Dodd ME, Silvertown J & Chase MW (1999) Phylogenetic analysis of trait evolution and
     species diversity variation among angiosperm families. *Evolution* 53:732–744
Ellstrand NC, Whitkus R & Rieseberg LH (1996) Distribution of spontaneous plant
     hybrids. *Proc Natl Acad Sci USA* 93:5090–5093
Fulton M & Hodges SA (1999) Floral isolation between *Aquilegia formosa* and *Aquilegia
     pubescens*. *Proc R Soc Lond B* 266:2247–2252
Galen C, Zimmer KA & Newport ME (1987) Pollination in floral scent morphs of
     *Polemonium viscosum*: a mechanism for disruptive selection on flower size.
     *Evolution* 41:599–606
Gill FB (1988) Trapline foraging by hermit hummingbirds: competition for an
     undefended, renewable resource. *Ecology* 69:1933–1942
Giurfa M, Núñez J, Chittka L & Menzel R (1995) Colour preferences of flower-naïve
     honeybees. *J Comp Physiol A* 177:247–259
Goldblatt P, Manning JC & Bernhardt P (1995) Pollination biology of *Lapeirousia*
     subgenus *Lapeirousia* (Iridaceae) in southern Africa: floral divergence and
     adaptation for long-tongued fly pollination. *Ann Missouri Bot Gard* 82:517–534
Goulson D & Jerrim K (1997) Maintenance of the species boundary between *Silene dioica*
     and *S. latifolia* (red and white campion). *Oikos* 79:115–126
Grant V (1949) Pollination systems as isolation mechanisms in angiosperms. *Evolution*
     3:82–97
Grant V (1971) *Plant speciation*. Columbia University Press, New York
Grant V (1994) Modes and origins of mechanical and ethological isolation in
     angiosperms. *Proc Natl Acad Sci USA* 91:3–10
Grimaldi D (1999) The co-radiations of pollinating insects and angiosperms in the
     Cretaceous. *Ann Missouri Bot Gard* 86:373–406
Heinrich B (1975) Bee flowers: a hypothesis on flower variety and blooming time.
     *Evolution* 29:325–334
Heinrich B (1976) The foraging specializations of individual bumblebees. *Ecol Monogr*
     46:105–128
Hodges SA, Burke JM & Arnold ML (1996) Natural formation of *Iris* hybrids:
     experimental evidence on the establishment of hybrid zones. *Evolution*
     50:2504–2509
Inouye DW (1978) Resource partitioning in bumble bees: experimental studies of
     foraging behavior. *Ecology* 59:672–678
Johnson SD (1996) Pollination, adaptation and speciation models in the Cape flora of
     South Africa. *Taxon* 45:59–66
Johnson SD & Steiner KE (2000) Generalization versus specialization in plant
     pollination systems. *Trends Ecol Evol* 15:140–143
Jones KN (1996) Fertility selection on a discrete floral polymorphism in *Clarkia*
     (Onagraceae). *Evolution* 50:71–79
Jones KN (1997) Analysis of pollinator foraging: tests for non-random behaviour. *Funct
     Ecol* 11:255–259

Jones KN & Reithel JS (2001) Pollinator-mediated selection on a flower color polymorphism in experimental populations of *Antirrhinum* (Scrophulariaceae). *Am J Bot*, in press.

Kadmon R (1992) Dynamics of forager arrivals and nectar renewal in flowers of *Anchusa strigosa*. *Oecologia* 92:552–555

Klips RA (1999) Pollen competition as a reproductive isolating mechanism between two sympatric *Hibiscus* species (Malvaceae). *Am J Bot* 86:269–272

Kondrashov AS & Kondrashov FA (1999) Interactions among quantitative traits in the course of sympatric speciation. *Nature* 400:351–354

Laverty TM & Plowright RC (1985) Competition between hummingbirds and bumble bees for nectar in flowers of *Impatiens biflora*. *Oecologia* 66:25–32

Leebens-Mack J & Milligan BG (1998) Pollination biology in hybridizing *Baptisia* (Fabaceae) populations. *Am J Bot* 85:500–507

Levin DA (1978) The origin of isolating mechanisms in flowering plants. *Evol Biol* 11:185–317

Levin DA (1983) Polyploidy and novelty in flowering plants. *Am Nat* 122:1–25

Lincoln R, Boxshall G & Clark P (1998) *A dictionary of ecology, evolution and systematics*. 2nd edn. Cambridge University Press, Cambridge

Lynch M & Walsh B (1998) *Genetics and analysis of quantitative traits*. Sinauer, Sunderland, MA

Milne RI, Abbott RJ, Wolff K & Chamberlain DF (1999) Hybridization among sympatric species of *Rhododendron* (Ericaceae) in Turkey: morphological and molecular evidence. *Am J Bot* 86:1776–1785

Ollerton J (1996) Reconciling ecological processes with phylogenetic patterns: the apparent paradox of plant–pollinator systems. *J Ecol* 84:767–769

Possingham HP (1988) A model of resource renewal and depletion: applications to the distribution and abundance of nectar in flowers. *Theor Popul Biol* 33:138–160

Raven P (1972) Why are bird-visited flowers predominantly red? *Evolution* 26:674

Rieseberg LH (1997) Hybrid origins of plant species. *Ann Rev Ecol Syst* 28:359–389

Rieseberg LH & Carney SE (1998) Plant hybridization. *New Phytol* 140:599–624

Schemske DW & Bradshaw HD (1999) Pollinator preference and the evolution of floral traits in monkeyflowers (*Mimulus*). *Proc Natl Acad Sci USA* 96:11910–11915

Segraves KA & Thompson JN (1999) Plant polyploidy and pollination: floral traits and insect visits to diploid and tetraploid *Heuchera grossulariifolia*. *Evolution* 53:1114–1127

Servedio MR (2000) Reinforcement and the genetics of nonrandom mating. *Evolution* 54:21–29

Soltis DE & Soltis PS (1993) Molecular data and the dynamic nature of polyploidy. *Crit Rev Plant Sci* 12:243–275

Thompson JD & Lumaret R (1992) The evolutionary dynamics of polyploid plants: origins, establishment and persistence. *Trends Ecol Evol* 7:302–307

Thomson JD (1996) Trapline foraging by bumblebees. I. Persistence of flight-path geometry. *Behav Ecol* 7:158–164

Turner ME, Stephens JC & Anderson WW (1982) Homozygosity and patch structure in plant populations as a result of nearest-neighbor pollination. *Proc Natl Acad Sci USA* 79:203–207.

Waser NM (1986) Flower constancy: definition, cause, and measurement. *Am Nat* 127:593–603

Waser NM (1998) Pollination, angiosperm speciation, and the nature of species boundaries. *Oikos* 82:198–201

Waser NM, Chittka L, Price MV, Williams N & Ollerton J (1996) Generalization in pollination systems, and why it matters. *Ecology* 77:279–296

Wilson P, Thomson JD (1996) How do flowers diverge? In: Lloyd DG & Barrett SCH (eds) *Floral biology: studies on floral evolution in animal-pollinated systems*, pp 88–111. Chapman & Hall, New York

Wolfe AD, Xiang QY & Kephart SR (1998) Diploid hybrid speciation in *Penstemon* (Scrophulariaceae). *Proc Natl Acad Sci USA* 95:5112–5115

## 14

# Behavioral responses of pollinators to variation in floral display size and their influences on the evolution of floral traits

The number of flowers open at any one time on a plant, i.e., floral display size, varies greatly among plant species. For example, some species flower during a brief period and have many open flowers, while others have extended flowering with only a few open flowers at one time (Gentry 1974; Bawa 1983). Also, floral display size often varies among individuals of the same plant species (e.g., Willson & Price 1977; Pleasants & Zimmerman 1990). The causes of such variations in floral display size are enduring interest to plant ecologists (reviewed by de Jong *et al.* 1992).

Numerous studies have reported that variation in floral display size produces marked alterations in pollinator behavior. Especially, two types of pollinator response to increased floral display size have been recognized from the perspective of their influences on pollen dispersal. First, larger floral displays attract more pollinators per unit of time (Fig 14.1A; reviewed by Ohashi & Yahara 1998). This will promote cross-pollination in terms of increased pollen receipt, removal, or potential mate diversity (Harder & Barrett 1996). Second, the number of flowers that individual pollinators probe per plant also increases with floral display size (Fig. 14.1B; also reviewed by Ohashi & Yahara 1998). This will increase self-pollination among flowers on the same plant ("geitonogamy"; Richards 1986; de Jong *et al.* 1993). Thus, variation in floral display size may lead to a substantial difference in pollen dispersal and, in turn, plant fitness.

To understand how plant fitness can be related to floral display size, we have to know the shapes of the functional relationships between floral display size and the two types of pollinator response, i.e., visitation rate per plant and the number of flowers probed per plant visit. One possible approach is to examine the actual pollinator behaviors empirically (e.g., Ohashi & Yahara 1998). Another is to consider how pollinators should

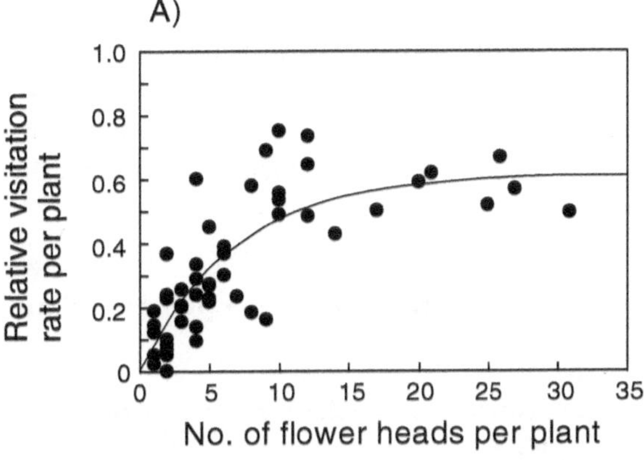



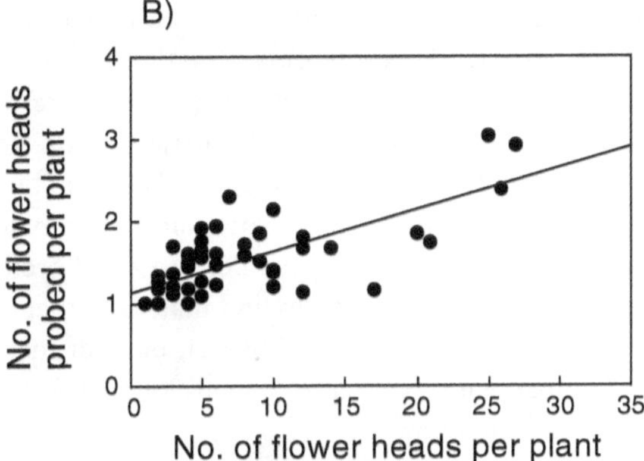



Fig. 14.1. The relationships between the number of flower heads on a *Cirsium purpuratum* plant and (A) the relative bumble bee visitation rate per plant ($y = 0.612[1 - \exp(-0.148x)]$, $n = 57$, $R^2 = 0.637$, $p < 0.0001$), and (B) the number of flower heads probed by a bee per plant ($y = 1.117 + 0.051x$, $n = 57$, $R^2 = 0.533$, $p < 0.0001$), obtained from six 1-day observations for 9–13 plants. Relative visitation rate per plant ($y$ axis of 1A) is calculated as [(number of bee visits to the focal plant)/(total number of bee visits to the observed plants)] for each of six 1-day observations. (Adapted from Ohashi & Yahara 1998.)

behave on plants to maximize their rates of energy gain. Pollinators have been often used as model animals for studying "optimal foraging." In particular, behavioral ecologists have regarded a plant or an inflorescence as a "patch" and have made intensive efforts to test the prediction of the optimal patch-use model, originally formulated by Charnov (1976), the so-called marginal-value theorem. Despite numerous relevant studies on pollinator behavior, however (reviewed by Orth & Waddington 1997), the functional response of pollinators to floral display remains to be solved theoretically. Bearing this in mind, here we provide a theoretical analysis of the optimal pattern of pollinator responses to variation in floral display size and its influences on pollen transfer within and between plants. Based on the results, we consider how pollinators could affect the evolution of floral display size and other floral traits.

## Optimal number of flowers probed per plant in relation to floral display size

### Possible factors causing patch depression during foraging on a plant

How many flowers does a pollinator probe after landing on a plant or inflorescence? As shown in Fig. 14.1B and Table 14.1, previous authors have found that pollinators tend to probe only a few flowers on a plant or inflorescence before leaving even when far more flowers are present. Furthermore, the proportion of flowers probed generally declines considerably with increasing display size. Reasons suggested by plant biologists for short visitation sequences include: (1) satiation of pollinators; (2) draining all floral rewards on a plant; (3) the need to avoid predator or (aggressive) competitors; and (4) the need to find mates or other types of food (Stephenson 1982; Snow *et al.* 1996). However, none of these ideas seems applicable to most plant–pollinator systems studied so far, especially to bumble bees and their flowers.

Instead, here we address this problem from the standpoint of optimal foraging. We use "plant" as a general term that may indicate either an individual plant or an inflorescence, which would in practice be regarded by pollinators as one flower patch. Because plants are distributed as discrete patches, moving between plants is more costly for a pollinator than moving within a plant in terms of time and energy expended (Heinrich 1975). Therefore, the pollinator should probe all available flowers on a plant unless the rate of energy gain declines as it stays longer on it. Thus, a pollinator's decision largely depends on the presence or absence of a gradual decrease in the rate of energy gain within a plant, i.e., "patch depression" (originally termed "depression" in Charnov *et al.* 1976).

Two major mechanisms could cause patch depression on a plant. One possibility is variation in nectar productivity coupled with pollinators' non-random flower choice. For example, plants that have flowers on simple vertical inflorescences often have a pattern of decreasing rate of

Table 14.1. *A tabulation of reports concerning how parameters of pollinator visitation (visitation rate per plant $V_p$, number of flowers probed on a plant $t_c$, proportion of flowers probed per plant $t_c/F$, and visitation rate per flower $V_f$) correlate with floral display size (F); the notations +, −, and ± indicate positive, negative, and statistically non-significant relationships found between these parameters and floral display size*

| Plant density and plant species | $V_p$ | $t_c$ | $t_c/F$ | $V_f$ | Pollinator | References |
|---|---|---|---|---|---|---|
| High ($\geq$ 0.1 plants m$^{-2}$) | | | | | | |
| *Aconitum columbianum* | ? | + | − | ? | bumble bee | Pyke (1982) |
| *Aconitum columbianum* | + | + | − | + | bumble bee | Pleasants & Zimmerman (1990) |
| *Asclepias exaltata* | + | ? | ? | ? | bumble bee, butterfly, honeybee | Broyles & Wyatt (1997) |
| *Asclepias syriaca* | + | + | − | ± | bumble bee, honeybee, etc. | Morse (1986) |
| *Cirsium purpuratum* | + | + | − | ± | bumble bee | Ohashi & Yahara (1998) |
| *Cynoglossum officinale* | + | + | − | ± | bumble bee | Vrieling *et al.* (1999) |
| *Delphinium nelsonii* | + | + | − | + | bumble bee | Pleasants & Zimmerman (1990) |
| *Eichhornia paniculata* | + | + | − | ± | bumble bee | Harder & Barrett (1995) |
| *Epilobium angustifolium* | + | + | − | ± | bumble bee | Schmid-Hempel & Speiser (1988) |
| *Glycine max* | + | ? | ? | ? | honeybee | Robacker & Erickson (1984) |
| *Mertensia ciliata* | + | + | − | ± | bumble bee | Geber (1985) |
| *Mimulus guttatus* | + | + | − | ± | bumble bee, honeybee | Robertson & Macnair (1995) |
| *Myosotis colensoi* | + | + | − | ± | tachnid fly | Robertson & Macnair (1995) |
| *Myosotis colensoi* | ? | + | − | ? | tachnid fly | Robertson (1992) |
| *Pulmonaria collina* | + | + | − | ? | bumble bee, beefly | Oberrath & Böning-Gaese (1999) |
| *Viscaria vulgaris* | + | + | − | ± | bumble bee | Dreisig (1995) |
| Low (<0.1 plants m$^{-2}$) | | | | | | |
| *Anchusa officinalis* | ± | + | ± | ± | bumble bee | Dreisig (1995) |

Table 14.1 (*cont.*).

| Plant density and plant species | $V_p$ | $t_c$ | $t_c/F$ | $V_f$ | Pollinator | References |
|---|---|---|---|---|---|---|
| **High and low** | | | | | | |
| *Cynoglossum officinale* | | | | | | |
| high (in population) | + (strong) | + (weak) | − | + | bumble bee | Klinkhamer *et al.* (1989) |
| low (isolated) | + (weak) | + (strong) | ± | ± | | |
| *Echium vulgare* | | | | | | |
| high (in population) | + | + (weak) | − (strong) | − | bumble bee | Klinkhamer & de Jong (1990) |
| low (isolated) | ± | + (strong) | − (weak) | − | | |
| **Unknown** | | | | | | |
| *Anchusa officinalis* | + | + | − | − | bumble bee | Andersson (1988) |
| *Aralia hispida* | + | + | − | − | bumble bee | Thomson (1988) |
| *Ipomopsis aggregata* | ? | + | − | ? | hummingbird | Pyke (1978*b*) |
| *Ipomopsis aggregata* | ± | + | − | ± | hummingbird | Brody & Mitchell (1997) |
| *Nepeta cataria* | + | + | − | + | honeybee | Sih & Baltus (1987) |
| | + | + | + | + | bumble bee | |
| | + | − | − | − | solitary bee | |
| *Sagittaria australis* | ? | ? | ? | ± | bumble bee, solitary bee | Muenchow & Delesalle (1994) |
| *Sagittaria latifolia* | ? | ? | ? | ± | bumble bee, solitary bee, etc. | Muenchow & Delesalle (1994) |
| *Symphytum officinale* | + | + | − | ± | bumble bee | Goulson *et al.* (1998*b*) |

*Source:* Adapted from Ohashi & Yahara (1999).

nectar production per flower from bottom to top (Pyke 1978*a*; Best & Bierzychudek 1982; but see Corbet *et al.* 1981). Pollinators moving vertically up each inflorescence, therefore, may experience a gradual decrease in gain per flower. Even without a spatial gradient of nectar productivity, pollinators that preferentially probe fresh flowers with higher nectar productivity (Gori 1989; Kevan *et al.* 1990; Oberrath & Böhning-Gaese 1999) may also experience patch depression, because they are increasingly likely to encounter an old, less-rewarding flower as they stay longer on a plant.

Another possibility is revisitation of flowers that it has previously probed, receiving little or no nectar. If the risk of revisiting flowers increases with the number of flowers probed, patch depression will result, even in the absence of variation in nectar productivity among flowers. In the few published studies, flower revisitation rates, defined as the fraction of flower visits that are to a flower previously probed while the pollinator was on the same plant, are low (2.9% in Pyke 1979; 3.5% in Pyke 1982; 0.2% and 3.0% in Galen & Plowright 1985). However, if pollinators avoid revisitation by curtailing their visit duration on plants, low frequency of total revisitation does not necessarily mean that it cannot cause patch depression. To evaluate flower revisitation, we must investigate whether the revisitation rate increases with the number of flowers probed on a plant. To our knowledge, only three field studies quantified this relationship (Pyke 1978*b*, 1981*a*, 1982), where revisitation rate increased with the number of flowers probed (up to approximately 10%–50% before all available flowers were probed). In an experimental study, Redmond & Plowright (1996) have also reported that revisitation rate increased with the number of flowers probed within a patch (up to approximately 25%).

To what degree can a pollinator remember flowers that it has previously probed on a plant? Pollinators often possess large long-term memory capacity for spatial information, such as the location of the home, the nest, and of flower patches, as well as their positions relative to surrounding landmarks (Heinrich 1976; Gould 1986; Menzel *et al.* 1996). However, the spatial scale on which they can use spatial long-term memory may be limited because a pollinator may probe hundreds of flowers during each foraging trip (Brown & Demas 1994). Even if a pollinator could use long-term memory, the need to browse its "library" at every flower may cause a time delay (Chittka *et al.* 1999). In fact, honeybees and vertebrates use short-term memory ("spatial working memory") for avoiding revisitation of food sources (Brown & Demas 1994; Brown *et al.*

1997). If short-term memory capacity is limited, the risk of flower revisitation would increase with the number of flowers probed. This scenario is rather speculative and clearly needs further exploration.

A pollinator's directionality in its foraging movement within a patch may help pollinators to avoid revisitation, irrespective of their spatial memory ability. For example, pollinators usually move in the same direction on plants with simple vertical inflorescences (Darwin 1876; Benham 1969), so that they seldom revisit flowers (Heinrich 1975; Pyke 1978a; Best & Bierzychudek 1982). In addition, the presence of landmarks may also promote movement directionality (Pyke & Cartar 1992; Redmond & Plowright 1996). Interestingly, Pyke & Cartar (1992) suggested that such a directional movement itself is partially constrained by pollinators' spatial memory ability for patch arrival direction. At present, we know very little about the perceptual and memory mechanisms underlying such effects.

Other potential mechanisms for avoiding flower revisitation are perceptual discrimination of nectar (Heinrich 1979a) or footprints (scent marks: Goulson *et al.* 1998a; electrostatic change: Erickson & Buchmann 1983). The availability of such mechanisms is likely to depend on pollinators' abilities to discriminate the cues and on pollinator visitation frequency (Dreisig 1995). Therefore, it will require careful investigation to determine how these mechanisms are effective for avoiding revisitation in the context of natural foraging.

### The effect of flower number

Recently we have developed a theoretical model to predict the relationship between the optimal number of flowers probed per plant and floral display size (Ohashi & Yahara 1999). We considered the increasing risk of revisitation to be the major cause of depression for a pollinator foraging on a plant, as it seems to be a very probable mechanism. To incorporate this effect into the model, we assumed that a pollinator remembers probing a maximum of $m$ flowers on the plant and avoids revisiting them. We referred to $m$ as "memory size", although it incorporates effects of both the actual memory capacity and of other factors discussed above. If the memory size ($m$) is limited, as we have noted, the risk of revisitation would increase with the number of flowers probed. Furthermore, larger displays offer more flowers from which to choose, which would decrease the risk of revisitation (e.g., see Table 14 in Pyke 1982). We also incorporated the cost of interplant movement by defining the mean discounting rate for visiting another plant as $k = $ [(flight time between flowers within a plant) + (handling time per flower)]/[(flight time between plants) + (handling time per flower)] ($0 \leq k \leq 1$). For simplicity, the rate of energy expenditure was assumed to be constant. Then, the following formula approximates the relationship between the optimal number of flowers per plant and the display size:

$$t_c = (1 - k)F + mk \qquad (F > m),$$

$$t_c = F \qquad (F \leq m) \qquad\qquad (14.1)$$

where $t_c$ is the number of flowers probed per plant and $F$ is the size of floral display.

The observed data (Table 14.1; Fig. 14.1B) agree well with the prediction of our model (Fig. 14.2B). First, pollinators probe fewer flowers than available. Second, the number of flowers probed increases with, but less rapidly than, the size of floral display (but see Sih & Baltus 1987). Third, the number of flowers per plant increased more rapidly with floral display size in low-density plant populations than it did in high-density populations (Klinkhamer *et al.* 1989; Klinkhamer & de Jong 1990).

It is worth noting that the aim of our model is not to demonstrate a patch-leaving rule that visitors actually follow, but simply to find the ideal point of plant departure for pollinators. In nature, it is well known that pollinators often leave the plant just after probing one to two flowers with little or no nectar (Fig. 14.3) (Pyke 1978*b*, 1982; Thomson *et al.* 1982; Hodges 1985). Pollinators usually do not have "complete information" on the nectar distribution on the current plant, so that such a simple probabilistic rule may provide a practical method for pollinators to approach the purely mathematical optimum (see also McNamara & Houston 1980; Iwasa *et al.* 1981). In a stochastic environment, probabilistic rules may work better than leaving plants deterministically following Eq. 14.1.

### Optimal visitation rate per plant in relation to floral display size

**What is "attractiveness" to pollinators in large floral displays?**
The model described above was based on the implicit assumptions that floral display size is invariable within a population and that pollinators arrive randomly at individual plants. In contrast, natural plant

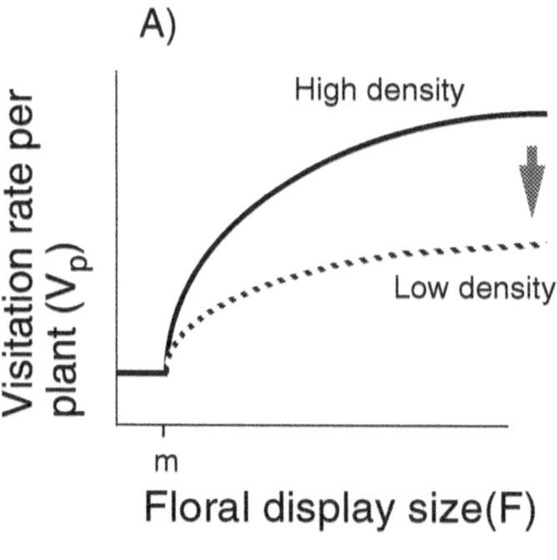


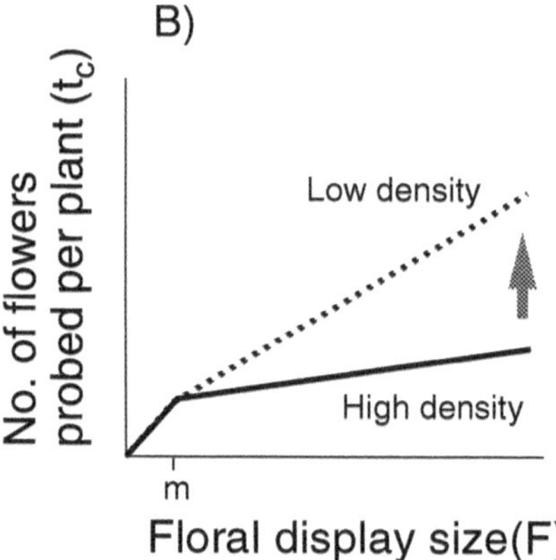


Fig. 14.2. Predicted relationships between floral display size (F) and (A) pollinator visitation rate per plant ($V_p$), and (B) the number of flowers probed per plant ($t_c$). (Adapted from Ohashi & Yahara 1999.)

populations usually include plants of various sizes (e.g., Pleasants & Zimmerman 1990). In such populations, pollinators prefer to visit large floral displays over smaller ones (Fig. 14.1A; Table 14.1).

Two major hypotheses have been proposed to account for the observed preference for large floral displays: (1) the detection-advertising hypothesis; (2) the flight-cost hypothesis. The detection-advertising hypothesis states that floral display size limits the distance from which it could be detected because of the insects' limited visual resolution (Dafni *et al.* 1997). For example, Giurfa *et al.* (1996) showed that the minimum display

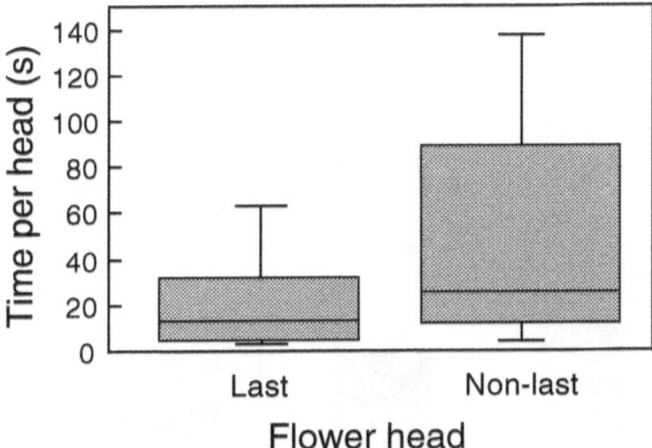

Fig. 14.3 Time spent by bumble bees on the last- and non-last-probed heads in each visit to a *Cirsium purpuratum* plant. The five horizontal lines of the plot indicate the 10th, 25th, 50th, 75th, and 90th percentiles of the data ($U = 824.5$, $p = 0.0056$; Mann–Whitney $U$-test). (Adapted from Ohashi 1998.)

size that would be detected by honeybees from 45-cm distances is about 5 cm in diameter in the best case. Moreover, color contrast (Lehrer & Bischof 1995; Giurfa *et al.* 1996) and relative motion speed against the background (Lehrer *et al.* 1990; Srinivasan *et al.* 1990) also influence the detectability of objects. Thus, this hypothesis may partially explain preference for large displays, especially when individual flowers are small. If large floral displays are infrequent in the population, however, this hypothesis may not hold; pollinators would more frequently choose small, but closer displays.

On the other hand, the flight-cost hypothesis emerges from an economic viewpoint. Harder & Cruzan (1990) and Harder & Barrett (1996) stated that pollinators visit large inflorescences because the proximity of many flowers reduces pollinator flight costs. Their statements have implied that pollinators prefer to visit larger displays on which they can probe more flowers, so that they can reduce the total movement costs required to probe a fixed number of flowers. That is, total movement costs required to probe N flowers is expressed as:

$$[(T + h) + (t + h)(i - 1)] N/i \tag{14.2}$$

where $i$ is the number of flowers probed per plant, $T$ is the flight time between plants, $t$ is the time between flowers on the same plant, and $h$ is the handling time per flower. As $T$ is usually longer than $t$, total movement costs would decrease with increasing $i$. This advantage is greater if pollinators could walk between flowers on plants or inflorescences, which

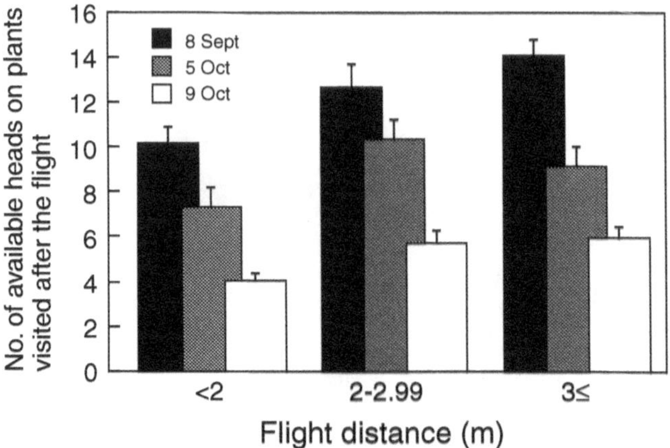


Fig. 14.4. Floral display size of plants visited by bumble bees after flying different distances. Bars indicate the average number of available heads per plant ($\pm$SE). The size of floral display visited after the flight differed significantly among distance categories (8 Sept $H = 12.04$, $p = 0.002$; 5 Oct $H = 13.00$, $p = 0.002$; 9 Oct $H = 12.69$, $p = 0.002$; Kruskal–Wallis test). (Adapted from Ohashi & Yahara 1998.)

requires about 90% less energy expenditure per time than flight (Heinrich 1975). As above, however, if large floral displays are infrequent in the population, this hypothesis also may not hold because pollinators would choose small but closer displays more frequently. In fact, some authors have found that pollinators often choose small displays when they are close (Fig. 14.4) (Pyke 1981$b$). Moreover, the visitation rate per flower rarely increases with floral display size (Table 14.1), which suggests that the reduced movement costs on large displays are of minor importance in determining pollinators' preferences for large displays. Thus, we have to reconsider our view about "attractiveness" of large floral displays.

### Incorporating the ideal free distribution into the model

Robertson & Macnair (1995) have suggested that, when plant density is relatively high, optimally foraging pollinators should visit flowers on all sizes of displays at equal rates, following Fretwell & Lucas's (1970) theorem of the "ideal free distribution." The ideal free distribution (IFD) is an equilibrium state that arises as a consequence of repetitive movements of competitors in search of more profitable local areas. In the case of plants and pollinators, the profitability of a plant (mean nectar crop) may decrease linearly with the average visitation rate per flower, because nectar crop per flower increases linearly with renewal time, at least at the scale of actual inter-visit times (Kadmon 1992). This situation corresponds to the simplest IFD model, i.e., the "continuous input" model (Parker & Sutherland 1986), which expects the average visitation rate per flower to be directly proportional to its nectar productivity. Since nectar production rate per flower often shows no significant correlation with floral display size (Harder & Cruzan 1990 and references therein), this model expects that the average visitation rate per flower would be equal between large and small displays (Dreisig 1995).

Based on this idea, Ohashi & Yahara (1999) have expanded the former model (Eq. 14.1) for the cases where display size is variable. The visitation rate per plant $(V_p)$ is expressed as:

$$V_p = V_f F / [(1 - k) F + m k] \qquad (F > m)$$

$$V_p = V_f \qquad\qquad (F \leq m) \qquad\qquad (14.3)$$

where $V_f$ is the average visitation rate per flower (constant under an IFD). As shown in Fig. 14.2A, pollinator visitation rate per plant $(V_p)$ is a decelerating function of floral display size $(F)$. This prediction agrees well with previous results (Fig 14.1A) (Iwasa *et al.* 1995 and references therein; but see Sih & Baltus 1987; Andersson 1988; Ohara & Higashi 1994). Moreover, visitation rate per plant $(V_p)$ increases more rapidly at higher plant density. This is because a reduction in the proportion of flowers probed per plant reduces the competition among pollinators on large floral displays. This prediction agrees with the observation that bumble bees visited large floral displays less preferentially at lower plant density (Klinkhamer *et al.* 1989; Klinkhamer & de Jong 1990; Dreisig 1995). Note that the prediction of our model is opposite to the intuitive prediction deduced from the previous two hypotheses. If the detectability of floral display is most important, the visitation rate per plant $(V_p)$ would increase less rapidly with display size $(F)$ at higher plant density because pollinators could detect smaller sizes of floral display. The same prediction will result when the flight cost is most important, because both the cost of interplant movement ($T$ in Eq. 14.2) and the proportion of flowers probed per plant $(t_c/F)$ would decrease with increasing plant density. Clearly, observation of pollinator behavior alone is not a sufficient proof for the relative importance of competition among pollinators. We emphasize the value of simultaneously exploring pollinator behavior and nectar availability in future studies. Moreover, functional responses of pollinators other than bumble bees (birds, honeybees, solitary bees, flies, butterflies, beetles, etc.) need to be explored more intensively.

The strategies that individual pollinators might use to achieve an IFD are still open to question. As Dreisig (1995) suggested, pollinators' preferences for large floral displays may partly explain the IFD. Furthermore, in the real world, nectar distribution among plants may fluctuate over time. If a pollinator could respond to such spatio-temporal variation, it would achieve an IFD more accurately. For example, pollinators are known to fly longer distances after encountering lower rewards ("area-restricted searching"; reviewed by Motro & Shmida 1995). By adopting this rule while foraging along its own "trapline" (Thomson *et al.* 1997 and references therein), pollinators may efficiently reduce the spatio-temporal bias in nectar distribution. Also, a trapline forager may occasionally sample new plants to detect and respond to temporal changes in nectar distribution among plants (Thomson *et al.* 1987). The spatial scale on which pollinators should adopt these strategies will depend on the spatial distribution of flowers, the frequency of revisitation, pollinators' energetic requirements, perceptual and memory constraints, and the number of competitors. Clearly too little is known at present to draw any conclusion about these issues.

### Evolutionary implications

### Effects of plant density on the evolution of floral display

Plants growing at low densities are said to experience some reproductive difficulties through alterations in pollinator behavior for at least three reasons. First, they may have trouble attracting pollinators away from competing resources because they are economically inefficient to exploit (Kunin 1997). Second, pollinators are more likely to behave as generalists on sparsely distributed plants and to lose pollen during interspecific flights or clog stigmas with foreign pollen (Kunin 1993). Third, pollinators may probe more flowers per plant at low density, which may increase geitonogamy (Bosch & Waser 1999 and references therein).

In addition to such population-level effects, lowered plant density may cause changes in pollen dispersal among different-sized displays. Ohashi & Yahara's (1999) model predicts that pollinators probe more flowers per plant with decreasing plant density particularly on larger floral displays. Moreover, the model predicts that pollinators would show a weaker preference for visiting large floral displays over small ones at lower plant density. Such an effect could aggravate the relative disadvantage of larger displays growing at low densities; it would reduce xenogamy and increase geitonogamy. To clarify these influences, we describe a model by incorporating pollinators' optimal behavior into the model of pollen transfer. We independently developed this model, but very similar theoretical ideas were developed by Iwasa $et\ al.$ (1995), who tried to explain the small number of flowers probed by a pollinator per plant as a plant's strategy to maximize pollen dispersal. We assume that: (1) pollen on a pollinator constitutes a single, homogeneous pool; (2) a pollinator deposits and picks up pollen in equal amounts at each flower ("pollination saturation"; de Jong $et\ al.$ 1993); and (3) this amount is a constant fraction of the amount of pollen held on a pollinator's body. This simple model (the exponential decay model with a constant pollen carryover) is the most commonly used theoretical description of pollen transfer (reviewed by Harder & Barrett 1996). Even when we adopt more realistic models such as the changing carryover model (Morris $et\ al.$ 1994) or models with pollen loss during transports (Rademaker $et\ al.$ 1997; Harder & Wilson 1998), the qualitative conclusion of the present analysis remains unchanged. We further assume that: (4) nectar production rate per flower is constant; (5) pollinators are always competing for floral resources; and (6) the plant is self-incompatible and the total number of pollen grains exported from a plant is a measure of its male fitness.

The number of pollen grains exported from a plant per pollinator per visit $(E)$ is expressed as:

$$E = A + A(1-d) + A(1-d)^2 + \ldots + A(1-d)^{t_c-1} = A(1-C^{t_c})/(1-C) \quad (14.4)$$

where $A$ is the amount of pollen held on a pollinator's body, $d$ is the fraction of pollen picked up or deposited at each flower, $t_c$ is the number of flowers probed per plant, and $C$ is pollen carryover $(C=1-d)$. Then, assuming that pollen dispersal is limited by pollinator visits, the total number of pollen grains exported from a plant with $F$ open flowers (male fitness, $W$) is found by combining Eqs. 14.3) and 14.4:

$$W = V_p E = V_f F A \left[1 - C^{(1-k)F+mk}\right] / \{[(1-k)F+mk](1-C)\}. \quad (14.5)$$

Figure 14.5 shows that under the assumption of the IFD (i.e., $V_f$ is constant), male fitness $(W)$ increases with floral display size $(F)$, but the average male fitness gain per flower $(W/F)$ decreases with display size.

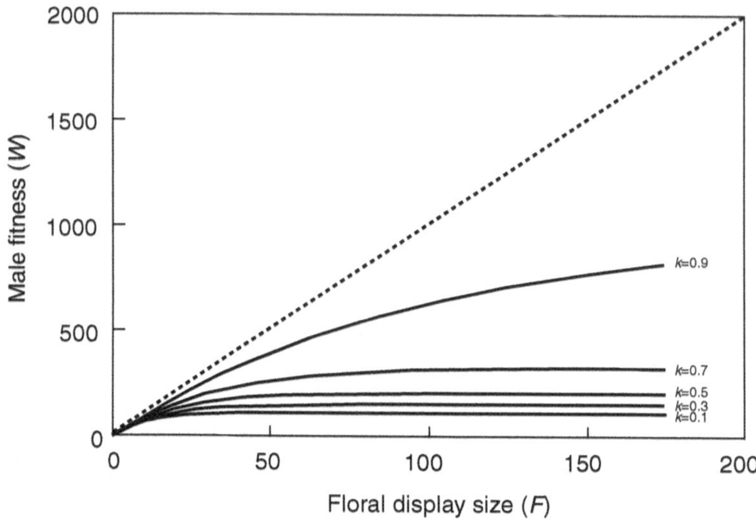


Fig. 14.5. Predicted relationship between floral display size ($F$) and male fitness ($W$). $W$ is determined by numerical solution of Eq. 14.5, for $V_f = 1$, $m = 1$, $A = 10$, and $C = 0.9$. The dashed line represents male fitness when $k = 1$.

Furthermore, male fitness gain per flower diminishes as the relative cost of interplant movement $(1 - k)$ increases, especially when $k$ is larger than 0.5.

Some authors have suggested that the costs associated with geitono-gamy may decrease with increasing display size because the proportion of flowers probed per plant declines on larger plants (Snow *et al.* 1996). However, our model reveals that the benefits of attracting more pollina-tors do not counteract the cost of increased geitonogamy, even if pollina-tor availability and pollen carryover is large (see also de Jong *et al.* 1993). In addition to this, we found that a small rise in the relative cost of inter-plant movement dramatically increases the cost of geitonogamy on larger displays. Based on this result, we can suggest that plants that typically grow at low densities (due to competition, predation, colonization to novel habitats, etc.) will be subject to strong natural selection favoring small displays or extended blooming. Both in warm-temperate and cool-temperate forest on Yaku Island, Yumoto (1987) found a suggestive pattern that climbers, epiphytes, and most of the understory shrubs, which typically grow at low density and are visited by specialist pollina-tors, exhibited extended blooming. At present, however, there are no empirical data on such a tendency in any particular plant–pollinator system.

Note that the optimal floral display size in actual plant–pollinator systems may be often larger than our models would predict because: (1) a plant population consisting of small displays cannot attract sufficient pollinators (Kunin 1997), so that among-population selection for larger displays may be strong enough to oppose individual selection for smaller displays; (2) flowering time can be constrained by biotic and/or abiotic factors such as frost, rainfall, or the availability of seasonal pollinators (Rathcke & Lacey 1985); and (3) the opportunity of geitonogamy can be reduced by spatio-temporal separation of sexes such as dioecy or gynodioecy (Thomson & Brunet 1990), synchronized dichogamy (Cruden 1988), and dichogamy coupled with vertical inflorescences (Pyke 1978*a*).

### Can plants manipulate pollinators to their own advantage? Some possibilities of plant traits that promote movements between plants

If plants can shorten pollinators' visit sequences, they can increase male fitness as a result of decreased geitonogamy. Iwasa *et al.* (1995) modeled this effect and found that pollinator behavior that maximizes pollen export (male fitness) is qualitatively similar to observed pollinator behavior. Is this agreement fortuitous, or a result of pollinator manipulation by plants in an evolutionary sense? Based on considerations of optimal foraging, we now discuss possible strategies by which plants can manipulate their pollinators to their own advantage.

(1) Low nectar reward. Many authors have reported that lower nectar rewards often cause pollinators to depart earlier from plants and promote interplant movements (e.g., Heinrich 1979*b*). Lower nectar reward might therefore be advantageous unless pollination is inefficient (Robertson 1992; Iwasa *et al.* 1995). Moreover, decreased investment in nectar production will allow plants to reallocate resources into ovules, which can improve fitness (Pyke 1991; Sakai 1993).

(2) Gradient of nectar production within a structured inflorescence. On plants with vertical inflorescences, spatial gradient in the nectar productivity (or crop) decreasing from bottom to top may be an important cause of patch depression in place of flower revisitation (see above). In fact, Orth & Waddington (1997) found that carpenter bees foraging on vertical inflorescences with no spatial gradient of nectar

rewards probed a larger proportion of flowers than reported in other studies where there was a nectar gradient.

(3) Within-plant variation in nectar productivity per flower. Rathcke (1992) stated that if within-plant variation in nectar per flower increases the likelihood of pollinators' encountering low-reward flowers, it might shorten visit sequences. Observed simple departure rules adopted by pollinators (e.g., Fig. 14.3) seem to support this idea. However, pollinators may alter their departure rules in response to changes in spatial distribution of nectar rewards (Iwasa *et al.* 1981). Moreover, different pollinator species may adopt different rules of plant departure (Collevatti *et al.* 1997). Thus, further empirical and theoretical studies are needed before generalizing this argument.

(4) Retention of old flowers coupled with floral color (or scent) change. After they have landed on plants, pollinators often avoid old, less-rewarding flowers by their color or scent, while they have little or no ability to discriminate between different-aged flowers at a distance (Oberrath & Böhning-Gaese 1999 and references therein). Therefore, some authors have suggested that the retention of old flowers – coupled with floral color or scent changes – may enable plants to increase the pollinator visitation rate per plant while simultaneously decreasing the proportion of flowers probed per plant (Gori 1983; Oberrath & Böhning-Gaese 1999). The adaptive value of this strategy may be greatest in plant species bearing small flowers, where the cost of retaining an old flower will be small, pollinators cannot discriminate different-colored flowers at a distance, and clustering of flowers may greatly improve the plant's long-distance attractiveness.

(5) Plant traits increasing the risk of flower revisitation. Spatial memory or directional movement may be affected by some plant traits. For example, Redmond & Plowright (1996) found that bumble bees revisited artificial "flowers" more often in irregular than in uniform configurations. They showed in addition that the presence of landmarks significantly reduced flower revisits when bees had to fly between flowers, but had no effect when bees could walk between flowers. Also, Brown *et al.* (1997) suggested that spatial working memory capacity of honeybees is limited by their ability to discriminate among locations in close proximity. Therefore, inflorescence architecture (complex or close-packed arrangements of flowers, the absence of bract leaves, etc.) may shorten pollinators' visit sequences, mediated through the increased risk of flower revisitation.

(6) Plant traits reducing the relative cost of interplant movement. If plants can reduce the relative cost of inter-plant movement, they can greatly improve their male fitness (Fig. 14.5). Based on our definition of the

mean discounting rate for visiting another plants ($k$; see above), we suggest two possible strategies. One is asynchronous flowering of adjacent flowers, which would increase the average flight distance within a plant. Another strategy is producing deep flowers – or veiling nectaries behind complex floral structures – which would increase handling time per flower (Harder 1983, 1986). This strategy may be efficient only when autogamy does not increase with handling time per flower (Zimmerman 1988). Deep flowers have been discussed in the context of nectar protection (Corbet 1990), evolutionary race with pollinator tongue (Darwin 1862; Nilsson 1988), the exclusion of generalists (Heinrich 1979$b$; Laverty 1980), and the promotion of flower constancy (Darwin 1876; Laverty 1994). In addition to these, we indicate a novel functional value of deep flowers, i.e., the promotion of pollen dispersal.

We must note that plants or populations adopting any strategies discussed above may sometimes increase the risk of pollinator deficiency because they could be economically inefficient to exploit (Zimmerman 1988). For example, pollinators often learn to avoid patches with low reward levels (Dreisig 1995) or high reward variation ("risk-aversive foraging"; Perez & Waddington 1996). Despite such possibilities, we feel that the available data indicate that plants can improve their pollen dispersal by altering pollinator behavior. The importance of these characteristics in improving pollen dispersal is totally hypothetical at this time, and again, awaits theoretical and empirical exploration.

## Acknowledgements

We thank Drs L. Chittka, Y. Iwasa, E. Kasuya, S. Sakai, and J. D. Thomson, and two anonymous reviewers for fruitful comments on our manuscript, and Y. Harada for assistance in model analyses. This work was supported by a research fellowship (no. 1597) of Japan Society for the Promotion of Science.

## References

Andersson S (1988) Size-dependent pollination efficiency in *Anchusa officinalis* (Boraginaceae): causes and consequences. *Oecologia* 76:125–130

Bawa KS (1983) Patterns in flowering in tropical plants. In: Jones CE & Little RJ (eds) *Handbook of experimental pollination biology*, pp 394–410. Scientific & Academic Editions, New York

Benham BR (1969) Insect visitors to *Chamaenerion angustifolium* and their behaviour in relation to pollination. *Entomologist* 102:222–228

Best LS & Bierzychudek P (1982) Pollinator foraging on foxglove (*Digitalis purpurea*): a test of a new model. *Evolution* 36:70–79

Bosch M & Waser NM (1999) Effects of local density on pollination and reproduction in *Delphinium nuttallianum* and *Aconitum columbianum* (Ranunculaceae). *Am J Bot* 86:871–879

Brody AK & Mitchell RJ (1997) Effects of experimental manipulation of inflorescence size on pollination and pre-dispersal seed predation in the hummingbird-pollinated plant *Ipomopsis aggregata*. *Oecologia* 110:86–93

Brown MF & Demas GE (1994) Evidence for spatial working memory in honeybees (*Apis mellifera*). *J Comp Psychol* 108:344–352

Brown MF, Moore JA, Brown CH & Langheld KD (1997) The existence and extent of spatial working memory ability in honeybees. *Anim Learn Behav* 25:473–484

Broyles SB & Wyatt R (1997) The pollen donation hypothesis revisited: a response to Queller. *Am Nat* 149:595–599

Charnov EL (1976) Optimal foraging, the marginal value theorem. *Theor Popul Biol* 9:129–135

Charnov EL, Orians GH & Hyatt K (1976) Ecological implications of resource depression. *Am Nat* 110:247 259

Chittka L, Waser NM & Thomson JD (1999) Flower constancy, insect psychology, and plant evolution. *Naturwiss* 86:313–329

Collevatti RG, Campos LAO & Schoereder JH (1997) Foraging behaviour of bee pollinators on the tropical weed *Triumfetta semitriloba*: departure rules from flower patches. *Insectes Soc* 44:345–352

Corbet SA (1990) Pollination and the weather. *Israel J Bot* 39:13–30

Corbet SA, Cuthill I, Fallows M, Harrison T & Hartley G (1981) Why do nectar-foraging bees and wasps work upwards on inflorescences? *Oecologia* 51:79–83

Cruden RW (1988) Temporal dioecism: systematic breadth, associated traits, and temporal patterns. *Bot Gaz* 149:1–15

Dafni A, Lehrer M & Kevan PG (1997) Spatial flower parameters and insect spatial vision. *Biol Rev Camb Phil Soc* 72:239–282

Darwin CR (1862) *On the various contrivances by which British and foreign orchids are fertilized by insects*. John Murray, London

Darwin CR (1876) *Cross and self fertilization in the vegetable kingdom*. John Murray, London

de Jong TJ, Klinkhamer PGL & van Staalduinen MJ (1992) The consequences of pollination biology for selection of mass or extended blooming. *Funct Ecol* 6:606–615

de Jong TJ, Waser NM & Klinkhamer PGL (1993) Geitonogamy: the neglected side of selfing. *Trends Ecol Evol* 8:321–325

Dreisig H (1995) Ideal free distributions of nectar foraging bumblebees. *Oikos* 72:161–172

Erickson EH & Buchmann SL (1983) Electrostatics and pollination. In: Jones CE & Little RJ (eds) *Handbook of experimental pollination biology*, pp 173–184. Scientific & Academic Editions, New York

Fretwell SD & Lucas HL (1970) On territorial behavior and other factors influencing habitat distribution in birds. *Acta Biotheor* 19:16–36

Galen C & Plowright RC (1985) The effects of nectar level and flower development on pollen carry-over in inflorescences of fireweed (*Epilobium angustifolium*) (Onagraceae). *Can J Bo* 63:488–491

Geber MA (1985) The relationship of plant size to self-pollination in *Mertensia cliata*. *Ecology* 66:762–772

Gentry AH (1974) Flowering phenology and diversity in tropical Bignoniaceae. *Biotropica* 6:64–68

Giurfa M, Vorobyev M, Kevan P & Menzel R (1996) Detection of coloured stimuli by honeybees: minimum visual angles and receptor specific contrasts. *J Comp Physiol A* 178:699–709

Gori DF (1983) Post-pollination phenomena and adaptive floral changes. In: Jones CE & Little RJ (eds) *Handbook of experimental pollination biology*, pp 31–49. Scientific & Academic Editions, New York

Gori DF (1989) Floral color change in *Lupinus argenteus* (Fabaceae): why should plants advertise the location of unrewarding flowers to pollinators? *Evolution* 43:870–881

Gould JL (1986) The locale map of honeybees do insects have cognitive maps? *Science* 232:861–863

Goulson D, Hawson SA & Stout JC (1998a) Foraging bumblebees avoid flowers already visited by conspecifics or by other bumblebee species. *Anim Behav* 55:199–206

Goulson DG, Stout JC, Hawson SA & Allen JA (1998b) Floral display size in comfrey, *Symphytum officinale* L. (Boraginaceae): relationships with visitation by three bumblebee species. *Oecologia* 113:502–508

Harder LD (1983) Flower handling efficiency of bumblebees: morphological aspects of probing time. *Oecologia* 57:274–280

Harder LD (1986) Effects of nectar concentration and flower depth on flower handling efficiency of bumblebees. *Oecologia* 69:309–315.

Harder LD & Barrett SCH (1995) Mating cost of large floral displays in hermaphrodite plants. *Nature* 373:512–515

Harder LD & Barrett SCH (1996) Ecology of geitonogamous pollination. In: Lloyd DG & Barrett SCH (eds) *Floral biology*, pp 140–190. Chapman & Hall, New York

Harder LD & Cruzan MB (1990) An evaluation of the physiological and evolutionary influences of inflorescence size and flower depth on nectar production. *Funct Ecol* 4:559–572

Harder LD & Wilson WG (1998) Theoretical consequences of heterogeneous transport conditions for pollen dispersal by animals. *Ecology* 79:2789–2807

Heinrich B (1975) Thermoregulation in bumblebees II. Energetics of warm-up and free flight. *J Comp Physiol* 96:155–166

Heinrich B (1976) The foraging specializations of individual bumblebees. *Ecol Monogr* 46:105–128

Heinrich B (1979a) Resource heterogeneity and patterns of movement in foraging bumblebees. *Oecologia* 40:235–245

Heinrich B (1979b) *Bumblebee economics*. Harvard University Press, Cambridge, MA

Hodges CM (1985) Bumblebee foraging: energetic consequences of using a threshold departure rule. *Ecology* 66:188 197

Iwasa Y, Higashi M & Yamamura N (1981) Prey distribution as a factor determining the choice of optimal foraging strategy. *Am Nat* 117:710–723

Iwasa Y, de Jong TJ & Klinkhamer PGL (1995) Why pollinators visit only a fraction of the open flowers on a plant: the plant's point of view. *J Evol Biol* 8:439–453

Kadmon R (1992) Dynamics of forager arrivals and nectar renewal in flowers of *Anchusa stigosa*. *Oecologia* 92:552–555

Kevan PG, Eisikowitch D, Ambrose JD & Kemp JR (1990) Cryptic dioecy and insect pollination in *Rosa setigera* Michx. (Rosaceae), a rare plant of Carolinian Canada. *Biol J Linn Soc* 40:229–244

Klinkhamer PG & de Jong TJ (1990) Effects of plant size, plant density, and sex differential nectar reward on pollinator visitation in the protandrous *Echium vulgare* (Boraginaceae). *Oikos* 57:399–405

Klinkhamer PGL, de Jong TJ & Bruyn G-J (1989) Plant size and pollinator visitation in *Cynoglossum officinale*. *Oikos* 54:201–204

Kunin WE (1993) Sex and the single mustard: population density and pollinator behavior effects on seed-set. *Ecology* 74:2145–2160

Kunin WE (1997) Population size and density effects in pollination: pollinator foraging and plant reproductive success in experimental arrays of *Brassica kaber*. *J Ecol* 85:225–234

Laverty TM (1980) The flower-visiting behaviour of bumblebees: floral complexity and learning. *Can J Zool* 58:1324–1335

Laverty TM (1994) Bumblebee learning and flower morphology. *Anim Behav* 47:531–545

Lehrer M & Bischof S (1995) Detection of model flowers by honeybees: the role of chromatic and achromatic contrast. *Naturwiss* 82:145–147

Lehrer M, Srinivasan MV & Zhang SW (1990) Visual edge detection in the honeybee and its chromatic properties. *Proc R Soc Lond B* 238:321–330

McNamara J & Houston A (1980) The application of statistical decision theory to animal behaviour. *J Theor Biol* 85:673–690

Menzel R, Geiger K, Chittka L, Joerges J, Kunze J & Müller U (1996) The knowledge base of bee navigation. *J Exp Biol* 199:141–146

Morris WF, Price MV, Waser NM, Thomson JD, Thomson B & Stratton DA (1994) Systematic increase in pollen carryover and its consequences for geitonogamy in plant populations. *Oikos* 71:431–440

Morse DH (1986) Inflorescence choice and time allocation by insects foraging on milkweed. *Oikos* 46:229–236

Motro U & Shmida A (1995) Near–far search: an evolutionarily stable foraging strategy. *J Theor Biol* 173:15–22

Muenchow G & Delesalle V (1994) Pollinator response to male floral display size in two *Sagittaria* (Alismataceae) species. *Am J Bot* 81:568–573

Nilsson LA (1988) The evolution of flowers with deep corolla tubes. *Nature* 334:147–149

Oberrath R & Böhning-Gaese K (1999) Floral color change and the attraction of insect pollinators in lungwort (*Pulmonaria collina*). *Oecologia* 121:383–391

Ohara M & Higashi S (1994) Effects of inflorescence size on visits from pollinators and seed set of *Corydalis ambigua* (Papaveraceae). *Oecologia* 98:25–30

Ohashi K (1998) Effects of floral display size on pollinator foraging behavior in *Cirsium purpuratum* (Asteraceae). PhD thesis, Kyushu University, Japan

Ohashi K & Yahara T (1998) Effects of variation in flower number on pollinator visits in *Cirsium purpuratum* (Asteraceae). *Am J Bot* 85:219–224

Ohashi K & Yahara T (1999) How long to stay on, and how often to visit a flowering plant? – a model for foraging strategy when floral displays vary in size. *Oikos* 86:386–392

Orth AI & Waddington KD (1997) The movement patterns of carpenter bees *Xylocopa micans* and bumblebees *Bombus pennsylvanicus* on *Pontederia cordata* inflorescences. *J Insect Behav* 10:79–86

Parker GA & Sutherland WJ (1986) Ideal free distributions when individuals differ in competitive ability phenotype limited ideal free models. *Anim Behav* 34:1222–1242

Perez SM & Waddington KD (1996) Carpenter bee (*Xylocopa micans*) risk indifference and a review of nectarivore risk-sensitivity studies. *Am Zool* 36:435–446

Pleasants JM & Zimmerman M (1990) The effect of inflorescence size on pollinator visitation of *Delphinium nelsonii* and *Aconitum columbianum*. *Collect Bot* 19:21–39

Pyke GH (1978*a*) Optimal foraging in bumblebees and coevolution with their plants. *Oecologia* 36:281–293

Pyke GH (1978*b*) Optimal foraging in hummingbirds: testing the marginal value theorem. *Am Zool* 18:739–752

Pyke GH (1979) Optimal foraging in bumblebees: rule of movement between flowers within inflorescences. *Anim Behav* 27:1167–1181

Pyke GH (1981*a*) Honeyeater foraging: a test of optimal foraging theory. *Anim Behav* 29:878–888

Pyke GH (1981*b*) Optimal foraging in hummingbirds: rule of movement between inflorescences. *Anim Behav* 29:889–896

Pyke GH (1982) Foraging in bumblebees: rule of departure from an inflorescence. *Can J Zool* 60:417 428

Pyke GH (1991) What does it cost a plant to produce floral nectar? *Nature* 350:58–59

Pyke GH & Cartar RV (1992) The flight directionality of bumblebees: do they remember where they came from? *Oikos* 65:321–327

Rademaker MCJ, de Jong TJ & Klinkhamer PGL (1997) Pollen dynamics of bumblebee visitation on *Echium vulgare. Funct Ecol* 11:554–563

Rathcke BJ (1992) Nectar distributions, pollinator behavior, and plant reproductive success. In: Hunter MD, Ohgushi T & Price PW (eds) *Effects of resource distribution on animal–plant interactions*, pp 113–138. Academic Press, New York

Rathcke B & Lacey EP (1985) Phenological patterns of terrestrial plants. *Ann Rev Ecol Syst* 16:179–214

Redmond D & Plowright CMS (1996) Flower revisitation by foraging bumblebees: the effects of landmarks and floral arrangement. *J Apic Res* 35:96–103

Richards AJ (1986) *Plant breeding systems*. Allen & Unwin, London

Robacker DC & Erickson EH (1984) A bioassay for comparing attractiveness of plants to honeybees. *J Apic Res* 23:199–203

Robertson AW (1992) The relationship between floral display size, pollen carryover and geitonogamy in *Myosotis colensoi* (Kirk) Macbride (Boraginaceae). *Biol J Linn Soc* 46:333–349

Robertson AW & Macnair MR (1995) The effects of floral display size on pollinator service to individual flowers of *Myosotis* and *Mimulus. Oikos* 72:106–114

Sakai S (1993) A model for nectar secretion in animal-pollinated plants. *Evol Ecol* 7:394–400

Schmid-Hempel P & Speiser B (1988) Effects of inflorescence size on pollination in *Epilobium angustifolium. Oikos* 53:98–104

Sih A & Baltus MS (1987) Patch size pollinator behavior and pollinator limitation in catnip. *Ecology* 68:1679–1690

Snow AA, Spira TP, Simpson R & Klips RA (1996) Ecology of geitonogamous pollination. In: Lloyd DG & Barrett SCH (eds) *Floral biology*, pp 191–216. Chapman & Hall, New York

Srinivasan MV, Lehrer M & Horridge GA (1990) Visual figure–ground discrimination in the honeybee: the role of motion parallax at boundaries. *Proc R Soc Lond B* 238:331–350

Stephenson AG (1982) When does outcrossing occur in a mass-flowering plant? *Evolution* 36:762–767

Thomson JD (1988) Effects of variation in inflorescence size and floral rewards on the visitation rates of traplining pollinators of *Aralia hispida*. *Evol Ecol* 2:65–76

Thomson JD & Brunet J (1990) Hypothesis for the evolution of dioecy in seed plants. *Trends Ecol Evol* 5:11–16

Thomson JD, Maddison WP & Plowright RC (1982) Behavior of bumblebee pollinators of *Aralia hispida* Vent. (Araliaceae). *Oecologia* 54:326–336

Thomson JD, Peterson SC & Harder LD (1987) Response of traplining bumblebees to competition experiments: shifts in feeding location and efficiency. *Oecologia* 71:295–300

Thomson JD, Slatkin M & Thomson BA (1997) Trapline foraging by bumblebees. II. Definition and detection from sequence data. *Behav Ecol* 8:199–210

Vrieling K, Saumitou-Laprade P, Cuguen J, van Dijk H, de Jong TJ & Klinkhamer PGL (1999) Direct and indirect estimates of the selfing rate in small and large individuals of the bumblebee pollinated *Cynoglossum officinale* L. (Boraginaceae). *Ecol Lett* 2:331–337

Willson MF & Price PW (1977) The evolution of inflorescence size in *Asclepias* (Asclepiadaceae). *Evolution* 31:495–511

Yumoto T (1987) Pollination systems in a warm temperate evergreen broad-leaved forest on Yaku Island Japan. *Ecol Res* 2:133–146

Zimmerman M (1988) Nectar production, flowering phenology, and strategies for pollination. In: Lovett Doust J & Lovett Doust L (eds) *Plant reproductive ecology*, pp 157–178. Oxford University Press, New York

LAWRENCE D. HARDER, NEAL M. WILLIAMS,
CRISPIN Y. JORDAN AND WILLIAM A. NELSON

## 15

# The effects of floral design and display on pollinator economics and pollen dispersal

Animal pollination is a mixed blessing for angiosperms. Animals carry pollen readily because they are mobile and large relative to pollen grains. Furthermore, animals learn to associate floral signals with the presence of food and so move between conspecific plants relatively consistently (Chittka *et al.*, this volume; Gegear & Laverty, this volume; Giurfa, this volume; Menzel, this volume). However, animals act in their own interests, which often conflict with successful pollen transport (e.g., only about 1% of a plant's pollen production reaches stigmas; Harder 2000). Consequently, manipulation of pollinator behavior to promote cross-pollination is a prevailing theme in the evolution of floral design (form, color, nectar, and fragrance production) and display (inflorescence size and architecture).

This chapter reviews three aspects of pollinator manipulation by plants and their effects on pollen dispersal. First, because pollen dispersal for most animal-pollinated plants depends on the general responses of feeding pollinators to their foraging environment, we consider the underlying economic principles that establish the opportunities for floral manipulation. Second, we outline influences on the typical pattern of pollen dispersal among flowers for plants with granular pollen, and summarize how flower design affects this pattern (for a review of dispersal of orchid pollen, see Harder 2000). Finally, because pollination and mating success are characteristics of entire plants, rather than individual flowers, we consider how floral display affects pollinator attraction and within-plant behavior to determine pollen dispersal.

## Pollinator economics

Most pollinators visit flowers to gather food. In general, foraging involves economic principles whereby a resource's utility depends ultimately on its relative contribution to the forager's fitness. However, animals probably cannot evaluate the fitness consequences of different foraging options; rather, they must assess opportunities based on the proximate benefits and costs associated with current physiological and ecological conditions. Often, the behavior of experienced feeding animals maximizes a single variable, or foraging currency, that integrates foraging benefits and costs (Stephens & Krebs 1986). Such behavior bears diverse consequences for pollination, because it affects a pollinator's choice of plant species (e.g., Rasheed & Harder 1997*a*), choice of individual plants (e.g., Heinrich 1979; Waser & Price 1983; Thomson 1988), and behavior on those plants (e.g., Galen & Plowright 1985; Hodges 1985; Rasheed & Harder 1997*b*).

Foraging pollinators typically visit flowers for nectar and/or pollen; these resources differ distinctly with respect to both foraging benefits and costs. Most pollinators visit flowers for the concentrated, easily digested energy in nectar. Because animals ingest nectar, they can readily determine their intake rate and whether a flower is empty (e.g., Dreisig 1989). The main handling cost of nectar collection involves the time and energy required to drink nectar from flowers. This cost depends primarily on the volume of nectar ingested, its depth within a flower, the animal's body size, and the length of its proboscis (see Montgomerie 1984; Harder 1986). Consequently, the choice of plants within and between plant species varies with pollinator size and morphology (e.g., Harder 1985, 1988). Furthermore, because the rate of flower manipulation increases with experience, foraging decisions can depend on an individual pollinator's learning ability (Gegear & Laverty, this volume).

Unlike nectar, pollen offers a rich source of protein, amino acids, lipids, and sterols, compared to most other plant tissues (Stanley & Linskens 1974). However, pollen use requires specific collecting and digestive abilities, given the small quantities of pollen available in individual flowers and the indigestibility of pollen exine. Consequently, only some pollinators (primarily non-parasitic bees, syrphid flies, and masarine wasps) satisfy their protein needs by feeding from flowers. Instead of ingesting pollen directly from anthers, most of these animals harvest it in three steps: (1) external removal on the animal's body; (2) grooming; and (3) either consumption or transfer to specialized, external carrying structures (scopae, including corbiculae) for transport to a nest (Holloway 1976; Michener *et al.* 1978; Thorp 1979). The removal step can be cost-free if it occurs passively during nectar collection, or it can require considerable effort, as when bees contract their flight muscles rapidly to buzz pollen from poricidal anthers (reviewed by Buchmann 1983; also see Harder & Barclay 1994; King & Buchmann 1996). Grooming also elevates the cost of pollen collection relative to nectar collection, especially because it typically occurs during flight (Holloway 1976; Harder 1990*a*; Michener *et al.* 1978), which increases metabolic effort 10-fold (Ellington *et al.* 1990). Even though most bees carry pollen externally, they detect variation in the amount and quality of pollen removed from individual flowers (Cane & Payne 1988; Buchmann & Cane 1989; Harder 1990*a*; Robertson *et al.* 1999), perhaps by setae on the scopae that are coupled to displacement sensors (Ford *et al.* 1981) or by assessment of grooming effort. In response to such variation, bees alter their behavior to promote pollen-collection profits (Rasheed & Harder 1997*a, b*). This behavior often results in individual bees not depleting flowers of pollen, even when they visit for no other resource (Harder 1990*b*; Harder & Barclay 1994).

In addition to handling costs, the relevant foraging currency must incorporate the time and energy expended on other activities. These additional costs always include travel within and between plants. For animals that visit flowers to provision offspring, transit costs between nests and foraging sites are also relevant, so that foraging costs equal the total expense of foraging. In contrast, pollinators that visit flowers to sustain other activities, such as defending a territory, finding mates, or searching out oviposition sites, must accommodate the additional costs associated with these activities (Montgomerie *et al.* 1984; Houston & Krakauer 1993).

Given the benefits ($B$), time costs ($T$), and energy costs ($E$) of nectar and pollen collection, what currencies do pollinators typically maximize? The behavior of nectar feeders usually maximizes either net intake rate ($[B-E]/T$: Hodges 1981; Gass & Roberts 1992; Hainsworth & Hamill 1993) or net foraging efficiency ($[B-E]/E$: Schmid-Hempel & Schmid-Hempel 1987; Tamm 1989), whereas that of pollen feeders maximizes gross efficiency ($B/E$: Rasheed & Harder 1997*a, b*). Maximization of foraging rate maximizes daily gains for animals that forage to satisfy their own needs, whereas animals that maximize efficiency while provisioning other individuals maximize the overall daily delivery of resources to their nests (Ydenberg *et al.* 1994). In addition, a provisioning forager that maximizes its foraging efficiency promotes its reproductive output when the chance of mortality increases with foraging effort (Houston *et al.* 1988). In general, energy costs influence efficiency more than rate, so that when pollinators maximize efficiency they limit expensive behaviors, especially flight. As a result, maximizing efficiency rather than rate requires pollinators to visit more flowers per inflorescence (Rasheed & Harder 1997b) and to work each flower longer. As we discuss below, both of these behavioral responses affect pollination, so that foraging currency will affect plant mating.

Most studies of the currencies that motivate pollinator behavior have considered animals feeding exclusively on either nectar or pollen. However, animals that rely on flowers for both energy and protein require a balanced diet of nectar and pollen for adequate nutrition (e.g., Haslett 1989; Camazine 1993; Plowright *et al.* 1993). To maintain this balance, flower-dependent animals may need to compromise the economic collection of either resource to optimize overall diet composition. Such compromises may be common, because flower-dependent animals often collect nectar and pollen from different plant species (Brian 1957; Liu *et al.* 1975; Teräs 1985) that differ in their relative availability of these resources. N.M. Williams and V.J. Tepedino (unpublished manuscript) proposed that, in such circumstances, solitary bees minimize the total time spent collecting all the pollen and nectar required to provision a single offspring. Given relatively constant mass and quality of offspring provisions, such behavior would maximize the gross rate of resource collection per provision. Williams and Tepedino observed that female *Osmia lignaria* (Megachilidae) divided their foraging effort between a rich nectar source (*Hydrophyllum capitatum*) and a rich pollen source (*Salix* spp.) in proportions expected from time minimization, even though these species were separated by 300 m. In contrast to solitary bees, social bees need not always compromise economic collection of nectar and pollen, because they can achieve a balanced input by varying the proportions of dedicated pollen- and nectar-foragers (Brian 1952; Cartar 1992; Camazine 1993; Plowright *et al.* 1993).

If a pollinator is to maximize some benefit–cost ratio between the beginnings of consecutive foraging bouts, how should it decide whether to continue its current behavior, such as a flower visit, or switch to a different behavior? Quite simply, an individual act serves the longer-term goal of currency maximization as long as its instantaneous benefit–cost ratio (i.e., marginal value) exceeds the average ratio expected by ending the current behavior and beginning anew (marginal value theorem: Charnov 1976). This principle underlies many aspects of pollinator behavior, including: whether to deplete individual flowers of nectar and/or pollen (Hodges & Wolf 1981); whether to move to another flower on the same plant, or to another plant (Pyke 1979; Hodges 1985; Kadmon & Shmida

1992; Rasheed & Harder 1997*b*); whether to move to a neighboring, or more distant plant (Cibula & Zimmerman 1984); whether to start feeding on a different plant species (Zimmerman 1981); and when to end a foraging bout and either return to the nest or transfer to another behavior (Schmid-Hempel *et al.* 1985). Three features of the involvement of a benefit–cost ratio in these decisions warrant notice. First, as will become apparent below, all of these behaviors influence the pattern of pollen dispersal, so that foraging currency defines the linkage between many floral characteristics and pollination success. Second, the consequences of a floral characteristic, such as nectar volume or concentration, for pollinator behavior depends on its influence on the relevant foraging currency, rather than its effects on benefits or costs alone (e.g., Harder & Real 1987). Finally, because the value of a particular behavior to a pollinator depends on the average currency in the environment, the details of pollinator behavior (and the associated pollination) are often context dependent (e.g., Harder & Barrett 1996; Kunin 1997; Smithson & Macnair 1997).

Notwithstanding the widespread occurrence of currency maximization by pollinators, some solitary bees restrict their pollen collection (but not necessarily nectar collection) to a few related plant species even when other species seemingly offer greater rewards (reviewed by Wcislo & Cane 1986). Based on the limited available data, such specialization seems to be genetically determined (Thorp 1969; Williams 1999). This innate specialization is sometimes associated with behavioral and morphological adaptations for harvesting resources from particular plant taxa, which may increase pollination effectiveness. Innate specialization can benefit pollination by promoting pollen transfer between conspecific plants, although in this respect it does not differ fundamentally from short-term specialization by a generalist pollinator that maximizes its current foraging returns (see Waser 1986). On the other hand, adaptations for collecting pollen from specific plant species can impair pollination if they enable pollen specialists to function more as pollen thieves than as pollinators (e.g., Eickwort 1967; Cane & Buchmann 1989; Williams & Thomson 2001).

### Flowers and pollen dispersal by individual pollinators

Individual flowers serve pollination by contributing to a plant's overall attractiveness to pollinators (including both signaling and energetics) and by controlling the transfer of pollen to and from each visitor. Because attraction typically involves all flowers open on a plant, we review this function when we discuss inflorescences (below). Here, we consider the role of individual flowers in controlling pollen exchange with pollinators. Floral characteristics mediate this exchange by determining which areas of pollinators' bodies contact a flower's pollen and stigma(s), and the intensity and duration of that contact. Most aspects of floral form, as well as the production of nectar and floral oils, contribute to these pollination functions. Before addressing the roles of specific floral features in pollen export and import, we review the general pattern of pollen dispersal by a single pollinator from a specific donor flower to recipient flowers.

For most angiosperms, individual pollinators transport donor pollen to several or many recipient flowers (reviewed by Morris *et al.* 1994; Harder & Wilson 1998), because each stigma receives only a fraction of the pollen carried by a pollinator from a specific donor flower. In general, successive recipient flowers receive progressively less pollen from a particular donor (see Fig. 15.1a) due to deposition on stigmas and transport losses (e.g., grooming). If the pollen carried by a pollinator behaved as a single, completely mixed population, receipt of donor pollen by successive recipient flowers would decline as expected for simple geometric decay – as in Fig. 15.1a. However, in reality the decline is more rapid among initial recipients and more gradual among later recipients (reviewed by Morris *et al.* 1994), suggesting that the pollen carried by a pollinator behaves as a subdivided population with heterogeneous transport conditions.

Such subdivision could arise either passively or actively (Harder & Wilson 1998). Passive segregation could arise through at least three mechanisms: differences in the ability of areas on a pollinator's body to carry pollen (e.g., hairy head versus smooth mouthparts); variation in the incidence and intensity of contact by pollinators with anthers and/or stigmas; and accumulation of pollen in layers on the pollinator. Active segregation of pollen on a pollinator arises from behaviors such as grooming or movements of the mouthparts. These behaviors affect pollen on some sites on a pollinator's body (exposed sites), but not on others (safe sites: e.g., Kimsey 1984; Thomson 1986). Such behaviors could also move pollen from exposed to safe sites, supplementing safe sites with pollen from flowers visited previously while depleting exposed sites. As a result, the proportion of pollen from a specific donor flower that is dispersed to stigmas via safe versus exposed sites increases steadily as the pollinator visits successive recipient flowers.

Variation in pollen removal from a donor flower by individual pollinators is modified by grooming and layering dynamics so that pollen export increases non-linearly as removal increases (Harder & Wilson 1997). When

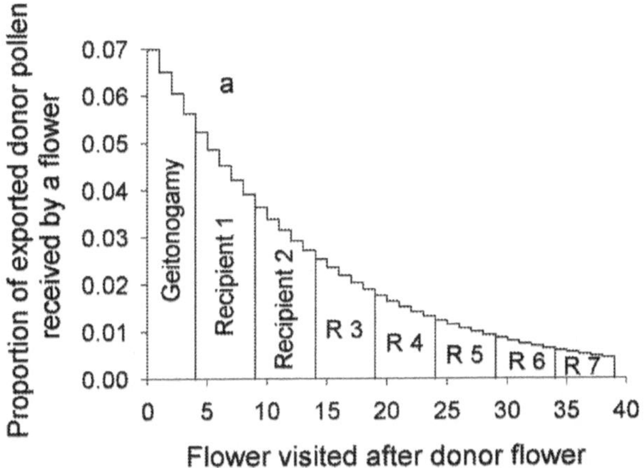


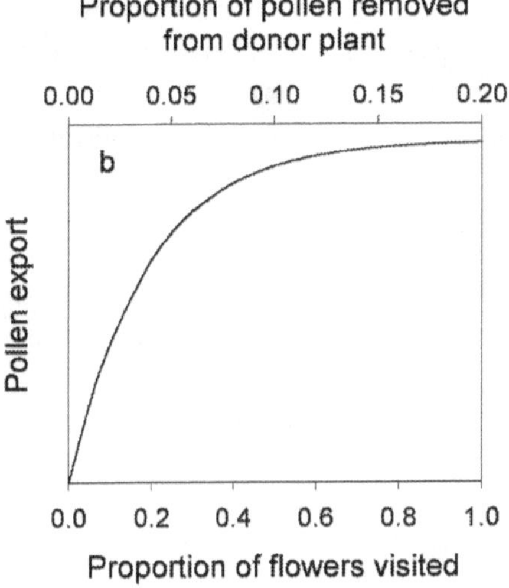


Fig. 15.1. Theoretical features of pollen dispersal by a single pollinator from (a) a single flower and (b) a 50-flowered plant (see Harder & Barrett 1996). Panel (a) considers the pollination fates of pollen removed from the first of five flowers (the donor flower) visited by a pollinator on a focal plant ($d_i = r\rho[1 - \rho]^{i-1}$, where $d_i$ is the proportion of donor pollen received by recipient flower $i$, $r$ is the proportion of available pollen removed from each flower, and $\rho$ is the proportion of pollen carried by the pollinator that is deposited on the stigma of each flower). Panel (b) illustrates how pollen export from a donor plant ($E$) to other plants varies with the proportion of flowers that a pollinator visits per inflorescence ($E = r[1 - \{1 - \rho\}^{v \cdot n}]/\rho$, where $v$ is the proportion of the $n$ open flowers visited by the pollinator). For both panels, $r = 0.2$, $\rho = 0.1$, and $n = 50$ flowers.

grooming intensity (and associated pollen loss) varies positively with the amount of pollen removed from a flower (e.g., Harder 1990*a*), enhanced removal increases subsequent pollen export at a decelerating rate (e.g., Thomson & Thomson 1989). This increased export arises both because each recipient flower receives more donor pollen and because donor pollen reaches more recipient flowers (Harder & Wilson 1998). With layering, total export by a single pollinator initially increases with pollen removal, as each recipient receives more donor pollen. However, greater increases in removal decrease total export, because pollen becomes buried more quickly and so does not reach distant recipients (Harder & Wilson 1998). As we discuss below, the diminishing returns associated with grooming and layering influence the evolution of attractiveness and floral control of pollen removal.

The importance of pollen exchange between flower and pollinator for pollen dispersal by individual pollinators should promote selection for floral features that mediate pollinator–flower interactions to a plant's advantage. Obvious features affecting pollen removal and deposition include corolla size and shape (Murcia 1990; Campbell *et al.* 1996; Kobayashi *et al.* 1999; but see Wilson 1995), the amount and schedule of pollen presentation (Harder & Thomson 1989; Harder & Wilson 1994), anther position (Harder & Barrett 1993), and stigma size, structure and position (e.g., Waser & Price 1984; Murcia 1990; Campbell *et al.* 1994; Conner *et al.* 1995). In addition, pollen exchange often varies with the duration of pollinator visits (Harder 1990*b*; Murcia 1990; Conner *et al.* 1995; Hurlbert *et al.* 1996; but see Mitchell & Waser 1992), which depends on the amount and quality of food (nectar or pollen) present in a flower (Montgomerie 1984; Harder 1986; Thomson 1986; Martínez del Río & Eguiarte 1987; Harder & Barclay 1994). Because food availability often increases as time elapses since the last pollinator visit (Waser & Mitchell 1990; Kadmon 1992; Williams 1997; Jones *et al.* 1998), flowers exchange more pollen with individual pollinators when they receive infrequent visits than when pollinators visit often (Harder & Thomson 1989; Klinkhamer & de Jong 1993; Harder & Wilson 1994).

### Inflorescences

#### Pollinator behavior on inflorescences

While foraging in the three-dimensional environment of an inflorescence, a pollinator must choose a starting flower, negotiate a route among visited flowers, and determine when to leave the plant. These decisions affect foraging benefits and costs by determining the number of flowers visited and the time and energy expended on flight. As we discuss in the next section, these decisions also establish the extent of between-flower self-pollination (geitonogamy) and pollen export to other plants.

Many pollinators visiting vertical spikes or racemes invariably start foraging on either lower (bees, wasps, and hawkmoths: e.g., Waddington & Heinrich 1979; Corbet *et al.* 1981; Dreisig 1985; Rasheed & Harder 1997*b*) or upper flowers (flies: Arista *et al.* 1999), thereby predetermining their subsequent movement direction within the inflorescence. For bumble bees, the proclivity to move upward apparently involves a functional constraint, as it persists on inverted inflorescences (Heinrich 1979) or when resources per flower increase or decrease along the inflorescence (Waddington & Heinrich 1979; Corbet *et al.* 1981). Given this constraint, bees respond to resource gradients in vertical inflorescences by altering their starting and leaving positions in ways that enhance their foraging economy (Pyke 1979; Waddington & Heinrich 1979; Rasheed & Harder 1997*b*). By generally moving upward, bumble bees seldom revisit flowers on vertical inflorescences (Pyke 1979; Galen & Plowright 1985). In contrast to insects, hummingbirds move less stereotypically on vertical inflorescences, starting on bottom or top flowers with roughly equal frequency (Wolf & Hainsworth 1986; Healy & Hurly, this volume).

Inflorescences with more three-dimensional structure than a raceme seem to complicate pollinator foraging. The only study to examine this effect (Hainsworth *et al.* 1983) compared the responses of hummingbirds to vertical, two-dimensional inflorescences and hemispheric, three-dimensional inflorescences. On three-dimensional inflorescences, birds probed fewer flowers, with proportionately fewer revisits than when they visited vertical inflorescences. In addition, flights between flowers lasted longer on hemispheric than on vertical inflorescences, even though the flowers were closer together. Hence, the spatial arrangement of surrounding flowers altered the cost of moving between two flowers separated by a specific distance. If this outcome applies more generally, different inflorescence architectures likely establish unique foraging environments for pollinators, and consequently influence pollen transfer within and between plants.

The effect of inflorescence architecture on pollinator behavior depends partly on whether pollinators modify foraging conditions predictably for subsequent visitors by depleting resources. On vertical inflorescences, the upward movement of bumble bees creates a positive correlation in nectar standing crop among flowers on the same inflorescence (Waddington 1981; Dreisig 1989), so the state of one flower provides information about that of higher flowers. In this environment, a bumble bee invariably continues feeding on its current inflorescence after visiting a rewarding flower, whereas it typically switches to another inflorescence after encountering a single empty flower (Dreisig 1989). In contrast, on the head-like inflorescences, such as *Monarda fistulosa*, bumble bees generally do not leave an inflorescence until encountering several empty flowers (Cresswell 1990). Presumably, this reduced responsiveness reflects the less stereotyped movement of bees on these less ordered inflorescences, thereby limiting correlation in reward availability among flowers (also see Kadmon & Shmida 1992; but see Wolf & Hainsworth 1986). The contrast between these responses indicates that the role of inflorescence architecture in modifying the economics of pollinator foraging (and pollen dispersal) extends beyond the effects of floral display on the actions of individual pollinators to include indirect interactions between all pollinators attracted to an inflorescence.

### Geitonogamy and outcross siring success

The presence of multiple flowers permits pollen transport between a plant's own flowers (reviewed by Harder & Barrett 1996; Snow *et al.* 1996; also see Brunet & Eckert 1998; Rademaker & de Jong 1998). In general, geitonogamy increases as a pollinator visits more flowers on a plant. For example, consider the destinations of pollen removed from the first of five flowers visited by a pollinator on a plant (Fig. 15.1a). Geitonogamous pollen transfer from this flower occurs during the pollinator's next four flower visits. If, instead, the pollinator visited eight additional flowers on the same plant, geitonogamy would claim a larger fraction of the total pollen dispersed from the donor flower and the plant as a whole. Because pollinators tend to visit more flowers on larger inflorescences (reviewed by Ohashi & Yahara, this volume), geitonogamy generally increases with display size (reviewed by Harder & Barrett 1996; Snow *et al.* 1996).

In addition to increasing the number of matings susceptible to inbreeding depression, geitonogamy reduces the pollen available for dispersal to other plants (pollen discounting: Kohn & Barrett 1994; Harder & Barrett 1995; Emms *et al.* 1997; Harder *et al.* 2000). Lloyd (1992) proposed that geitonogamy always diminishes outcrossing opportunities when pollen transport between flowers on the same plant involves the same processes as transport between plants. Because pollen dispersal between a donor and recipient flower varies non-linearly with the number of flower visits that separate them, each additional flower that a pollinator visits adds a progressively smaller increment to total pollen export (see Fig. 15.1b)(Harder & Barrett 1996; Emms *et al.* 1997; Rademaker & de Jong 1998).

The significant mating cost of pollen discounting probably favors inflorescence designs that restrict geitonogamy, including limits on the number of flowers displayed simultaneously, segregation of the sex roles among flowers within inflorescences, and heterostyly (Kohn & Barrett 1994; Harder & Barrett 1995; Harder *et al.* 2000). However, the benefits of specific anti-geitonogamy mechanisms depend on pollinator characteristics. For example, presentation of male flowers above female flowers limits pollen discounting for vertical inflorescences when bees move upward (Harder *et al.* 2000), but would aggravate discounting if pollinators move downward, and would have little effect if pollinators move unpredictably among flowers within inflorescences. Hence, plant species pollinated predominately by animals with different movement patterns should exhibit different patterns of sexual segregation.

### Attraction

Many features that distinguish animal-pollinated species from abiotically pollinated species serve the signaling and reward functions that govern a plant's attractiveness to pollinators, including nectar and pollen availability, a showy perianth, and fragrance (reviewed by Mitchell 1993). Attractiveness generally increases with the number of flowers displayed simultaneously (reviewed by Ohashi & Yahara, this volume), and so depends on the aggregate signal perceived by pollinators and their expected foraging returns from a plant's entire floral display (e.g., Weiss 1991). Consequently, the signals and rewards of individual flowers must be considered in the context of their collective contributions to a plant's overall reproductive success.

Given the expense of attraction (reviewed by Morgan 1992), the benefits must be significant. Obviously, a plant must attract enough pollinators to engage much of its pollen in dispersal and to bring in enough pollen to fertilize most of its ovules. Furthermore, participation of many individual animals in pollination increases a plant's mate diversity when different pollinators follow different foraging paths. A less obvious, but significant, benefit of attracting many pollinators arises because of the diminishing returns associated with increased pollen removal by individual pollinators which accompany pollinator grooming, pollen layering, and geitonogamous pollen discounting. Because of diminishing returns, a pollinator that removes half of a plant's pollen will export more than half as much as another pollinator that removes it all (see Fig. 15.1b). Consequently, plants enhance their pollen export by restricting pollen removal by individual pollinators and involving many pollinators in dispersal (Harder & Thomson 1989; Iwasa *et al.* 1995). Indeed, optimal restriction of pollen removal could increase siring success by more than an order of magnitude when pollinators are abundant, although time-dependent processes, such as loss of pollen viability and competition among male gametophytes for access to ovules, counteract the benefits of restricted removal (Harder & Wilson 1994).

To appreciate the benefits of restricting pollen removal, consider the relation of total pollen export to the proportion of a plant's flowers visited by each pollinator (see Iwasa *et al.* 1995 for mathematical details). Because of the diminishing returns caused by geitonogamous pollen discounting (see Fig. 15.1b), two pollinators that each visited half of a plant's flowers would export more pollen overall than a single pollinator that visited all the flowers, *even though the number of visits per flower is identical.* Attraction of many pollinators further enhances pollen export, as long as each pollinator visits only a fraction of a plant's open flowers, thereby limiting pollen discounting (see Fig. 15.2). However, if pollinators visit too few flowers, pollen can remain in anthers (pollen-removal failure), thereby reducing the plant's total pollen export. Hence, as the solid curve in Fig. 15.2 illustrates, maximization of pollen export occurs when the proportion of flowers visited by each pollinator balances the risk of pollen-removal failure against the mating cost of geitonogamous pollen discounting. The appropriate balance depends on the number of pollinators attracted. Deviation from this optimum reduces total pollen export, particularly when many pollinators visit. However, total pollen export declines asymmetrically on either side of the optimum, so that plants lose less from erring towards too much pollen discounting than from having pollen left in anthers (Fig. 15.2).

Given that enhanced attractiveness increases pollen export only if each pollinator removes a limited amount of pollen, how do plants restrict pollen removal? Two types of mechanisms serve this purpose: *packaging*

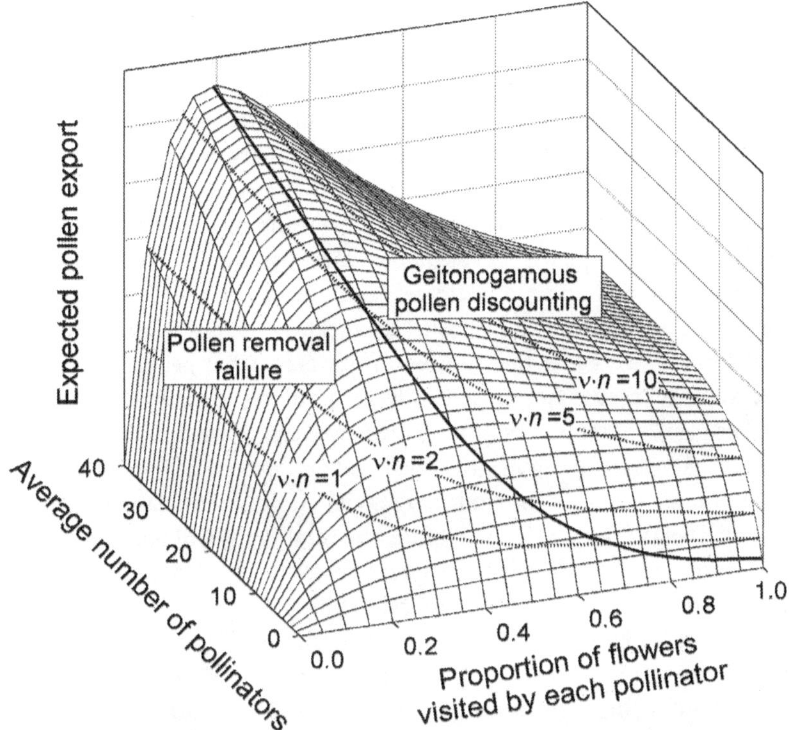



Fig. 15.2. Relation of expected pollen export by all pollinators to average pollinator attraction and the proportion of available flowers visited by each pollinator (based on Eq. 1 in Iwasa *et al.* 1995). The bold solid line depicts the proportion of flowers visited that maximizes expected pollen export for a specific average pollinator availability. Depression of pollen export around this optimum results from pollen-removal failure or geitonogamous pollen discounting, as indicated. The dashed lines illustrate expected pollen export for fixed expected intensities of visits per flower ($vn$). This example involves the same parameter values as Fig. 15.1.

*mechanisms* control the amount of pollen exposed at one time, whereas *dispensing mechanisms* limit the amount of exposed pollen removed by each pollinator (Harder & Thomson 1989). Packaging mechanisms can be implemented in individual flowers through staggered anther dehiscence, or on the entire plant through staggered opening of flowers. These mechanisms enable strict management of pollen removal because they are completely under a plant's control. In contrast, many dispensing mechanisms adjust pollen removal to a plant's prevailing frequency of pollinator visits (see Harder & Wilson 1994; Harder & Barclay 1994). Floral mechanisms that serve as dispensing mechanisms include anther position (Harder & Barrett 1993), poricidal anthers (Harder & Barclay 1994; King & Buchmann 1996), secondary pollen presentation (Yeo 1993; Harder & Wilson 1994), anther tripping (Armstrong 1992; Lebuhn & Anderson 1994), and nectar production.

Nectar production provides a unique means of dispensing pollen, because it allows plants to counteract diminishing returns on pollen removal from both individual flowers and inflorescences. Nectar volume influences pollen removal by positively affecting the duration of visits to individual flowers and the number of flowers visited per inflorescence (see above). Because nectar generally accumulates steadily (Búrquez & Corbet 1991), individual pollinators remove less pollen from individual flowers (Jones *et al.* 1998) and visit fewer flowers per inflorescence (Kadmon & Shmida 1992; Hodges 1995) when pollinators are abundant and visits occur frequently. Therefore, the combination of nectar production rate and visit frequency enables restricted pollen removal in a manner that responds to pollinator availability.

In seeming contradiction to this proposal, the rate of nectar production varies considerably among flowers within inflorescences, with some flowers producing little nectar (Feinsinger 1978; Brink 1982; Marden 1984; Gilbert *et al.* 1991). Bell (1986) proposed that empty flowers allowed plants to save some of the expense of nectar production without forfeiting much pollinator service. However, nectar may not be expensive for many plants (reviewed by Harder & Barrett 1992), so that empty flowers may do little to reduce the cost of attraction. Instead, we propose that by maintaining a fraction of flowers that produce little nectar, plants encourage pollinators to leave inflorescences after visiting only a fraction of their open flowers, thereby restricting pollen removal per pollinator and enhancing the aggregate pollen dispersal provided by all pollinators that visit. According to this hypothesis, empty flowers should be most common in species pollinated by abundant pollinators, because restricted removal is most beneficial when pollinators visit frequently.

The preceding discussion of attraction focused on the benefits for male success, rather than female success, through pollen receipt. We adopted this emphasis because the needs of pollen receipt are often realized with fewer pollinator visits than are those of pollen dispersal (e.g., Young & Stanton 1990; Mitchell & Waser 1992; Aizen & Basilio 1998; Bell & Cresswell 1998). This asymmetry arises from the dissimilarity in mating opportunities through female and male roles. The opportunities for paternal success depend on the number of available ovules in the population as a whole. As a result, outcross siring success increases continuously with a plant's relative contribution of pollen to stigmas. In contrast, each pistil contains a limited number of ovules, so that female outcross success levels off as stigmas receive an increasing share of exported pollen (e.g., Snow 1982; Shore & Barrett 1984; Galen 1992). Indeed, receipt of too much pollen can cause interference between pollen tubes and reduce seed production (reviewed by Young & Young 1992). Because of this asymmetry, the considerable effort expended on attraction by many animal-pollinated plants seems to benefit male success more than female success, even though pollen export must equal pollen import at the population level.

We conclude by emphasizing two essential features of the selection of floral design and display. The first feature arises from recognition that floral and inflorescence characteristics create the environment within which pollinators tend to maximize a specific foraging currency. Because of this role, plant evolution could improve foraging benefits or alleviate costs; however, it will do so only to the extent that such changes promote plant mating (e.g., Harder & Cruzan 1990; Harder & Barclay 1994). Therefore, the evolutionary relevance of specific floral or inflorescence traits must extend beyond their impact on pollinator behavior to realized mating outcomes. The second feature deserving emphasis is that, despite the key role of individual flowers in controlling pollen exchange with pollinators, mating fundamentally involves entire plants. For example, contrary to the expectation that a plant's pollen export increases monotonically with the number of pollinator visits received by each flower (e.g., Harder & Thomson 1989; Harder & Wilson 1994), increasing visits *per flower* with no change in the average number of visits *per plant* eventually reduces export (Fig. 15.2). Because of such non-monotonic effects, selection of floral traits will often optimize pollination of individual flowers to maximize a plant's mating success.

### Acknowledgements

We thank S.C.H. Barrett, L. Chittka, K.N. Jones, and J.D. Thomson for helpful comments on an earlier version of the manuscript. This review is based on research supported by the Natural Sciences and Engineering Research Council of Canada (LDH) and the Izaak Walton Killam Fund for Advanced Studies (NMW).

# References

Aizen MA & Basilio A (1998) Sex differential nectar secretion in protandrous *Alstroemeria aurea* (Alstroemeriaceae): is production altered by pollen removal and receipt? *Am J Bot* 85:245–252

Arista M, Ortiz PL & Talavera S (1999) Apical pattern of fruit production in the racemes of *Ceratonia siliqua* (Leguminosae: Caesalpinioideae): role of pollinators. *Am J Bot* 86:1708–1716

Armstrong JE (1992) Lever action anthers and the forcible shedding of pollen in *Torenia* (Scrophulariaceae). *Am J Bot* 79:34–40

Bell G (1986) The evolution of empty flowers. *J Theor Biol* 118:253–258

Bell SA & Cresswell JE (1998) The phenology of gender in homogamous flowers: temporal change in the residual sex function of flowers of oil-seed rape (*Brassica napus*). *Funct Ecol* 12:298–306

Brian AD (1952) Division of labour and foraging in *Bombus agrorum* Fabricius. *J Anim Ecol* 21:223–240

Brian AD (1957) Differences in the flowers visited by four species of bumble-bees and their causes. *J Anim Ecol* 26:71–98

Brink D (1982) A bonanza-blank pollinator reward schedule in *Delphinium nelsonii* (Ranunculaceae). *Oecologia* 52:292–294

Brunet J & Eckert CG (1998) Effects of floral morphology and display on outcrossing in blue columbine, *Aquilegia caerulea* (Ranunculaceae). *Funct Ecol* 12:596–606

Buchmann SL (1983) Buzz pollination in angiosperms. In: Jones CE & Little RJ (eds) *Handbook of experimental pollination biology*, pp 73–113. Van Nostrand Reinhold, New York

Buchmann SL & Cane JH (1989) Bees assess pollen returns while sonicating *Solanum* flowers. *Oecologia* 81:289–294

Búrquez A & Corbet SA (1991) Do flowers reabsorb nectar? *Funct Ecol* 5:369–379.

Camazine S (1993) The regulation of pollen foraging by honeybees: how foragers assess the colony's need for pollen. *Behav Ecol Sociobiol* 32:265–272

Campbell DR, Waser NM & Price MV (1994) Indirect selection of stigma positions in *Ipomopsis aggregata* via a genetically correlated trait. *Evolution* 48:55–68

Campbell DR, Waser NM & Price MV (1996) Mechanisms of hummingbird-mediated selection for flower width in *Ipomopsis aggregata*. *Ecology* 77:1463–1472

Cane JH & Buchmann SL (1989) Novel pollen-harvesting behavior by the bee *Protandrena mexicanorum* (Hymenoptera: Andrenidae). *J Insect Behav* 2:431–436

Cane JH & Payne JA (1988) Foraging ecology of the bee *Habropoda laboriosa* (Hymenoptera: Anthophoridae), an oligolege of blueberries (Ericaceae: *Vaccinium*) in the southeastern United States. *Ann Entomol Soc Am* 81:419–427

Cartar RV (1992) Adjustment of foraging effort and task switching in energy-manipulated wild bumblebee colonies. *Anim Behav* 44:75–87

Charnov EL (1976) Optimal foraging, the marginal value theorem. *Theor Popul Biol* 9:129–136

Cibula DA & Zimmerman M (1984) The effect of plant density on departure decisions: testing the marginal value theorem using bumblebees and *Delphinium nelsonii*. *Oikos* 43:154–158.

Conner JK, Davis R & Rush S (1995) The effect of wild radish floral morphology on pollination efficiency by four taxa of pollinators. *Oecologia* 104:234–245

Corbet SA, Cuthill I, Fallows M, Harrison T & Hartley G (1981) Why do nectar-foraging bees and wasps work upwards on inflorescences? *Oecologia* 51:79–83

Cresswell JE (1990) How and why do nectar-foraging bumblebees initiate movements between inflorescences of wild *Monarda fistulosa* (Lamiaceae)? *Oecologia* 82:450–460

Dreisig H (1985) Movement patterns of a clear-wing hawkmoth, *Hemaris fuciformis*, foraging at red catchfly, *Viscaria vulgaris*. *Oecologia* 67:360–366

Dreisig H (1989) Nectar distribution assessment by bumblebees foraging at vertical inflorescences. *Oikos* 55:239–249

Eickwort GC (1967) Aspects of the biology of *Chilicola ashmeadi* in Costa Rica (Hymenoptera: Colletidae). *J Kans Entomol Soc* 40:42–73

Ellington CP, Machin KE & Casey TM (1990) Oxygen consumption of bumblebees in forward flight. *Nature* 347:472–473

Emms SK, Stratton DA & Snow AA (1997) The effect of inflorescence size on male fitness: experimental tests in the andromonoecious lily, *Zigadenus paniculatus*. *Evolution* 51:1481–1489

Feinsinger P (1978) Ecological interactions between plants and hummingbirds in a successional tropical community. *Ecol Monogr* 48:269–287

Ford DM, Hepburn HR, Moseley FB & Rigby RJ (1981) Displacement sensors in the honeybee pollen basket. *J Insect Physiol* 27:339–346

Galen C (1992) Pollen dispersal dynamics in an alpine wildflower, *Polemonium viscosum*. *Evolution* 46:1043–1051

Galen C & Plowright RC (1985) Contrasting movement patterns of nectar-collecting and pollen-collecting bumble bees (*Bombus terricola*) on fireweed (*Chamaenerion angustifolium*) inflorescences. *Ecol Entomol* 10:9–17

Gass CL & Roberts WM (1992) The problem of temporal scale in optimization: three contrasting views of hummingbird visits to flowers. *Am Nat* 140:829–853

Gilbert FS, Haines N & Dickson K (1991) Empty flowers. *Funct Ecol* 5:29–39

Hainsworth FR & Hamill T (1993) Foraging rules for nectar: food choices by painted ladies. *Am Nat* 142:857–867

Hainsworth FR, Mercier T & Wolf LL (1983) Floral arrangements and hummingbird feeding. *Oecologia* 58:225–229

Harder LD (1985) Morphology as a predictor of flower choice by bumble bees. *Ecology* 66:198–210

Harder LD (1986) Effects of nectar concentration and flower depth on flower handling efficiency of bumble bees. *Oecologia* 69:309–315

Harder LD (1988) Choice of individual flowers by bumble bees: interaction of morphology, time and energy. *Behaviour* 104:60–77

Harder LD (1990*a*) Behavioral responses by bumble bees to variation in pollen availability. *Oecologia* 85:41–47

Harder LD (1990*b*) Pollen removal by bumble bees and its implications for pollen dispersal. *Ecology* 71:1110–1125

Harder LD (2000) Pollen dispersal and the floral diversity of Monocotyledons. In: Wilson KL & Morrison D (eds) *Monocots: systematics and evolution*, pp 243–257. CSIRO Publishing, Melbourne

Harder LD & Barclay RMR (1994) The functional significance of poricidal anthers and buzz pollination: controlled pollen removal from *Dodecatheon*. *Funct Ecol* 8:509–517

Harder LD & Barrett SCH (1992) The energy cost of bee pollination for *Pontederia cordata*. *Funct Ecol* 6:226–233

Harder LD & Barrett SCH (1993) Pollen removal from tristylous *Pontederia cordata*: effects of anther position and pollinator specialization. *Ecology* 74:1059–1072

Harder LD & Barrett SCH (1995) Mating cost of large floral displays in hermaphrodite plants. *Nature* 373:512–515

Harder LD & Barrett SCH (1996) Pollen dispersal and mating patterns in animal-pollinated plants. In: Lloyd DG & Barrett SCH (eds) *Floral biology: studies on floral evolution in animal-pollinated plants*, pp 140–190. Chapman & Hall, New York

Harder LD & Cruzan MB (1990) An evaluation of the physiological and evolutionary influences of inflorescence size and flower depth on nectar production. *Funct Ecol* 4:559–572

Harder LD & Real LA (1987) Why are bumble bees risk averse? *Ecology* 68:1104–1108

Harder LD & Thomson JD (1989) Evolutionary options for maximizing pollen dispersal of animal-pollinated plants. *Am Nat* 133:323–344

Harder LD & Wilson WG (1994) Floral evolution and male reproductive success: optimal dispensing schedules for pollen dispersal by animal-pollinated plants. *Evol Ecol* 8:542–559

Harder LD & Wilson WG (1997) Theoretical perspectives on pollination. *Acta Hort* 437:83–101

Harder LD & Wilson WG (1998) Theoretical consequences of heterogeneous transport conditions for pollen dispersal by animals. *Ecology* 79:2789–2807

Harder LD, Barrett SCH & Cole WW (2000) The mating consequences of sexual segregation within inflorescences of flowering plants. *Proc R Soc Lond B* 267:315–320

Haslett JR (1989) Adult feeding by holometabolous insects: pollen and nectar as complementary nutrient sources for *Rhingia campestris* (Diptera: Syrphidae). *Oecologia* 81:361–363

Heinrich B (1979) Resource heterogeneity and patterns of movement in foraging bumblebees. *Oecologia* 40:235–245

Hodges CM (1981) Optimal foraging in bumblebees: hunting by expectation. *Anim Behav* 29:1166–1171

Hodges CM (1985) Bumble bee foraging: the threshold departure rule. *Ecology* 66:179–187

Hodges CM & Wolf LL (1981) Optimal foraging in bumblebees: why is nectar left behind in flowers? *Behav Ecol Sociobiol* 9:41–44

Hodges SA (1995) The influences of nectar production on hawkmoth behavior, self-pollination, and seed production in *Mirabilis multiflora* (Nyctaginaceae). *Am J Bot* 82:197–204

Holloway BA (1976) Pollen-feeding in hover-flies (Diptera: Syrphidae). *NZ J Zool* 3:339–350

Houston AI & Krakauer DC (1993) Hummingbirds as net rate maximisers. *Oecologia* 94:135–138

Houston AI, Schmid-Hempel P & Kacelnik A (1988) Foraging strategy, worker mortality and the growth of the colony in social insects. *Am Nat* 131:107–114

Hurlbert AH, Hosoi SA, Temeles EJ & Ewald PW (1996) Mobility of *Impatiens capensis* flowers: effect on pollen deposition and hummingbird foraging. *Oecologia* 105:243–246

Iwasa Y, de Jong TJ & Klinkhamer PGL (1995) Why pollinators visit only a fraction of the open flowers on a plant: the plant's point of view. *J Evol Biol* 8:439–453

Jones KN, Reithel JS & Irwin RE (1998) A trade-off between the frequency and the duration of bumble bee visits to flowers. *Oecologia* 117:161–168

Kadmon R (1992) Dynamics of forager arrivals and nectar renewal in flowers of *Anchusa strigosa*. *Oecologia* 92:552–555

Kadmon R & Shmida A (1992) Departure rules used by bees foraging for nectar: a field test. *Evol Ecol* 6:142–151

Kimsey LS (1984) The behavioural and structural aspects of grooming and related activities in euglossine bees (Hymenoptera: Apidae). *J Zool* 204:541–550

King MJ & Buchmann SL (1996) Sonication dispensing of pollen from *Solanum laciniatum* flowers. *Funct Ecol* 10:449–456

Klinkhamer GL & de Jong TJ (1993) Attractiveness to pollinators: a plant's dilemma. *Oikos* 66:180–184

Kobayashi S, Inoue K & Kato M (1999) Mechanism of selection favoring a wide tubular corolla in *Campanula punctata*. *Evolution* 53:752–757

Kohn JR & Barrett SCH (1994) Pollen discounting and the spread of a selfing variant in tristylous *Eichhornia paniculata*: evidence from experimental populations. *Evolution* 48:1576–1594

Kunin WE (1997) Population biology and rarity: on the complexity of density-dependence in insect–plant interactions. In: Kunin WE & Gaston KJ (eds) *The biology of rarity: causes and consequences of rare–common differences*, pp 150–173. Chapman & Hall, London

Lebuhn G & Anderson GJ (1994) Anther tripping and pollen dispensing in *Berberis thunbergii*. *Am Midl Nat* 131:257–265

Liu HJ, Macfarlane RP & Pengelly DH (1975) Relationships between flowering plants and four species of *Bombus* (Hymenoptera: Apidae) in southern Ontario. *Can Entomol* 107:577–588

Lloyd DG (1992) Self- and cross-fertilization in plants. II. The selection of self-fertilization. *Int J Plant Sci* 153:370–380

Marden JH (1984) Intrapopulation variation in nectar secretion in *Impatiens capensis*. *Oecologia* 64:418–422

Martínez del Río C & Eguiarte LE (1987) The effect of nectar availability on the foraging behavior of the stingless bee *Trigona testacea*. *Southwestern Nat* 32:313–319

Michener CD, Winston ML & Jander R (1978) Pollen manipulation and related activities and structures in bees of the family Apidae. *Univ Kans Sci Bull* 51:575–601

Mitchell RJ (1993) Adaptive significance of *Ipomopsis aggregata* nectar production: observation and experiment in the field. *Evolution* 47:25–35

Mitchell RJ & Waser NM (1992) Adaptive significance of *Ipomopsis aggregata* nectar production: pollination success of single flowers. *Ecology* 73:633–638

Montgomerie RD (1984) Nectar extraction by hummingbirds: response to different floral characters. *Oecologia* 63:229–236

Montgomerie, RD, Eadie JM & Harder LD (1984) What do foraging hummingbirds maximize? *Oecologia* 63:357–363

Morgan MT (1992) Attractive structures and the stability of hermaphroditic sex expression in flowering plants. *Evolution* 46:1199–1213

Morris WF, Price MV, Waser NM, Thomson JD, Thomson BA & Stratton DA (1994) Systematic increase in pollen carryover and its consequences for outcrossing in plant populations. *Oikos* 71:431–440

Murcia C (1990) Effect of floral morphology and temperature on pollen receipt and removal in *Ipomoea trichocarpa*. *Ecology* 71:1098–1109

Parker FD & Frohlich DR (1983) Hybrid sunflower pollination by a manageable composite specialist: the sunflower leafcutter bee (Hymenoptera: Megachilidae). *Environ Entomol* 12:576–581

Plowright RC, Thomson JD, Lefkovitch LP & Plowright CMS (1993) An experimental study of the effect of colony resource level manipulation on foraging for pollen by worker bumble bees (Hymenoptera: Apidae). *Can J Zool* 71:1393–1396

Pyke GH (1979) Optimal foraging in bumblebees: rule of movement between flowers within inflorescences. *Anim Behav* 27:1167–1181

Rademaker MCJ & de Jong TJ (1998) Effects of flower number on estimated pollen transfer in natural populations of three hermaphroditic species: an experiment with fluorescent dye. *J Evol Biol* 11:623–641

Rasheed SA & Harder LD (1997*a*) Economic motivation for plant-species preferences of pollen-collecting bumble bees (Apidae, Hymenoptera, *Bombus*). *Ecol Entomol* 22:209–219

Rasheed SA & Harder LD (1997*b*) Foraging currencies for non-energetic resources: pollen collection by bumblebees. *Anim Behav* 54:911–926

Robertson AW, Mountjoy C, Faulkner BE, Roberts MV & Macnair MR (1999) Bumble bee selection of *Mimulus guttatus* flowers: the effects of pollen quality and reward depletion. *Ecology* 80:2594–2606.

Schmid-Hempel P & Schmid-Hempel R (1987) Efficient nectar-collecting by honeybees. II. Response to factors determining nectar availability. *J Anim Ecol* 56:219–227

Schmid-Hempel P, Kacelnik A & Houston AI (1985) Honeybees maximize efficiency by not filling their crop. *Behav Ecol Sociobiol* 17:61–66.

Shore JS & Barrett SCH (1984) The effect of pollination intensity and incompatible pollen on seed set in *Turnera ulmifolia* (Turneraceae). *Can J Bot* 62:1298–1303

Smithson A & Macnair MR (1997) Density-dependent and frequency-dependent selection by bumblebees *Bombus terrestris* (L.) (Hymenoptera: Apidae). *Biol J Linn Soc* 60:401–417

Snow AA (1982) Pollination intensity and potential seed set in *Passiflora vitifolia*. *Oecologia* 55:231–237

Snow AA, Spira TP, Simpson R & Klips RA (1996) The ecology of geitonogamous pollination. In: Lloyd DG & Barrett SCH (eds), *Floral biology: studies on floral evolution in animal-pollinated plants*, pp 191–216. Chapman & Hall, New York

Stanley RG & Linskens HF (1974) *Pollen biology, biochemistry, and management.* Springer-Verlag, Heidelberg

Stephens DW & Krebs JR (1986) *Foraging theory.* Princeton University Press, Princeton, NJ

Tamm S (1989) Importance of energy costs in central place foraging by hummingbirds. *Ecology* 70:195–205

Tepedino VJ (1981) The pollination efficiency of the squash bee (*Peponapis pruinosa*) and the honeybee (*Apis mellifera*) on summer squash (*Cucurbita pepo*). *J Kans Entomol Soc* 54:359–377

Teräs I (1985) Food plants and flower visits of bumblebees (*Bombus*: Hymenoptera, Apidae) in southern Finland. *Acta Zool Fennica* 179:1–120

Thomson JD (1986) Pollen transport and deposition by bumble bees in *Erythronium*: influences of floral nectar and bee grooming. *J Ecol* 74:329–341

Thomson JD (1988) Effects of variation in inflorescence size and floral rewards on the visitation rates of traplining pollinators of *Aralia hispida*. *Evol Ecol* 2:65–76

Thomson JD & Thomson BA (1989) Dispersal of *Erythronium grandiflorum* pollen by bumblebees: implications for gene flow and reproductive success. *Evolution* 43:657–661

Thorp RW (1969) Systematics and ecology of bees of the subgenus *Diandrena* (Hymenoptera: Andrenidae). *Univ Calif Publ Entomol* 52:1–178

Thorp RW (1979) Structural, behavioral, and physiological adaptations of bees (Apoidea) for collecting pollen. *Ann Missouri Bot Gard* 66:788–812

Waddington KD (1981) Factors influencing pollen flow in bumblebee-pollinated *Delphinium virescens. Oikos* 37:153–159

Waddington KD, Heinrich B (1979) The foraging movements of bumblebees on vertical "inflorescences": an experimental analysis. *J Comp Physiol A* 134:113–117

Waser NM (1986) Flower constancy: definition, cause, and measurement. *Am Nat* 127:593–603

Waser NM & Mitchell RJ (1990) Nectar standing crops in *Delphinium nelsonii* flowers: spatial autocorrelation among plants? *Ecology* 71:116–123

Waser NM & Price MV (1983) Pollinator behaviour and natural selection for flower colour in *Delphinium nelsonii. Nature* 302:422–424

Waser NM & Price MV (1984) Experimental studies of pollen carryover: effects of floral variability in *Ipomopsis aggregata. Oecologia* 62:262–268

Wcislo WT & Cane JH (1996) Floral resource utilization by solitary bees (Hymenoptera: Apoidea) and exploitation of their stored foods by natural enemies. *Ann Rev Entomol* 41:257–286

Weiss MR (1991) Floral colour changes as cues for pollinators. *Nature* 354:227–229

Williams CS (1997) Nectar secretion rates, standing crops and flower choice by bees on *Phacelia tanacetifolia. J Apic Res* 36:23–32

Williams NM (1999) The evolution and ecology of diet specialization in two Osmiine bees. PhD thesis, State University of New York, Stony Brook, NY

Williams NM & Thomson JD (2001) Pollinator quality in native bees and honeybees: comparing pollen removal and deposition on *Phacelia tanacetifolia. Thomas Say Publ Entomol*, in press

Wilson P (1995) Selection for pollination success and the mechanical fit of *Impatiens* flowers around bumblebee bodies. *Biol J Linn Soc* 55:355–383

Wolf LL & Hainsworth FR (1986) Information and hummingbird foraging at individual inflorescences of *Ipomopsis aggregata. Oikos* 46:15–22

Ydenberg RC, Welham CVJ, Schmid-Hempel R, Schmid-Hempel P & Beauchamp G (1994) Time and energy constraints and the relationships between currencies in foraging theory. *Behav Ecol* 5:28–34

Yeo PF (1993) *Secondary pollen presentation.* Springer-Verlag, New York

Young HJ & Stanton ML (1990) Influences of floral variation on pollen removal and seed production in wild radish. *Ecology* 71:536–547

Young HJ & Young TP (1992) Alternative outcomes of natural and experimental high pollen loads. *Ecology* 73:639–647

Zimmerman M (1981) Optimal foraging, plant density and the marginal value theorem. *Oecologia* 49:148–153

16

## Pollinator behavior and plant speciation: looking beyond the "ethological isolation" paradigm

> Floral isolating mechanisms consist of barriers to interspecific
> pollination in angiosperms imposed by structural contrivances ...
> [and] by the constancy of the pollinators to one kind of flower...
>
> Grant (1949), p 93

> Ist die Pollen-übertragung durch Insekten geeignet, die zur
> Artbildung nötige (mechanische) Isolierung zu fördern? (Is pollen
> transfer by insects suitable for promoting the mechanical isolation
> needed for speciation?)
>
> Werth (1955), p 163

> Another very obvious deficiency of observations indispensable to be
> made on the subject ... resulted ...[from] ...the fertilisation of flowers
> by insects being studied by botanists but little acquainted with insects.
>
> Müller (1873), p 187

It often is claimed that Darwin had little to say about the evolution of species, in spite of the title of his 1859 book. This is not strictly true: a close reading of the *Origin of Species* reveals that Darwin envisioned speciation for the most part as the eventual extension of a process of divergence beginning at a much smaller scale within a single species, and driven for the most part by natural selection. What is true, however, is that a detailed understanding of speciation in its many forms remains an elusive and desirable prize: speciation is, so to speak, the holy grail of evolutionary biology. Many questions confront us still. How often is speciation a simple extension of microevolution as Darwin proposed? When it is not, what new mechanisms of evolution come into play? What role is played by processes other than natural selection? Do different traits (e.g., floral phenotype, vegetative ecology, reproductive isolation) evolve together during speciation? What are the spatial elements of the process;

in particular, how common is speciation in near or complete sympatry? What are the genetics of the process? What is the time course; is speciation slow or fast? These questions exemplify those that have been discussed vigorously since the Evolutionary Synthesis of the 1930s and 1940s (see e.g., Endler 1977; Felsenstein 1981; Templeton 1981; Maynard Smith 1983; Otte & Endler 1989; Harrison 1990; Howard & Berlocher 1998; Dieckmann & Doebeli 1999).

Since the Synthesis, the flowering plants have been put forth as a group in which speciation and macroevolution are straightforward. The reason for this optimism is easy to see. The remarkable evolutionary radiation of the angiosperms over the last 100 million years, which makes them the dominant land plants, was matched in broad outline by radiations in several animal groups, including birds and those insect taxa comprising most pollinators (Crepet 1983; Grimaldi 1999). Furthermore, radiation appears to have been much more rapid in families of animal-pollinated plants than in those whose members are pollinated abiotically (Dodd *et al.* 1999). The traits by which taxonomists recognize angiosperm species with complex flowers are disproportionately reproductive ones, suggesting that these traits have been centrally involved in speciation (Grant 1949). These observations, combined with apparent specialization of many plants to a specific type of pollinator (e.g., a specific insect order or vertebrate class; Pijl 1961), suggest a cohesive and attractive view of the mechanics of angiosperm speciation, and of the role of animal pollinators.

This view is as follows: plants tend to evolve toward specialization in their attraction of specific pollinators, and the pollinators themselves exhibit "fidelity", i.e., are specialized to visit only (or mostly) the plant species in question; or, if not specialized as species, exhibit individual "constancy" in flower visits, meaning that an individual pollinator exhibits fidelity at least over the shorter term of a single foraging bout, a single day, or several days. Because of these behaviors, pollinators contribute to two linked processes that comprise angiosperm speciation (Grant 1949, 1952, 1994; Straw 1956; Pijl 1960; Baker 1963; Grant & Grant 1964, 1965, 1967; Free 1966; Stebbins 1970; Macior 1971; Jones 1978; Levin 1978; Crepet 1983, 1984; Wells *et al.* 1983; Hodges & Arnold 1994; Bradshaw *et al.* 1995; Schemske & Bradshaw 1999). As *agents of selection*, pollinators foster divergence in floral traits, because different pollinators select in different directions. Simultaneously, as *agents of gene flow*, their fidelity causes a great reduction or complete cessation of gene exchange between incipient species, as a pleiotropic side-product of floral divergence.

With reference to Grant's (1949) concept of pollinator-mediated reproductive isolation (see the quotation above), I refer to this as the "ethological isolation paradigm." Some evidence has accumulated to support the paradigm, even in its boldest claims of cospeciation of plants and pollinators (e.g., Powell 1992) and of speciation in sympatry (Hiesey *et al.* 1971; Vickery 1995; Ippolito & Holtsford 1999). But this positive evidence is surprisingly limited so far. In part this proves how difficult it is to accumulate empirical evidence that will satisfy a skeptical inquirer (as opposed to accumulating "plausibility arguments" for the paradigm). In addition there are reasons to suppose that the paradigm will be far from universal, stemming from recent empirical studies of the aforementioned skeptical enquirers which suggest that the situations fostering disruptive selection by pollinators in sympatry, and allowing cospeciation, will be special ones (e.g., Patel *et al.* 1993; Herrera 1996; Wcislo & Cane 1996; Wilson & Thomson 1996). Further doubts arise from contemplating how pollinators are expected to behave in choosing flowers.

It is my purpose in this chapter to argue that closer ties between botanists and zoologists, in particular those studying the behavior of foraging animals, will enrich our understanding of pollinator-mediated speciation in flowering plants. In so doing I will develop a different scenario for the role of pollinators in plant speciation. I contend that faster progress will be made if we search for the grail armed with a range of scenarios, in particular ones that include the perspective of foraging animals.

## The foraging behavior of pollinators

One of the fathers of pollination biology, Hermann Müller, clearly saw the danger of studying pollination without knowledge of insect biology, as the quotation at the head of this chapter shows. Unfortunately, Müller's (1873) warning remains apropos. Relatively few of the systems that have served as models in pollination biology have been studied with equal emphasis on botany and zoology. Too many studies include only a superficial treatment of pollinator behavior, relying instead on long-standing truisms about what the animals do, and why. Pollination biology will benefit greatly if it can replace this casual approach with a tradition that is more rigorous and includes quantitative and experimental study of behavior where appropriate, along with an appreciation of recent advances in behavior and cognitive biology of insects and other pollinators.

With regard to the ethological isolation paradigm, it will help to step back and examine the central assumptions of specialization and flower constancy. A reasonable place to begin is the theory of foraging behavior (see Pyke 1984; Stephens & Krebs 1986; Parker & Maynard Smith 1990). In particular, the theory of optimal diets considers a forager searching an environment that contains food items of different values for fitness. The value or "utility" of each food item is taken as its expected mean return in terms of calories or some other nutrient; the variance around this expectation also may constitute part of the value (Kacelnik & Bateson 1996; Smallwood 1996). The forager is assumed to have accurate knowledge of the values of different items and the search costs associated with them. The optimal behavior is to specialize on the most valuable item if this is encountered sufficiently often, but if not, to expand the diet to include the second-ranked item. Depending on the rate of encounter with first- and second-ranked items, the diet will be expanded further to the third-ranked item, and so on. These predictions can be modified for conspicuous (rather than cryptic) food items, in which case the forager should choose a path through the environment that yields a maximum rate of caloric or nutrient return. Depending on the spatial arrangement of food items this optimal path may involve visits to a single item or a mixture (Mitchell 1989).

This summary shows that a continuum from complete specialization to complete generalization can be expected of the same animal, depending on properties of the "prey" community. This conclusion assumes that constraints of cognition (and of physiology or morphology) are not absolute. Available evidence suggests that this is a reasonable assumption for pollinating animals (Waser *et al.* 1996). Even specialized oligolectic solitary bees appear usually to be constrained to particular plant species only in terms of pollen collection (due apparently to digestive physiology of larvae), and not in terms of nectar collection, which they tend to undertake at many flowers. Hence it is not surprising to find that pollinators such as nectar-collecting bees often behave in reasonable agreement with predictions of first-generation diet models (Pyke 1984, and references therein). Still, it has become clear that the models are incomplete, for example because foraging behavior is often moderately (if not absolutely) constrained by features of pollinator morphology, physiology, or cognition (for more details see Weiss, this volume; Menzel, this volume; Dukas, this volume). The behavior of flower constancy, properly defined as a propensity to visit the same type of flower as last visited irrespective of the value of alternatives, in particular suggests cognitive constraints such as those involving memory retrieval (for further development see Chittka *et al.* 1999). Several second-generation foraging models incorporate such constraints (e.g., Hughes 1979; McNair 1981; Dukas & Ellner 1993; Kunin & Iwasa 1996). These models predict for example that constancy will break down at low flower density.

The discussion so far concerns small temporal and spatial scales, e.g., a single foraging bout or single day or single meadow. What is predicted on larger scales? Generalization is predicted as an optimal strategy so long as the abundances of preferred plant species fluctuate sufficiently in time and space (Waser *et al.* 1996). For example, if the pollinator lifespan exceeds the flowering of a single plant species, generalization must ensue over an individual's lifetime. Similarly, if the pollinator's habitat affinities and geographic range incompletely overlap those of a single plant species, generalization must occur across pollinator populations (e.g., Herrera 1988, 1995).

Several conclusions follow. We should be unsurprised to find that many pollinators are opportunistic and generalized in choice of flowers; selection has molded pollinator physiologies and behaviors in floral environments that often vary substantially in time and space, where strict specialization may be disadvantageous. Conversely, an observation of short-term specialization does not necessarily indicate a morphological, physiological, or cognitive constraint. A strong constraint *is* indicated if the specialization is fixed throughout the lifetime of a forager and across individuals. In addition, the intriguing behavior of flower constancy is not predicted of an unconstrained forager, and should not be absolute even in a constrained forager, but rather should depend on ecological context. The overall conclusion from animal behavior is this: oligolecty and constancy are not necessary directional outcomes of evolution. Any view of these as traits of "advanced" pollinators (e.g., Crepet 1984) should be replaced by curiosity about features of the immediate floral environment, and of past environments, that affect the behavioral machinery of foraging animals to cause generalization vs. various forms of specialization (Waser 1986; Waser & Price 1998; Wcislo & Cane 1996).

Nature is the final arbiter; what do we see from direct observation of foraging pollinators? The evidence to date is that generalization and opportunism are common in pollinating insects of several orders (most studies have focused on Hymenoptera, and on Lepidoptera; Weiss, this volume) and in pollinating vertebrates (such as birds). To the references assembled by Waser *et al.* (1996), limited space allows me to add only a few recent ones. Memmott (1999) showed that the interactions between flowers and insect visitors in a British meadow form a highly connected web. Although some insects are specialist visitors, the plants they visit often attract more generalized insects as well, and these often connect the plants to the rest of the web (see also Jordano 1987). Careful study by Cotton (1998) showed that hermit hummingbirds are as generalized as non-hermits, contrary to the usual assumption. Momose *et al.* (1998) presented an impressive community-wide analysis of pollination in a lowland tropical forest, showing that many pollinators are generalized and opportunistic in flower use. Fleming & Holland (1998) found that the obligate mutualism between senita cactus and senita moth in northern México is supplemented by pollination from generalist solitary bees. Similarly, Kwak & Velterop (1997) documented pollination by generalists to an endangered plant species in Holland and France, in addition to pollination by a specialist bee; they also showed that the faunal composition of generalists changed through time and space. Olsen (1997) showed that the most effective pollinators of a native composite in Texas (those producing the most seed from a single visit) were among the *least* abundant of 10 different pollinators, seemingly at odds with Stebbins's (1970) "most effective pollinator" principle. Finally, Leebens-Mack *et al.* (1998) found that yucca moths are not so specialized to two sympatric yucca species as to prevent hybridization between them (see also Webber 1960).

This last example brings us directly back to the topic of the chapter. Even a relatively small proportion of incompletely constant generalist pollinators may begin to genetically connect related plant species that grow in sympatry (compare May & Anderson 1987), so long as the species are otherwise interfertile. The transition from an unconnected set of plant–pollinator interactions to a web of genetic connection may be non-linear (compare Green 1994). Such genetic connection will hinder progress toward complete sympatric speciation of taxa that otherwise might so diverge. Similarly, if secondary contact occurs between sister species that are newly diverged in parapatry or allopatry, increasing gene flow from generalist pollinators may foster hybridization and possibly fusion into a single species (e.g., Arnold 1997).

## Examples from pollination interactions

The phlox family (Polemoniaceae) takes center stage in the ethological isolation paradigm. Grant & Grant (1965) argued that specialization to different pollinators has been critical to adaptive radiation within the family, by leading to ethological isolation. In addition these authors argued that divergence in floral phenotype would affect mechanical isolation (placing of pollen on separate parts of the same pollinator's body), the second facet of Grant's (1949, 1952, 1994) "floral isolation".

Grant proposed the congeners *Ipomopsis aggregata* and *I. tenuituba* as a good example of floral isolation. These species occupy broadly overlapping geographic ranges in the western USA. Although they often exhibit different habitat affinities, situations of contact or near contact (i.e., parental populations growing within a few km of each other) are not uncommon. In some but not all of these contact situations one finds obvious hybrids (Grant & Wilken 1988). Normally, hybridization is considered to be precluded or limited by floral isolation (along with habitat differences), in the form of predominant hummingbird pollination of the red, trumpet-shaped tubular flowers of *I. aggregata* and predominant hawkmoth pollination of the longer, narrower, pale flowers of *I. tenuituba* (Grant & Grant 1965; Grant 1994).

My colleagues and I are studying a hybrid zone in Colorado, and find a more complex role of pollinators. At our site, parental populations of *I. aggregata* and *I. tenuituba* are separated by about 3 km and 300 m elevation, with hybrids in between displaying clinal variation in floral phenotypes. The predominant pollinators are hummingbirds, which select for shorter, wider, more darkly pigmented flowers (Campbell *et al.* 1997). In experimental mixtures, hummingbirds undervisit but do not absolutely shun *I. tenuituba*, relative to *I. aggregata* and hybrids. Further experiments (Meléndez-Ackerman *et al.* 1997) show that this discrimination stems mostly from inferior nectar rewards of the former relative to the latter types of flowers. In other words, hummingbirds are making an "economic" choice as predicted by foraging theory. The birds exhibit no detectable flower constancy. Aviary experiments indicate that mechanical isolation is weak; i.e., hummingbirds transfer substantial pollen between flowers of the two parental species, and the presence of hybrids facilitates this gene flow (Campbell *et al.* 1998). Finally, we have found that hawkmoths are rare; when present they visit all flower types, but are most common in populations toward the *I. tenuituba* side of the hybrid zone.

When hawkmoths are present, they and the hummingbirds select different floral phenotypes that most resemble the parent species (Campbell *et al.* 1997).

In years when hawkmoths visit, then, pollinators appear to cause disruptive selection on floral phenotype in sympatry, and divergent selection in parapatry. But there is a major caveat: hawkmoths are absent or very rare in most years. This situation may conceivably be ancestral, or may be a result of recent anthropogenic change in the western USA. Whatever the explanation, the present selection regime exerted by pollinators in most years strongly favors floral phenotypes that resemble those of *I. aggregata*, and does so throughout the hybrid zone (Campbell *et al.* 1997).

In terms of pollinator-mediated gene flow, furthermore, the *Ipomopsis* system fails to conform to a strict construction of the ethological isolation paradigm. Although hummingbirds and hawkmoths each prefer different floral phenotypes, the preferences are far from absolute. Hence both pollinators affect some genetic connection between the parental plant species, rather than isolating them completely. Independent genetic evidence agrees (Wolf *et al.* 1991, 1993; Wolf & Soltis 1992) in suggesting past gene flow between these taxa in various locations (and between other species in the genus as well), and also in suggesting multiple origins of *I. tenuituba*, perhaps from surrounding *I. aggregata* populations (i.e., contact situations just as likely represent primary divergence as they do secondary contact; Wolf *et al.* 1997).

Notice that this turns the classical question on its head. Rather than ask, "How does hybridization occur given the barriers imposed by pollinators?", we instead might profit from asking, "What keeps hybridization from happening more often, given overlapping pollinator preferences?" (Leebens-Mack & Milligan 1998 raise the same issue). In the *Ipomopsis* system we are just beginning to explore this latter question. For example, Alarcón & Campbell (2000) have shown that competitive superiority of conspecific pollen does not occur in our hybrid zone. But such a block to hybridization might occur in other *Ipomopsis* contact situations as it does in other plant species (e.g., Arnold *et al.* 1993; Riesberg *et al.* 1995); indeed there is evidence for strong reproductive barriers, based in part on conspecific pollen precedence, in a contact situation between *I. aggregata* and another species, *I. arizonica* (Wolf *et al.* 2001). Similarly, known hybrid individuals grown from seeds transplanted across our hybrid zone survive well on average (Campbell & Waser 2001). But this does not preclude the possibility of low or zero hybrid viability in other contact situations.

Unfortunately we have few detailed studies in hand of hybrid zones or other situations in which evidence for the ethological isolation paradigm might be obtained. However, those few studies now in hand do suggest that the *Ipomopsis* system is representative of many pollination systems. For example, Werth (1955), whose title is quoted at the beginning of this chapter, showed that the behavior of a few generalist insects suffices to connect several pairs of related species in the German flora, and so answered his own question in the negative. Leebens-Mack & Milligan (1998) found that bumble bees and carpenter bees exhibited preference and constancy in experimental mixtures of two species of *Baptisia* and their hybrids, but these insects still caused substantial gene flow, which was enhanced by the presence of hybrids. Galen (1996) showed that bumble bees select for large flowers of *Polemonium viscosum*, resembling the phenotypes at high elevation where they are the main pollinators. At lower elevations both flies and bumble bees pollinate, and flowers are smaller, suggesting possible disruptive selection in parapatry, for which there is some evidence (Galen *et al.* 1987). However, visits by bumble bees at all sites suggest the likelihood of substantial gene flow. Goulson & Jerrim (1997) used allozymes, pollinator observations, and the movement of fluorescent dye powders to determine that floral isolation between two species of *Silene* in Britain is insufficient to prevent hybridization, and that hybrid individuals serve as a bridge for gene flow. Similarly, Wesselingh & Arnold (2000) found that neither hummingbirds nor bumble bees were strongly specific to different parental phenotypes in mixtures of two *Iris* species and their hybrids. Bradshaw *et al.* (1995) conversely implied that differences in flower color and morphology would confer virtually total reproductive isolation on two interfertile species of monkeyflower (*Mimulus*) via distinct preferences of bumble bee and hummingbird pollinators. However, Hiesey *et al.* (1971) and Sutherland & Vickery (1993) had earlier shown that floral traits of monkeyflowers in no case appear to erect absolute barriers to either bumble bees or hummingbirds. Recent experiments by Schemske & Bradshaw (1999) with mixtures of parental species and hybrids do indicate that bumble bees and hummingbirds prefer different trait expressions, but these preferences are not absolute, in keeping with the previous findings. Hence some ethological isolation exists, but it does not seem to be enough to explain the absence of observed hybrids between the parental species.

These examples are not meant to suggest that the ethological isolation paradigm will never apply. For example, Fulton & Hodges (1999) reported that hummingbirds show strong fidelity to *Aquilegia formosa*, and hawkmoths show complete fidelity to *A. pubescens*, in experimental arrays of the two species. Our experience with *Ipomopsis* suggests that these studies should be extended across years and sites to determine how consistent and complete the apparent ethological isolation is, and to carefully examine the role of other flower visitors such as bees and flies. Furthermore it would be interesting to know how pollinators respond to arrays that include hybrids between the two *Aquilegia* species, because this presumably mimics an initial stage of evolution within a single ancestral population (whereas the use only of parental species in arrays may mimic a situation of secondary contact after differentiation in allopatry). In this regard, the experiments of A. Ippolito & T. Holtsford (1999; personal communication), which yield preliminary evidence for distinct specialization by hummingbirds vs. hawkmoths within single hybrid populations of Neotropical *Nicotiana*, may be the best example of pollinators imposing virtually complete ethological isolation within a unimodal set of floral phenotypes.

## The paradigm revisited

The foregoing examples argue that we often will need to look beyond the strict ethological isolation paradigm in attempting to understand how animal pollinators contribute to angiosperm speciation. So long as the pollinator fauna contains even a minority of generalists, disruptive selection on floral phenotype in sympatry is likely to be weakened. If individual plant species are served by a diversity of pollinators, the distribution of pollinator phenotypes may be uni- rather than multi-modal at any site; sympatric divergence of phenotype is still theoretically possible (via an "adaptive dynamics" process; Dieckmann & Doebeli 1999), although this is unexplored to date. In allopatry or parapatry, relatively generalized pollination often may mean that divergent selection, if it occurs, stems from quantitative differences among populations in the relative abundances of different pollinators, rather than from qualitative turnover in pollinators. This view of selection by a pollinator *fauna* rather than a single pollinator *species* is not new (e.g., Grant & Grant 1965, pp 162–163), but has not been stressed sufficiently (Waser 1998; Dilley *et al.* 2000).

Opportunistic and generalized pollinators pose a more central challenge to the ethological isolation paradigm, however. Even if selection on floral phenotype is disruptive (in sympatry) or divergent (in parapatry), there may be enough pollinator infidelity to prevent strong reproductive isolation from this source alone. In this case, the pleiotropic connection between the evolution of floral diversity and of reproductive isolation is weakened. If this proposal is correct – if pollinators do not automatically provide strong or complete reproductive barriers via behavioral responses that by definition are *extrinsic* to the plants – then the focus in angiosperm speciation should expand to include more study of barriers that are clearly *intrinsic* to the plants. These barriers are those expressed in the success of parental crosses, and in the viability and fertility of hybrids produced from them. Several questions arise immediately: on what scale of physical, ecological, genetic, or taxonomic separation do such intrinsic reproductive barriers first arise within angiosperm species? Are incipient intrinsic barriers elaborated eventually into complete reproductive barriers, and by what stages? Or is final reproductive isolation a mixture of ethological isolation and what I term intrinsic barriers, and if so in what proportions? What forms do intrinsic barriers take, and what does this imply about their genetic and ecological mechanisms? Finally, what role do animal pollinators play, however indirectly?

Insect biologists appear to be far ahead of botanists in answering such questions (Oliver 1972; Coyne & Orr 1997). Still, botanists can offer tentative answers, based on studies of crossing relationships that were popular until only a few decades ago (and should be revived). For example, Kruckeberg's (1957) studies of *Streptanthus*, Grant & Grant's (1960) of *Gilia*, and Vickery's (1978) of *Mimulus*, all indicate that intrinsic crossing barriers are scattered at all taxonomic levels both within single species and among species within a genus (see also Levin 1978). Conversely, morphologically and ecologically distinct species within an angiosperm genus often are interfertile (indeed, distinct genera are sometimes interfertile). Strong crossing barriers are sometimes found over scales of tens of km or less, which can be assumed to correspond to an early stage of genetic differentiation among populations (see also Waser *et al.* 2000). The nature of the barriers is variable, encompassing sterility or reduced productivity of the parental cross, inviability or reduced viability in $F_1$ hybrids, sterility of the hybrids, and similar breakdown in later-generation hybrids (see also Levin 1978). In these regards partial barriers within species resemble stronger barriers among species or genera. Finally, the distribution of barriers, and their strength, appears somewhat haphazard within species, correlating weakly if at all with geographic, ecological, or phenotypic separation of populations, or other estimators of their degree of evolutionary differentiation such as subspecific status (e.g., Hughes & Vickery 1974, 1975). These observations are concordant with the view that intrinsic reproductive barriers are evolving partly independently of the phenotypic traits, for example floral traits, by which angiosperm taxa are usually recognized.

The role of pollinators in this scenario of the evolution of reproductive isolation is much more passive than in the strict ethological isolation paradigm. The converse of predicted opportunism of a relatively unconstrained forager in a mixture of flower species is that this forager should not travel farther than necessary between food items, especially if travel is energetically costly. Flying pollinators endure high costs of travel and tend to move short distances between flowers (e.g., Pyke 1984; Wolf *et al.* 1989). This behavior sets up conditions of local mating and genetic isolation by distance which foster divergence of populations as a function of their physical separation, due either to adaptation to local environments or to random genetic drift. Crosses between such populations will ultimately begin to exhibit reproductive barriers (Hughes & Vickery 1974; Waser *et al.* 2000), and these barriers may be reinforced upon secondary contact (e.g., Paterniani 1969; Crosby 1970; Levin 1978; Waser & Price 1993) or may strengthen as a pleiotropic effect of further genetic divergence. Thus pollinators may indirectly facilitate microevolutionary differentiation and incipient reproductive isolation which eventually becomes elaborated into complete isolation between higher angiosperm taxa.

A very different role of animal pollinators must be given a brief mention. By their very act of generalization and opportunism in flower visits, these animals appear certain to contribute to another important source of genetic and phenotypic novelty in plants (see also Webber 1960). There is evidence both old and new for an important role of hybridization in angiosperm macroevolution (e.g., Lewis & Epling 1959; Riesberg 1995). By fostering hybridization, animal pollinators stand to play a central role in the "instantaneous" generation of species via allopolyploidy and homoploid hybrid speciation (see Arnold 1997). From this, and by causing gene exchange among taxa even when no new species arise, they should contribute centrally to patterns of reticulate evolution in angiosperms.

### Future directions

The view developed above is that floral divergence often will arise not from qualitatively different specificities for pollinators, but rather from quantitatively different selection regimes imposed by different suites of visitors; and that reproductive divergence often will arise not from different absolute specificities for pollinators, but rather from combinations of partial pollinator preferences and of (potentially adaptive) genetic divergence in plant populations, itself facilitated by the area-restricted foraging of pollinators. This scenario is directly tied to the observed foraging behavior of many pollinators. By applying an animal perspective, phenomena that once were considered rare and pathological, such as foraging "mistakes" and hybridization, come into new focus as common and normal events that are important to plant reproductive ecology and evolution.

By extension this suggests that the chance for new insights in pollination biology will be greatly accelerated if we can build lasting bridges to zoology – not the least because of the accelerated pace of progress in animal behavior and cognitive biology. The way to construct such bridges is for specialists to be encouraged to broaden their training and to form collaborations that span a broader range of expertise (Waser & Price 1998).

Indeed, it is my conviction that a closer marriage with behavioral biology will allow pollination biology to develop more realistic organizing principles to supplant the various versions of the "pollination syndromes" now in common use (Waser & Chittka 1998; Waser & Price 1998). As pollination biology moves beyond typologies, its practitioners may more easily be encouraged to expand their focus from the *central tendencies* in plant–pollinator relationships to include *variation* in those relationships through time and space, and to focus on mechanisms underlying variation. Nothing could be more important as we strive to conserve pollination systems in the face of increasing anthropogenic change (Kearns *et al.* 1998).

### Acknowledgements

For discussions over the years and comments on the manuscript I thank Diane Campbell, Lars Chittka, Reuven Dukas, Carlos Herrera, Scott Hodges, Jeff Ollerton, Mary Price, Graham Pyke, James Thomson, Bob Vickery, Paul Wilson, and Paul Wolf. Many of our studies of *Ipomopsis* were supported by the United States National Science Foundation.

## References

Alarcón R & Campbell DR (2000) Absence of conspecific pollen advantage in the dynamics of an *Ipomopsis* (Polemoniaceae) hybrid zone. *Am J Bot* 87:819–824

Arnold ML (1997) *Natural hybridization and evolution*. Oxford University Press, New York

Arnold ML, Hamrick JL & Bennett BD (1993) Interspecific pollen competition and reproductive isolation in Iris. *J Hered* 84:13–16

Baker HG (1963) Evolutionary mechanisms in pollination biology. *Science* 139:877–883

Bradshaw HDJ, Wilbert SM, Otto KG & Schemske DW (1995) Genetic mapping of floral traits associated with reproductive isolation in monkeyflowers (*Mimulus*). *Nature* 376:762–765

Campbell DR & Waser NM (2001) Genotype by environment interaction and the fitness of plant hybrids in the wild. *Evolution* in press

Campbell DR, Waser NM & Meléndez-Ackerman EJ (1997) Analyzing pollinator-mediated selection in a plant hybrid zone: hummingbird visitation patterns on three spatial scales. *Am Nat* 149:295–315

Campbell DR, Waser NM & Wolf PG (1998) Pollen transfer by natural hybrids and parental species in an *Ipomopsis* hybrid zone. *Evolution* 52:1602–1611

Chittka L, Thomson JD & Waser NM (1999) Flower constancy, insect psychology, and plant evolution. *Naturwiss* 86:361–377

Cotton PA (1998) Coevolution in an Amazonian hummingbird-plant community. *Ibis* 140:639–646

Coyne JA & Orr HA (1997) "Patterns of speciation in *Drosophila*" revisited. *Evolution* 51:295–303

Crepet WL (1983) The role of insect pollination in the evolution of the angiosperms. In: Real L (ed) *Pollination biology*, pp 29–50. Academic Press, Orlando, FL

Crepet WL (1984) Advanced (constant) insect pollination mechanisms: pattern of evolution and implications vis-à-vis angiosperm diversity. *Ann Missouri Bot Gard* 71:607–630

Crosby JL (1970) The evolution of genetic discontinuity: computer models of the selection of barriers to interbreeding between species. *Heredity* 25:253–297

Dieckmann U & Doebeli M (1999) On the origin of species by sympatric speciation. *Nature* 400:354–357

Dilley JD, Wilson P & Mesler MR (2000) The radiation of *Calochortus*: generalist flowers moving through a mosaic of potential pollinators. *Oikos* 89:209–222

Dodd ME, Silvertown J & Chase MW (1999) Phylogenetic analysis of trait evolution and species diversity variation among angiosperm families. *Evolution* 53:732–744

Dukas R & Ellner S (1993) Information processing and prey detection. *Ecology* 74:1337–1346

Endler JA (1977) *Geographic variation, speciation, and clines*. Princeton University Press, Princeton, NJ

Felsenstein J (1981) Skepticism towards Santa Rosalia, or why are there so few kinds of animals? *Evolution* 35:124–138

Fleming TH & Holland JN (1998) The evolution of obligate pollination mutualisms: senita cactus and senita moth. *Oecologia* 114:368–375

Free JB (1966) The foraging behavior of bees and its effect on the isolation and speciation of plants. In: Hawkes JG (ed) *Reproductive biology and taxonomy of vascular plants*, pp 76–91. Pergamon Press, Oxford

Fulton M & Hodges SA (1999) Floral isolation between *Aquilegia formosa* and *A. pubescens*. *Proc R Soc Lond B* 266:2247–2252

Galen C (1996) Rates of floral evolution: adaptation to bumblebee pollination in an alpine wildflower, *Polemonium viscosum*. *Evolution* 50:120–125

Galen C, Zimmer KA & Newport ME (1987) Pollination in floral scent morphs of *Polemonium viscosum*: a mechanism for disruptive selection on flower size. *Evolution* 41:599–606

Goulson D & Jerrim K (1997) Maintenance of the species boundary between *Silene dioica* and *S. latifolia* (red and white campion). *Oikos* 79:115–126

Grant KA & Grant V (1964) Mechanical isolation of *Salvia apiana* and *Salvia mellifera* (Labiatae). *Evolution* 18:196–212

Grant KA & Grant V (1967) Effects of hummingbird migration on plant speciation in the California flora. *Evolution* 21:457–465

Grant V (1949) Pollination systems as isolating mechanisms in angiosperms. *Evolution* 3:82–97

Grant V (1952) Isolation and hybridization between *Aquilegia formosa* and *A. pubescens*. *Aliso* 2:341–360

Grant V (1994) Modes and origins of mechanical and ethological isolation in angiosperms. *Proc Natl Acad Sci USA* 91:3–10

Grant V & Grant A (1960) Genetic and taxonomic studies in *Gilia*. XI. Fertility relationships of the diploid cobwebby gilias. *Aliso* 4:435–481

Grant V & Grant KA (1965) *Flower pollination in the phlox family*. Columbia University Press, New York

Grant V & Wilken DH (1988) Natural hybridization between *Ipomopsis aggregata* and *I. tenuituba* (Polemoniaceae). *Bot Gaz* 149:213–221

Green DG (1994) Connectivity and complexity in landscapes and ecosystems. *Pacif Cons Biol* 3:194–200

Grimaldi D (1999) The co-radiations of pollinating insects and angiosperms in the Cretaceous. *Ann Missouri Bot Gard* 86:373–406

Harrison RC (1990) Hybrid zones: windows on evolutionary processes. *Oxf Surv Evol Biol* 7:69–128

Herrera CM (1988) Variation in mutualisms: the spatiotemporal mosaic of a pollinator assemblage. *Biol J Linn Soc* 35:95–125

Herrera CM (1995) Microclimate and individual variation in pollinators: flowering plants are more than their flowers. *Ecology* 76:1516–1524

Herrera CM (1996) Floral traits and plant adaptation to insect pollinators: a devil's advocate approach. In: Lloyd DG & Barrett SCH (eds) *Floral biology: studies on floral evolution in animal-pollinated plants*, pp 65–87. Chapman & Hall, New York

Hiesey WM, Nobs MS & Björkman O (1971) Experimental studies on the nature of species. V. Biosystematics, genetics and physiological ecology of the Erythranthe section of *Mimulus*. *Carnegie Institution of Washington, Washington DC, Publication 628*

Hodges SA & Arnold ML (1994) Columbines: a geographically widespread species flock. *Proc Natl Acad Sci USA* 91:5129–5132

Howard DJ & Berlocher SH (1998) *Endless forms: species and speciation*. Oxford University Press, New York

Hughes KW & Vickery RK Jr (1974) Patterns of heterosis and crossing barriers from increasing genetic distance between populations of the *Mimulus luteus* complex. *J Genet* 61:235–245

Hughes KW & Vickery RK Jr (1975) Evolutionary divergence in closely related populations of *Mimulus guttatus* (Scrophulariaceae). *Great Basin Nat* 35:240–244

Hughes RN (1979) Optimal diets under the energy maximization premise: the effects of recognition time and learning. *Am Nat* 113:209–221

Ippolito A & Holtsford T (1999) Selection of *Nicotiana* (Solanaceae) floral morphology in the wild. *Abstr XVI Int Bot Congr* 442

Jones CE (1978) Pollinator constancy as a pre-pollination isolating mechanism between sympatric species of Cercidium. *Evolution* 32:189–198

Jordano P (1987) Patterns of mutualistic interactions in pollination and seed dispersal: connectance, dependence asymmetries, and coevolution. *Am Nat* 129:657–677

Kacelnik A & Bateson M (1996) Risky theories – the effects of variance on foraging decisions. *Am Zool* 36:402–434

Kearns CA, Inouye DW & Waser NM (1998) Endangered mutualisms: the conservation of plant–pollinator interactions. *Ann Rev Ecol Syst* 29:83–112

Kruckeberg A (1957) Variation in infertility of hybrids between isolated populations of the serpentine species *Streptanthus glandulosus* Hook. *Evolution* 11:185–211

Kunin WE & Iwasa Y (1996) Pollinator foraging strategies in mixed floral arrays: density effects and floral constancy. *Theor Popul Biol* 49:232–263

Kwak MM & Velterop O (1997) Flower visitation by generalists and specialists: analysis of pollinator quality. *Proc Exp Appl Entomol Neth Entomol Soc* 8:85–89

Leebens-Mack J & Milligan BG (1998) Pollination biology in hybridizing *Baptisia* (Fabaceae) populations. *Am J Bot* 85:500–507

Leebens-Mack J, Pellmyr O & Brock M (1998) Host specificity and the genetic structure of two yucca moth species in a yucca hybrid zone. *Evolution* 52:1376–1382

Levin DA (1978) The origin of isolating mechanisms in flowering plants. *Evol Biol* 11:185–317

Lewis H & Epling C (1959) *Delphinium gypsophilum*, a diploid species of hybrid origin. *Evolution* 13:511–525

Macior LW (1971) Co-evolution of plants and animals – systematic insights from plant–insect interactions. *Taxon* 20:17–28

May RM & Anderson RM (1987) Transmission dynamics of HIV infection. *Nature* 326:137–142

Maynard Smith J (1983) Current controversies in evolutionary biology. In: Grene M (ed) *Dimensions of Darwinism*, pp 273–286. Cambridge University Press, Cambridge

McNair JN (1981) A stochastic foraging model with predator training effects. II. Optimal diets. *Theor Popul Biol* 9:147–162

Meléndez-Ackerman EJ, Campbell DR & Waser NM (1997) Hummingbird behavior and mechanisms of selection on flower color in *Ipomopsis*. *Ecology* 78:2532–2541

Memmott J (1999) The structure of a plant–pollinator food web. *Ecol Lett* 2:276–280

Mitchell WA (1989) Informational constraints on optimally foraging hummingbirds. *Oikos* 55:145–154

Momose K, Yumoto T, Nagamitsu T, Kato M, Nagamasu H, Sakai S, Harrison RD, Itioka T, Hamid AA & Inoue T (1998) Pollination biology in a lowland dipterocarp forest in Sarawak, Malaysia. I. Characteristics of the plant–pollinator community in a lowland dipterocarp forest. *Am J Bot* 85:1477–1501

Müller H (1873) On the fertilisation of flowers by insect and on the reciprocal adaptation of both. *Nature* 8:187–189, 205–206

Oliver CG (1972) Genetic and phenotypic differentiation and geographic distance in four species of Lepidoptera. *Evolution* 26:221–241

Olsen KM (1997) Pollination effectiveness and pollinator importance in a population of
    *Heterotheca subaxillaris* (Asteraceae). *Oecologia* 109:114–121
Otte D & Endler JA (1989) *Speciation and its consequences.* Sinauer, Sunderland, MA
Parker GA & Maynard Smith J (1990) Optimality theory in evolutionary biology. *Nature*
    348:27–33
Patel A, Hossaert-McKey M & McKey D (1993) *Ficus*-pollinator research in India: past,
    present and future. *Curr Sci* 65:243–253
Paterniani E (1969) Selection for reproductive isolation between two populations of
    maize, *Zea mays* L. *Evolution* 23:534–547
Pijl L van der (1960) Ecological aspects of flower evolution. I. Phyletic evolution.
    *Evolution* 14:403–416
Pijl L van der (1961) Ecological aspects of flower evolution. II. Zoophilous flower classes.
    *Evolution* 15:44–59
Powell JA (1992) Interrelationships of yuccas and yucca moths. *Trends Ecol Evol* 7:10–15
Pyke GH (1984) Optimal foraging theory: a critical review. *Ann Rev Ecol Syst* 15:523–575
Riesberg LH (1995) The role of hybridization in evolution: old wine in new skins. *Am J
    Bot* 82:944–953
Riesberg LH, Desrochers AM & Youn SJ (1995) Interspecific pollen competition as a
    reproductive barrier between sympatric species of *Helianthus* (Asteraceae). *Am J
    Bot* 82:515–519
Schemske DW & Bradshaw HD Jr (1999) Pollinator preference and the evolution of floral
    traits in monkey flowers (*Mimulus*). *Proc Natl Acad Sci USA* 96:11910–11915
Smallwood PD (1996) An introduction to risk sensitivity: the use of Jensen's inequality
    to clarify evolutionary arguments of adaptation and constraint. *Am Zool*
    36:392–401
Stebbins GL (1970) Adaptive radiation of reproductive characteristics in angiosperms. I.
    Pollination mechanisms. *Ann Rev Ecol Syst* 1:307–326
Stephens DW & Krebs JR (1986) *Foraging theory.* Princeton University Press, Princeton, NJ
Straw RM (1956) Floral isolation in *Penstemon. Am Nat* 90:47–53
Sutherland SD & Vickery RK Jr (1993) On the relative importance of floral color, shape,
    and nectar rewards in attracting pollinators to *Mimulus. Great Basin Nat*
    53:107–117
Templeton AR (1981) Mechanisms of speciation – a population genetic approach. *Ann
    Rev Ecol Syst* 12:23–48
Vickery RK Jr (1978) Case studies in the evolution of species complexes in *Mimulus. Evol
    Biol* 11:405–507
Vickery RK Jr (1995) Speciation in *Mimulus*, or, can a simple flower color mutant lead to
    species divergence? *Great Basin Nat* 55:177–180
Waser NM (1986) Flower constancy: definition, cause, and measurement. *Am Nat*
    127:593–603
Waser NM (1998) Pollination, angiosperm speciation, and the nature of species
    boundaries. *Oikos* 82:198–201
Waser NM & Chittka L (1998) Bedazzled by flowers. *Nature* 394:835–836
Waser NM & Price MV (1993) Crossing distance effects on prezygotic performance in
    plants: an argument for female choice. *Oikos* 68:303–308
Waser NM & Price MV (1998) What plant ecologists can learn from zoology. *Persp Plant
    Ecol Evol Syst* 1:137–150
Waser NM, Chittka L, Price MV, Williams N & Ollerton J (1996) Generalization in
    pollination systems, and why it matters. *Ecology* 77:279–296

Waser NM, Price MV & Shaw RG (2000) Outbreeding depression varies among cohorts of *Ipomopsis aggregata* planted in nature. *Evolution* 54:485–491

Wcislo WT & Cane JH (1996) Floral resouce utilization by solitary bees (Hymenoptera: Apoidea) and exploitation of their stored foods by natural enemies. *Ann Rev Entomol* 41:257–286

Webber JM (1960) Hybridization and instability of *Yucca*. *Madroño* 15:187–192

Wells H, Wells PH & Smith DM (1983) Ethological isolation of plants. I. Colour selection by honeybees. *J Apic Res* 22:33–44

Werth E (1955) Ist die Pollen-übertragung durch Insekten geeignet, die zur Artbildung nötige (mechanische) Isolierung zu fördern? *Ber Deut Bot Ges* 68:163–166

Wesselingh RA & Arnold ML (2000) Pollinator behaviour and the evolution of Louisiana iris hybrid zones. *J Evol Biol* 13:171–180

Wilson P & Thomson JD (1996) How do flowers diverge? In: Lloyd DG & Barrett SCH (eds) *Floral biology: studies on floral evolution in animal-pollinated systems*, pp 88–111. Chapman & Hall, New York

Wolf PG & Soltis PS (1992) Estimates of gene flow among populations, geographic races, and species in the *Ipomopsis aggregata* complex. *Genetics* 130:639–647

Wolf PG, Soltis PS & Soltis DE (1991) Genetic relationships and patterns of allozymic divergence in the *Ipomopsis aggregata* complex and related species (Polemoniaceae). *Am J Bot* 78:515–526

Wolf PG, Soltis PS & Soltis DE (1993) Phylogenetic relationships of the *Ipomopsis aggregata* complex and related species (Polemoniaceae): evidence from chloroplast DNA restriction site variation. *Syst Bot* 18:652–662

Wolf PG, Murray RA & Sipes SD (1997) Species-independent, geographical structuring of chloroplast DNA haplotypes in a montane herb *Ipomopsis* (Polemoniaceae). *Mol Ecol* 6:283–291

Wolf PG, Campbell DR, Waser NM, Sipes SD, Toler TR & Archibald JK (2001) Tests of pre- and post-pollination barriers to hybridization between sympatric species of *Ipomopsis* (Polemoniaceae). *Am J Bot*, in press

Wolf TJ, Schmid-Hempel P, Ellington CP & Stevenson RD (1989) Physiological correlates of foraging efforts in honey-bees; oxygen consumption and nectar load. *Func Ecol* 3:417–424

# Index


*Acacia*, 229
*Acer*, 114
achromatic cues, 63–5, 68–9
*Acyrthosiphon pisum*, 220
*Adalia bipunctata*, 220
*Agave*, 166
agents of selection, plant speciation and, 319
alpha-cellulose powder, 47–8
*Amaranthus*, nectar availability, 230
andrenid bees, 95
ant colonies, assessment of predation risk, 219(fig.)
antennae, odor signals, 87–8
anthomyiids, innate color preferences, 175
anthophilous plants, 114
*Anthurium*, 85
*Antirrhinum majus*, 267
ants, 218
*Apis mellifera*, 203, 216, 224
  bilateral symmetry, 76(fig.)
  color vision system, 62–3
  dance, *see* dance, bee
  decision-making process, 45(fig.)
  evolution of flower-color preference, 119
  foraging cycle in, 27(fig.)
  life span, 225
  memory storage and retrieval, 28(fig.)
  nectar concentration and, 52
  odors and, 90
  pattern vision, 61–79
  photoreceptors, 62(fig.)
  predation by crab spiders, 217(fig.)
  proboscis extension response, 49
  sugar concentration and, 54
  temporal dynamics of choice behavior, 30(fig.), 31(fig.)
  training patterns, 78
  vision and learning, 179(fig.)
*Aquilegia formosa*, 327

*Aquilegia pubescens*, 327
*Aralia hispida*, 205
  pollinator individuality, 194
*Archilochus alexandri*, 129
*Archilochus anna*, 129
*Archilochus colubris*, 129
arthropod evolution, color vision and, 116
artificial flowers
  bumble bees on, 241–3
  frequency-dependent visits to, 242(table), 243–5
artificial inflorescences, bumble bee behavior on, 245–6
*Asclepias syriaca* (milkweed), 216, 226
associative induction, 30–2
associative learning, spatial memory and, 26
assortative mating
  evolutionary significance of, 260–2
  pollinator-mediated, 259–70
avoidance learning, predation risk assessment, 218–20

*Baptisia*, 269
  pollination interactions, 326
Bateman's Index, 11(fig.), 111
Batesian mimicry, as frequency-dependent selection, 253–4
bats, 86, 148–67, 216
  daily energy expenditure, 150–2, 151(fig.)
  foraging energetics, 148–67
  energy costs of flight, 153–5
  flight speeds, 153
  glossophagine, 152(table)
  hovering flights, 154–5
  responses to fragrance, 91
  vision, 160–2
  *see also individual species*
*Bauhinia*, 160, 165

bees
    andrenid, 95
    assessment of predation risk and, 218
    choice behavior, 41–57
    color vision system, 62–3
    crab spiders and, 228(fig.)
    dance, *see* dance, bee
    euglossine, 85, 86
    evolution of flower-color preference,
        119–23
    flower color and, 111
    green contrast, 68
    movement patterns, 198(table)
    pattern vision, 61–79
    pollinator individuality, 194
    search times, 108
    subjective evaluation by, 41–57
    training patterns, 78
    trait variability hypothesis, 7–8
    ultraviolet vision, 115
    *see also individual species*
bee-eaters, 215
bee-wolf (*Philanthus triangulum*), 216–17
beetles, vision and learning in, 171–84
bilateral symmetry, bee pattern vision,
        76(fig.)
biosynthesis
    fragrance and, 83–5
    volatile production in plants, 84(fig.)
black-and-white stimuli, achromatic spatial
        cues in, 68–9
blowflies (*Calliphora*), 89
blue jays (*Cyanocitta cristata*), search image
        hypothesis, 5
*Bombacopsis quinatum*, 160
*Bombus*, 216, 224
    behavior on artificial inflorescences,
        245–6
    behavior on single artificial flowers, 241–3
    captured, 223(fig.)
    color preferences of, 120(fig.)
    color vision, 118
    Darwin's interference hypothesis, 2–4
    evolution of flower-color preference,
        119–23
    flower color and, 107–10, 109(fig.)
    frequency-dependent visits by, 243–5
    nectar concentration and, 51
    photoreceptor values, 117(fig.)
    phylogeny, 120(fig.)
    pollination interactions, 326
    pollinator individuality, 194
    predation by crab spiders, 217(fig.)
    scent-marking, 206
    sugar concentration and, 54
    trait variability hypothesis, 7–8

    trapline behavior, 196–7(fig.)
    vision and learning, 178–9
    wing wear, 222
*Bombus agrorum*, 193
*Bombus appositus*, 267
*Bombus bifarius*, 195
*Bombus ephippiatus*, 193
*Bombus fervidus*, 55
*Bombus flavifrons*, 192
*Bombus hypocrita*, 119
*Bombus ignitus*, 119
*Bombus impatiens*
    color vision, 116
    pollinator individuality, 194
    sugar concentration and, 55
    trait variability hypothesis, 7–8
    visits to low-variance flowers, 56(fig.)
*Bombus lapidarius*
    color vision, 116
    evolution of flower-color preference, 119
*Bombus lucorum*, 119
*Bombus monticola*, 116
*Bombus morio*, 116
*Bombus occidentalis*
    color vision, 121
    evolution of flower-color preference, 119
    flower color and, 110
*Bombus pratorum*, 119
*Bombus ternarius*, 194
*Bombus terrestris*
    biogeography of color preferences,
        122(fig.)
    color vision, 116, 121
    flower color and, 108
    on artificial inflorescences, 245–6
*Bombus terrestris canariensis*, 120, 121
*Bombus terrestris dalmatinus*, 119
*Bombus terrestris sassaricus*, 120
*Bombus terrestris terrestris*, 119
*Bombus terrestris xanthopus*, 120
*Bombus vagans*, 216
bombyliids, 173
    innate color preferences, 175
bugs, predaceous, 215
bumble bee, *see Bombus*
bumblebee-wolf, 217, 222
butterflies, 224
    Darwin's interference hypothesis, 2–4
    fragrance and, 92
    noxious, 220
    spectral range, 176
    vision and learning in, 171–84, 179(fig.)
    *see also individual species*

*Caladenia*, 95
*Calibri t. thalassinus*, 129

*Calliandra*, 160
calliphorid flies, 175, 181
*Capparis*, 160
carabid beetle, innate color preferences, 175
*Carassius auratus*, 129
carpenter bees (*Xylocopa micans*), 54
carvone oxide, 86
*Cepea* snails, 238
chiropterophily, 149, 159
*Choeronycteris*, 165
choice behavior, 41–57
    expected rewards, 49–55
    memory dynamics and, 26–30
    variation in nectar concentration, 54–5
    visual angle and, 65(fig.)
chromatic cues, 63–8
*Cimicifuga simplex*, 95
*Cirsium palustre*, 248, 249
*Cirsium purpuratum*, 283(fig.), 275(fig.)
*Clarkia*, 267
*Clarkia breweri*, 85
*Clarkia gracilis*, 248
*Cleome moritziana*, 165, 164
*Cobaea*, bats and, 166
cognitive capacities, bee pattern vision, 77–8
coleopterans, 172
*Colias* butterflies, as solitary insects, 181
color, flower, *see* flowers, color
color blindness, 120
color contrast, 62–3
color vision
    as adapted to flower color, 115–18
    evolution of, 106–23
    innate color preferences, 175–6
conditioned proboscis extension (CPE), 89,
    91
conditioned stimuli (CS), 29, 32, 91
conopid flies, 216, 217
continuous input model, 285
corolla size, pollen dispersal and, 304
*Corydalis cava*, 199
costs
    cutting, 159–65
    nectar and, 44–7
crab spiders, 215, 216, 221
    bees and, 228(fig.)
    predation by, 217(fig.), 219, 229–30
*Crithidia bombi*, 202
cryptic prey types, search image hypothesis,
    5
*Cryptostylis*, 95
cue preference, schematic test for, 135(fig.)
*Cycnoches*, 85

*Dactylanthus taylorii*, 95
*Dactylorhiza sambucina*, 251, 252

morph frequency and, 251(fig.)
*Dalechampia*, 86, 114
dance, bee
    nectar concentration and, 51(fig.)
    pollen quality and, 48(fig.)
    subjective evaluation of, 44–8
    sucrose concentration and, 46(fig.),
        47(fig.)
Darwin, Charles, 2–4, 10, 191, 318
*Daucus*, nectar availability and, 230
decision-making process, bees and, 29,
    43–4, 45(fig.)
decision rules, foraging in hummingbirds,
    128–9
*Delphinium nelsonii*, 248
detection, versus recognition, 65–8
detection-advertising hypothesis, 282–3
dimethylsulfide, 160
dioecy, 289
dipterans, 172
dispensing mechanisms, attraction and,
    309
diurnally foraging butterflies, 174
dragonflies, 215
drongos, 216
drosophilid flies
    innate color preferences, 175
    learning ability, 177

echolocation, flower shape and, 162–4
ecotypes, assortative mating and, 261
electroantennogram (EAG), 88–9
electroretinogram (ERG), 175
electrostatic change, 280
energy requirements
    flight in bats, 153–5
    flower-visitation patterns, 181–2
    glossophagine bats, 150, 152(table)
*Epilobium angustifolium*, 205
epiphytic orchid, 248
*Erigeron*, nectar availability and, 230
*Erythrina glauca*, 160
ethological isolation paradigm, 318–30
*Eugenes fulgens*, memory for flower location,
    133
euglossine bees, 85, 86
*Euphydryas editha*, learning ability, 177
eusocial hymenopterans, 172
evolution, floral, 237–56
Evolutionary Synthesis, plant speciation
    and, 319
exaptation, 106, 113–15, 116

fireflies, innate color preferences, 175
flavonoid coloration, 113–14
flesh fly, 225

flies
   vision and learning in, 171–84, 179(fig.)
   *see also individual species*
flight
   distance, floral display size and,
      284(fig.)
   energy costs of, in bats, 153–5
   horizontal forward, 153
   hovering, in bats, 154–5
   speeds, in bats, 153
flight-cost hypothesis, 283
floral…, *see also* flowers
floral adaptations, bats and, 148–67
floral displays
   attractiveness in, 281–4, 307–11
   color change, 180
   pattern vision and, 61–79
   plant density and, 286–9
   pollen dispersal and, 297–311
   searching image hypothesis, 5–6
   size, 274–91
floral diversity
   in communities, 15–16
   selective foraging behavior, 13–14
floral polymorphisms, 237
   evolutionary dynamics of, 255–6
floral scent, 83–97
   emissions, 94, 96(fig.)
   odor space, 86–7
   perception of, 87–91
   pollinator-mediated selection, 95–7
   response to, 91–7
   variation, 85–6
   *see also* olfaction
flower(s)
   choice, mechanistic model of, 36–8
   color
      change, 180
      color vision and, 115–18
      diversity, 16
      evolution of, 106–23, 113–15
      and pollination syndrome, 107–10
   evolution, 183–4, 237–56
   individual pollinators and, 301–4
   location of
      cues, 174–5
      memory of, 132–40
   multiple use of, 182
   scent-marking at, 206–8
   shape
      echolocation and, 162–4
      evolution, 183–4
      predation and, 229–30
   specialization, constancy and, 226–8
   switching, Darwin's interference
      hypothesis, 3–4

visitation patterns, energy requirements
      and, 181–2
   *see also* floral …
flower constancy, 12–13, 174
   bees, 112(fig.)
   determinants of, 21–38
   flower color and, 110–13, 174
   flower specialization and, 226–8
   plant speciation and, 322
   switching costs and, 4
flycatchers, 216
food quality, assessment of predation risk
      and, 219(fig.)
foraging
   areas, small, 192–3
   bats, 148–67
   costs, cutting, 159–65
   energetics, 148–67
   in field conditions, 246–8
   honeybees, 27(fig.), 28(fig.), 29(fig.)
   hummingbirds, 127–45
   memory dynamics, 34–6
   mode, variation in, 195–9, 205
   parasite-induced changes, 202
   patch depression, 276–80
   patterns, as frequency-dependent, 239–43
   plant speciation and, 320–3
   scent-driven, 83–97
   strategies, 92–5, 203–5
foraging–mortality tradeoff, 220
fragrance, *see* floral scent
frequency-dependent selection (FDS),
      237–56
   analyzing data, 239–41
   as fiction, 252–6
   in field conditions, 246–8
   pollinator choice and, 240–1(fig.),
      247(table)
   predicted fitness relationships and, 253–5
   in rewarding plant populations, 248–9
   in unrewarding plant populations,
      249–52
fruit-bat flowers, 149

gas chromatographic-
      electroantennographic detection
      (GC-EAD), 88–9
geitonogamy, 274, 286, 287, 288, 289
   inflorescences and, 305
   outcross siring success, 306–7
gene flow, plant speciation and, 319
general landscape memory (GLM), 23–6, 37
general odorant-binding proteins (GOBPs),
      88
genetic drift, flower color and, 106
genotype, effects of, 201–2

genotypic variation
  bees and, 42
  dance and, 44–8
  proboscis extension response and, 49
*Gilia*, 328
glomeruli, scent-driven foraging behavior
    and, 89–90
*Glossophaga*, responses to fragrance, 91
*Glossophaga commissarisi*, 160
  flight speeds, 153–5
  minimum nectar energy densities and,
    155
*Glossophaga soricina*
  hovering flights, 154(fig.)
  olfactory attractiveness, 161(fig.)
  spatial memory, 164
glossophagine flowers,148–9, 159
  daily energy expenditure (DEE) on,
    151(fig.)
  energy relations and, 150, 152(table)
goldenrod (*Solidago juncea*), 226
green contrast, bees and, 68
groundcherry (*Physalis pubescens*), 226,
    227(fig.)
gynodioecy, 289

handling times, Darwin's interference
    hypothesis, 3–4
harmonic radar tracking technique, 24
hawkmoths, 86, 90, 173, 174, 216
  long-tongued, 230
  pollination interactions, 325
  responses to fragrance, 91
*Helianthus annuus*, 226, 227(fig.)
*Heliconius*, 92, 224, 178
*Heliothis virescens*, 226
hermit hummingbirds, 92
  plant speciation and, 323
heterostyly, 237
*Heuchera grossulariifolia*, 268
honeybee, *see Apis mellifera*
hoverflies, 177
hummingbirds, 107, 137–45, 216
  assortative mating, 264
  feeder patterns, 141(fig.)
  flower rewards, 142(fig.)
  focal flowers, 139(fig.)
  fragrance and, 92
  plant speciation and, 323, 324
  pollination interactions, 324, 326
  return visits to flowers, 134(fig.)
  search times, 108
  spatial learning, 127–45
  *see also individual species*
hybridization, pollination interactions,
    325

*Hydrophyllum capitatum*, 300
*Hyles*, 91
*Hylonycteris underwoodi*, 158(fig.)
*Hymenoptera*, 217

individuality
  learning-related, 199–201
  in pollinators, 191–210
inflorescences, 304–11
  bumble bee behavior on, 245–6
innate preferences, multiple flower use and,
    182
insects
  non-hymenopteran, 172–4
  qualities of, 182–3
interference hypothesis, Darwin's, 2–4
*Ipomoea purpurea*, 248, 249
*Ipomopsis*
  assortative mating, 261
  pollination interactions, 326
*Ipomopsis aggregata*, 140, 268
  foraging, 128
  pollination interactions, 324–5
  visual cues and, 130, 131
*Ipomopsis arizonica*, 325
*Ipomopsis tenuituba*, 268
  pollination interactions, 324–5
  visual cues and, 130, 131

jacamars, 216

*Keckiella*, 107

laboratory behavior, and pollinator field
    behavior, 252–3
ladybird larvae, 220
*Lampornis clemenciae*, memory for flower
    location, 133
lampyrids, innate color preferences, 175
*Lapeirousia silenoides*, 266
*Lasius pallitarsis* ant colonies, assessment of
    predation risk, 219(fig.)
late short-term memory (lSTM), 32–3
*Lavandula*, 183
learning
  ability, 176–8, 180–1
  assessment of predation risk, 218–20
  beetles, 176–8
  fly taxa, 177
  hummingbirds, 127–45
  mechanisms of, 21–38
  neglected pollinators, 171–84
  *see also* memory
*Leiophron pallipes*, nectar availability and,
    230
lekking hummingbirds, 132

lepidopterans, 172
  learning ability, 177
  spectral range, 176
*Leptonycteris*, 165
long-term memory (LTM), 33–4, 37
long-tongued flies, 173
*Lonicera*, 91
*Lygus lineolaris*, nectar availability and,
    230

*Macrocarpaea*, 166
*Macroglossum stellatarum*, 177, 178
*Magnolia*, 87
*Manduca*, 90
marginal-value theorem, 275
megachiropterans, 149
memory
  associative learning and, 26
  dynamics of, 26–31, 34–6
  early short-term, 29–32
  of flower locations, 132–40
  formation, 29–30, 200
  late short-term, 32–3
  long-term, 33–4, 37
  mechanisms of, 21–38
  mid-term, 33
  models, 28(fig.)
  navigational, 22–6
  phases, 30–4
  retrieval, 29–30, 29(fig.)
  short-term, 31, 279–80
  spatial, 164–5, 279
  working, 29
*Merops apiaster*, 217
*Michelia*, 87
microchiropterans, 149
mid-term memory (MTM), 33
milkweed (*Asclepius syriaca*), 216, 226
mimicry, Müllerian, 220
*Mimulus*, 328, 326
*Mimulus cardinalis*, 264
*Mimulus lewisii*, 264
*Misumena vatia* (crab spiders), 216
*Monarda fistulosa*, 306
monkeyflower, 326
moths, 216
  vision and learning in, 171–84
*Mucuna holtonii*, 162–4, 163(fig.)
Müllerian mimicry, 220
*Musa*, 160
muscid flies
  innate color preferences, 175
  learning ability, 177
mushroom bodies, memory formation and,
    200

navigational memories, structure of, 22–6
nectar
  availability of, 230
  concentration of, 50–3, 54–5
  costs, 44–7
  energy densities, 155–9
  low reward, 289
  perceptual discrimination of, 280
  pollinator economics, 298
*Nemestrinidae*, 173
*Neobellieria bullata*, 225
Neotropical orchids, 85
*Nephila clavipes*, 221
nested visual signals, 92
neural mechanisms, flower constancy and,
    21–38
*Nicotiana attenuata*, 96
  fragrance emissions, 96(fig.)
*Nigella arvensis*, 115
non-hymenopteran insects, 172–4
nymphalids, spectral range of, 176

objective information, subjective
    evaluation of, 50–4
odor coding, 89–91
odor signal detection, 87–8
odor space, multidimensional, 86–7
*Oenothera*, nectar availability and, 230
olfaction, 83–97
  attractiveness, relative, 161(fig.)
  cues, flower location and, 174–5
  cutting foraging costs, 159–60
  *see also* floral scent
olfactory conditioning, 30
  flies and, 177
olfactory receptors
  peripheral, 88–9
  tuned to plant volatiles, 89
*Ophrys*, 95
orb-weaving spiders, 215
orchids, 85
  spur length and, 230–1
*Osmia lignaria*, pollinator economics and,
    300

pallitarsis ant colonies, 219(fig.)
paper wasps, 226
parasitoids, 216
  bumble bees and, 226
  of flower-visiting animals, 215
  foraging behavior and, 202
pasture rose (*Rosa carolina*), 226
patch depression, 276–80
pea aphids, 220
*Pelargonium sericifolium*, 266
*Penstemon*, 107

*Penstemon strictus*
  movement patterns of bees, 198(table)
  pollinator individuality, 194
  small foraging areas, 192–3
  variation in foraging mode, 195
  visit frequency, 193(fig.)
*Petunia*, 86
*Phallus impudicus*, 92
*Philanthus bicinctus*, 217, 222
*Philanthus triangulum*, 216–17
  predation by, 215
phlox family, plant speciation and, 324
*Phlox pilosa*, 249
phorid flies, 216
photoreceptors
  bumble bees, 117(fig.)
  honeybees, 62(fig.)
*Physalis pubescens*, 226, 227(fig.)
*Pieris* butterflies
  interindividual variance, 209
  spectral range, 176
  vision and learning, 179
*Pieris rapae*, as solitary insects, 181
plants
  as adaptive units, 26
  assortative mating and, 264–9
  density, evolution of floral display and,
    286–9
  effects of predation risk on, 214–31
  manipulating pollinators of, 289–91
  pollinator familiarity with
    characteristics, 205–6
  populations, frequency-dependent
    selection in 248–52
  speciation, pollinator behavior and,
    318–30
  *see also* flowers
pleiotropy, 113–15, 106
*Polemonium*, 261
*Polemonium viscosum*, 95
  pollination interactions, 326
*Polistes arizonensis*, 226
  tobacco budworms and, 227(fig.)
pollen, 47–8, 53–4
  dispersal, 297–311
  export, attraction and, 309(fig.)
  movement, fragrance and, 92–5
  targeting, assortative mating and, 261
pollination interactions, 324–7
pollination saturation, 287
pollination syndromes, flower color and,
    107–10
pollinator(s)
  assortative mating, 259–70
  attack rates on, 216–18
  economics, pollen dispersal and, 297–311

field behavior, and laboratory behavior,
    252–3
flower constancy, *see* flower constancy
individuality, 191–210
inflorescences and, 304–6
plant speciation and, 318–30
predation, *see* predation of pollinators
preference, 237–56
response to perceived predation, 220–2
sensory physiology, interindividual
    variance, 208–9
single, pollen dispersal and, 303(fig.)
taxa, 264–9
pollinator–plant interactions, 229
polymorphisms
  color vision and, 120
  evolutionary dynamics of, 255–6
praying mantids, 215
predaceous bugs, 215
predation of pollinators, 214–31
  behavioral adaptations to, 225–8
  and floral traits, 229–31
  and flower shape, 229–30
  indirect effects of, 218
  levels, 214–15
  perceived, 220–2
  and pollinator behavior, 218–25
  rates, 216–18
proboscis extension response (PER), 49, 57
*Prosoeca*, 266
protein kinase C pathway, 33
*Pseudobombax septenatum*, 160
*Psiguria*, 92
pteropodid bats, 152
*Ptiloglossa*, 53
*Puya*, 166

radar tracking technique, 24
radiation, plant speciation and, 319
rainforest bats, 158
receptor-specific contrasts, colored targets
    and, 63
recognition
  colored targets and, 62–5
  versus detection, 65–8
red flowers, 107–10
  hummingbirds and, 129
  visual cues and, 130
red preferences, color vision and, 121–3
retention index, 87
robber flies, 215
*Rosa carolina*, 226

*Salix*, 300
sampling error, choice behavior and, 42
sarcophagid flies, as solitary insects, 181

satyrids, spectral range of, 176
*Sauromatum guttatum*, 92
scarlet gilia, foraging and, 128
scent, floral, *see* floral scent
scent-driven foraging behavior, 83–97
scent-marking, at flowers, 206–8, 280
search image hypothesis, 4–6
    field experiments, 6
    selective foraging behavior patterns, 10
*Selasphorus, see also* hummingbirds
*Selasphorus platycercus*, 130
*Selasphorus rufous*, 130
*Senecio*, flower-visitation patterns, 182
sequential choice behavior, memory
        dynamics, 34–6
sheep blowflies, 177
short-term memory
    associative induction and, 30–2
    late, 32–3
    spatial working memory, 279–80
shrikes, 216
*Silene*, 92
    pollination interactions, 326
snapdragons, 267
*Solanum*, 53
*Solidago*
    nectar availability, 230
    pollinator individuality, 194
*Solidago juncea*, 226
solitary insects, 180–1
spatial cues, in black-and-white stimuli,
        68–9
spatial frequencies, discrimination among,
        71(fig.)
spatial memory
    associative learning and, 26
    bats, 164–5
    long-term, 279
specialized route memory (SRM), 23–6
speciation, pollinator behavior and, 318–30
spectral range, 176
sphingids, spectral range of, 176
spider mites, assessment of predation risk
        and, 219–20
*Stanhopea*, 87
stereo olfaction, 87–8
*Streptanthus*, 328
subjective evaluation, bees, 41–57
sulfur-containing compounds, 160
sunflowers, 226
swallows, 216
swifts, 216

switching costs
    flower constancy and, 4
    Darwin's interference hypothesis, 3–4,
        4(fig.)
*Syconycteris australis*, 152
symmetry
    cues and, 75(fig.)
    role of, 74–7
synchronized dichogamy, 289
syrphid flies, 174, 175

tarsal glands, scent-marking and, 206–7
templates, pattern recognition by, 70–3
tephritid flies
    innate color preferences, 175
    learning ability, 177
*Tetranychus urticae*, 220
thynnine wasps, 94–5
tobacco budworms (*Heliothis virescens*), 226,
        227(fig.)
*Tolumnia variegata*, 248, 249
*Tragopogon mirus*, 268
trait variability hypothesis, 6–10, 9(table)
transponders (PIT-tags), minimum nectar
        energy densities and, 155
trapline behavior, 224
    as adaptive, 202
    bumble bees, 196–7(fig.)
    fragrance and, 92
    holdover, 206
    memory for flower location and, 132
    pollinator individuality, 194
trichromatic color vision, 62–3
*Turdus*, foraging, 129

ultrasound echoes, from *Mucuna holtonii*,
        163(fig.)
ultraviolet vision, in bees, 115
unconditioned stimulus (US), 29, 32

vision
    as adapted to flower color, 115–18
    floral displays and, 61–79, 106–23
visual angle, correct choice and, 65(fig.)
visual signals, nested, 92
*Vriesea*, 164
*Vriesea gladioliflora*, 153–5

wasps, 94–5, 215

xenogamy, 287
*Xylocopa micans*, 54